CHALLENGES IN CORROSION

WILEY SERIES IN CORROSION

R. Winston Revie, Series Editor

CHALLENGES IN CORROSION

Costs, Causes, Consequences, and Control

V. S. SASTRI
Sai Ram Consultants, Ottawa, Ontario, Canada

Published by John Wiley & Sons, Inc., Hoboken, New Jersey
Published simultaneously in Canada

For general information on our other products and services or for technical support, please contact our Customer Care Department within the United States at (800) 762-2974, outside the United States at (317) 572-3993 or fax (317) 572-4002.

Wiley also publishes its books in a variety of electronic formats. Some content that appears in print may not be available in electronic formats. For more information about Wiley products, visit our web site at www.wiley.com.

Library of Congress Cataloging-in-Publication Data:

Sastri, V. S. (Vedula S.), 1935-
  Challenges in corrosion : costs, causes, consequences and control / V.S. Sastri.
    pages cm
  Includes index.
  ISBN 978-1-118-52210-3 (cloth)
1. Corrosion and anti-corrosives–Industrial applications. 2. Corrosion and anti-corrosives–Costs.
3. Machine parts–Failures–Economic aspects. I. Title.
  TA462.S31485 2015
  620.1'1223–dc23
                                                    2015000694

Typeset in 10pt/12pt Times by SPi Global, Chennai, India

Printed in the United States of America

10 9 8 7 6 5 4 3 2 1

1   2015

*Dedicated to Sri Vighneswara, Sri Venkateswara, Sri Anjaneya,
Sri Satya Sai Baba, my parents and teachers, my revered wife Bonnie,
and children, Anjali Eva Sastri and Martin Anil Kumar Sastri.*

CONTENTS

PREFACE

This book is an attempt to present a subject which affects everyone of us in our daily lives in a simple form. Corrosion manifests itself in many forms. Without the use of metals, our society would not have advanced, but we allow our most valuable natural resources to be wasted through corrosion.

Corrosion is inevitable, but it can be controlled. Scientists have a major responsibility by their contributions and efforts to reduce unnecessary levels of corrosion, and it is the responsibility of everyone who uses metals to the best advantage. An understanding of corrosion and its control is important for everybody, if only for the obvious reason that it saves money. Studies of the effects of corrosion on society have shown that we seem unable to learn from our mistakes. Thus, the aim and scope of this book is to address corrosion problems, the resulting costs and methods of preventing and controlling the corrosion problems to achieve reduced corrosion costs, save lives and lessen environmental damage.

This book is centered around five facets of corrosion science: (i) introduction and forms of corrosion; (ii) corrosion costs in various economic sectors; (iii) causes of various forms of corrosion; (iv) various methods of corrosion control and prevention in various sectors and (v) consequences of corrosion.

The first chapter constitutes an introduction to corrosion and various forms of corrosion such as general or uniform or quasi-uniform corrosion, galvanic corrosion, stray current corrosion, localized corrosion, such as pitting and crevice corrosion, metallurgically influenced and microbiologically influenced corrosion, mechanically assisted corrosion and environmentally induced cracking.

The second chapter deals with corrosion costs in various countries such as the United States of America, the United Kingdom, Australia, Kuwait, West Germany, Finland, Sweden, India and China in various industry sectors such as highway bridges, gas and liquid transmission pipelines, waterways and ports, hazardous materials transport and storage, airports, railroads, utilities, gas distribution, drinking water and sewer systems, electrical utilities, telecommunications, transportation, motor vehicles, ships, aircraft, railroad cars, production and manufacturing, oil and gas exploration and production, mining, petroleum refining, chemical, petrochemical and pharmaceutical production, pulp and paper, agricultural production, food processing, electronics, home appliances, government, defense, nuclear waste storage and tribology.

The third chapter discusses causes of corrosion in a variety of industry sectors such as concrete and steel bridges, underground pipelines, waterways and ports, hazardous materials storage, corrosion problems in airports, railroads, gas pipelines, drinking water and sewer systems, electrical utilities, telecommunications, automobiles, ships, aircraft, mining industry, petroleum refining, chemical, petrochemical and pharmaceutical industries, pulp and paper, agricultural production, food processing, electronics, home appliances and nuclear waste storage.

The fourth chapter is concerned with corrosion control and prevention, including topics such as protective coatings, metals and alloys, corrosion inhibitors, engineering composites and plastics, cathodic and anodic protection, corrosion control of bridges, mitigating corrosion of steel in underwater tunnels, gas and liquid pipelines, waterways and ports, hazardous waste storage, storage tanks, railroads, drinking water and sewer systems, electric utilities, telecommunications, automobiles, ships, aircraft, oil and gas industry, mining industry, hazardous materials transport, petroleum refining, chemical and petrochemical industry, pulp and paper industry, agricultural products, food industry, electronics, home appliances, defense and other preventive strategies.

The final chapter discusses the consequences of corrosion such as economic losses, loss in production, fatal accidents resulting in injuries and loss of lives and damage to our living environment by polluting the environment.

This monograph will be useful to students in engineering and applied chemistry as a prescribed book in both undergraduate and graduate courses. The book may also be used by students in general arts as a general interest elective course.

V. S. SASTRI

ACKNOWLEDGMENTS

I wish to express my deep gratitude to my wife, Bonnie, for her painstaking efforts in transcribing the text. I also wish to thank my son-in-law, Gerry Burtenshaw, and my daughter, Anjali, for imaging the figures and my son, Martin, for his moral support.

I wish to thank the following organizations for granting permission to reproduce figures:

American Society of Metals, Canadian Institute of Mining, Metallurgy and Petroleum, Elsevier, National Association of Corrosion Engineers, Society of Petroleum Engineers, John Wiley & Sons. Thanks are also due to David Raymond, City of Ottawa (retired), for using his photographs.

Finally, I wish to express my gratitude to the editorial and production staff of John Wiley & Sons for their kind support throughout the writing, reviewing and production of the manuscript.

V. S. Sastri
Sai Ram Consultant

Ottawa, Ontario, Canada

1

INTRODUCTION AND FORMS OF CORROSION

Corrosion is basically a combination of technical and economic problems. To understand the economics of corrosion, it is necessary that one is proficient in both the science of corrosion and the fundamental principles of economics. There are many forms of corrosion, which can be deleterious in a variety of ways. It is logical to discuss the various forms of corrosion of metallic structures occurring in different corrosive environments.

1.1 GENERAL OR UNIFORM OR QUASI-UNIFORM CORROSION

General corrosion is the most common form of corrosion. This can be uniform (even), quasi-uniform, or uneven. General corrosion accounts for the greatest loss of metal or material. Electrochemical general corrosion in aqueous media can include galvanic or bimetallic corrosion, atmospheric corrosion, stray current dissolution, and biological corrosion (Table 1.1).

Dissolution of steel or zinc in sulfuric or hydrochloric acid is a typical example of uniform electrochemical attack. Uniform corrosion often results from exposure to polluted industrial environments, exposure to fresh, brackish, and salt waters, or exposure to soils and chemicals. Some examples of uniform or general corrosion are the rusting of steel, the green patina on copper, tarnishing silver, and white rust on zinc on atmospheric exposure. Tarnishing of silver in air, oxidation of aluminum in air, attack of lead in sulfate-containing environments results in the formation of thin protective films and the metal surface remains smooth. Oxidation, sulfidation,

Challenges in Corrosion: Costs, Causes, Consequences, and Control, First Edition. V. S. Sastri.
© 2015 John Wiley & Sons, Inc. Published 2015 by John Wiley & Sons, Inc.

TABLE 1.1   Forms of Corrosion[1]

1. General corrosion	Uniform, quasi-uniform, nonuniform corrosion, galvanic corrosion
2. Localized corrosion	Pitting corrosion, crevice corrosion, filiform corrosion
3. Metallurgically influenced corrosion	Intergranular corrosion, sensitization, exfoliation, dealloying
4. Microbiologically influenced corrosion	
5. Mechanically assisted corrosion	Wear corrosion, erosion–corrosion, corrosion fatigue
6. Environmentally induced cracking	Stress-corrosion cracking; hydrogen damage, embrittlement; hydrogen-induced blistering; high-temperature hydrogen attack; hot cracking, hydride formation; liquid metal embrittlement; solid metal-induced embrittlement

[1] ASM Metals Handbook, Corrosion, Vol. 13, 9th ed., Craig and Pohlman, pp. 77–189.

carburization, hydrogen effects, and hot corrosion can be considered as types of general corrosion[16].

Liquid metals and molten salts at high temperatures lead to general corrosion[1]. Microelectrochemical cells result in uniform general corrosion. Uniform general corrosion can be observed during chemical and electrochemical polishing and passivity where anodic and cathodic sites are physically inseparable. A polished surface of a pure active metal immersed in a natural medium (atmosphere) can suffer from galvanic cells. Most of the time, the asperities act as anodes and the cavities as cathodes. If these anodic and cathodic sites are mobile and change in a continuous dynamic manner, uniform or quasi-uniform corrosion is observed. If some anodic sites persist and are not covered by protective corrosion products, or do not passivate, localized corrosion is observed (1).

Some macroelectrochemical cells can cause a uniform or near-uniform general attack of certain regions. General uneven or quasi-uniform corrosion is observed in natural environments. In some cases, uniform corrosion produces a somewhat rough surface by the removal of a substantial amount of metal that either dissolves in the environment or reacts with it to produce a loosely adherent, porous coating of corrosion products. After careful removal of rust formed because of general atmospheric corrosion of steel, the surface reveals an undulated surface, indicating nonuniform attack of different areas (1) as shown in Figure 1.1.

In natural atmospheres, the general corrosion of metals can be localized. The corrosion morphology is dependent on the conductivity, ionic species, temperature of the electrolyte, alloy composition, phases, and homogeneity in the microstructure of the alloy, and differential oxygenation cell. The figure also shows high-temperature attack that is generally uniform. It is also possible to observe subsurface corrosion films within the matrix of the alloy because of the film formation at the interface of certain microstructures in several alloys at high temperatures (3).

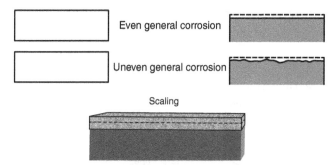

Figure 1.1 Even and uneven general corrosion and high-temperature attack. (Reproduced by permission, Elsevier Ltd. (2).)

The main factors governing general corrosion are: (i) agitation, (ii) pH of the medium, (iii) temperature, and (iv) protective passive films.

(i) The agitation of the medium has a profound influence on the corrosion performance of the metals as agitation accelerates corrosion performance of the metals, accelerates the diffusion of corrosive species, or destroys the passive film mechanically.

(ii) Low pH (acidic) values accelerate the rate of corrosion as for an active metal such as iron or zinc, the cathodic reaction controls the rate of reaction in accordance with the equation

$$E = E^\circ - 0.0592 \, \text{pH}$$

The plot of electrode potential against the logarithm of current density gives rise to a Tafel plot shown in Figure 1.2. From this plot, a logarithm of corrosion current density can be obtained. The Evans diagrams obtained by the extrapolation of Tafel slopes for the cathodic and anodic polarization curves shown in Figure 1.2 can also been seen in Figures 1.3 and 1.4. In general, the cathodic Tafel slopes are reproducible and reliable for evaluation of corrosion rates as they represent noncorroded original surface of the metal. It is obvious that the corrosion current is greater in acidic solution. The influence of pH also depends on the composition of the alloy as seen in Figure 1.4. When the zinc is present with mercury amalgam, the corrosion current is lower than when the metal is zinc alone. When zinc is present along with platinum, high corrosion rates are observed as platinum provides effective cathodic sites for hydrogen evolution. In addition to this, the stability of the passive film in acid, neutral, or alkaline pH is a contributing factor. Some examples are the stability of magnesium fluoride in alkaline medium and the amphoteric nature of aluminum oxide in pH of 4–8 solutions.

(iii) The difference in temperature can create a corrosion cell in the case of copper tubes. In general, increase in temperature results in increased corrosion rate. The corrosion rate of steel in acid solutions doubles for an increase of 10 °C

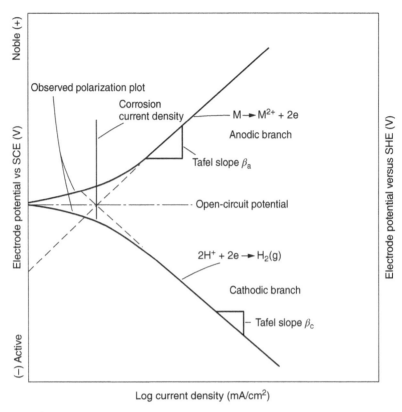

Figure 1.2 Theoretical Tafel plots. (Reproduced by permission, ASM International (4).)

between 15 and 70 °C. At temperatures above 70 °C, the solubility of oxygen in aqueous solutions is low, and the rate of reaction cannot be doubled.

(iv) Protective passive films similar to that of stainless steels result in uniform corrosion because of the mobility of the active sites that passivate readily. Corrosion products and/or passive films are characteristic of numerous electrochemical reactions of the alloys. The film is protective depending on coverage capacity, conductivity, partial pressure, porosity, toughness, hardness, and resistance to chemicals and gases. Rust, oxides of iron, and zinc oxide (white rust) are not protective, while patina (CuO), Al_2O_3, MgO, and Cr_2O_3 are protective in certain environments. Corrosion is generally controlled by diffusion of active species through the film.

1.2 GALVANIC CORROSION

When a metal or alloy is electrically coupled to another metal or conducting non-metal in the same electrolyte, a galvanic cell is formed. The electromotive force and

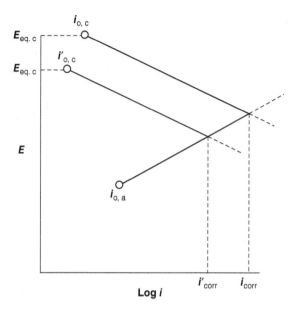

Figure 1.3 Evans diagram for corrosion of zinc as a function of pH. (Reproduced by permission, Elsevier Ltd., (2).)

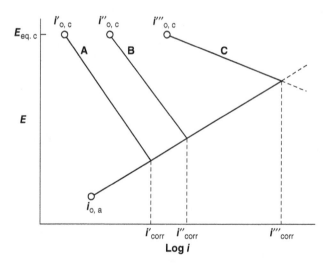

Figure 1.4 Evans diagram for corrosion of zinc alloys. (Reproduced by permission, Elsevier Ltd., (2).)

the current of the galvanic cell depend on the properties of the electrolyte and the polarization characteristics of the anodic and the cathodic reactions. Galvanic corrosion is caused by the contact of two metals or conductors with different potentials. The galvanic corrosion is also called as dissimilar metallic corrosion or bimetallic corrosion where the metal is the conductor material.

Galvanic corrosion can lead to general corrosion, localized corrosion, and sometimes both. Although the dissolution of active metals in acids is because of the numerous galvanic cells on the same metallic surface, it is generally referred to as general corrosion. In less aggressive media such as natural media consisting of dissimilar electrode cells, galvanic corrosion can start as general corrosion that can lead to localized corrosion because of different microstructures or impurities in several cases. Localized galvanic attack depends on the distribution and morphology of metallic phases, solution properties, agitation, and temperature. Localized galvanic corrosion can result in the perforation or failure of the structure.

Galvanic corrosion occurs when two metals with different electrochemical potentials are in contact in the same solution (Figs. 1.5 and 1.6). In both cases[5,7] the corrosion of iron/steel is exothermic, and the cathodic reaction controls the rate of corrosion. The more noble metal, copper, increases the corrosion rate through the cathodic reaction of hydrogen ion reduction and hydrogen evolution. A passive oxide film on stainless steel can accelerate hydrogen reduction reaction.

In engineering design, a junction of two different metals is seldom recommended. It is possible to use alloys with close values of potential in a certain medium. Some of the factors that can be deleterious are mechanical shaping, bending, or lamination of part of the metal, thermal treatment of metallic structure, welding, and cooking coils in vessels and heat exchangers can create galvanic cells of the same metal. These cells are known as macrogalvanic cells, which are different from microgalvanic cells present even in pure metals in a corrosive medium (5, 7).

In general, the galvanic cell is influenced by (i) the difference in potential between two metals/materials, (ii) the nature of the medium or environment, (iii) polarization of the metals, and (iv) the geometry of the cathodic and anodic sites such as shape, relative surface areas, distance.

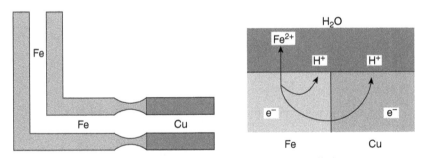

Figure 1.5 Galvanic corrosion of mild steel elbow fixed to a copper pipe. (Reproduced by permission, National Association of Corrosion Engineers International (5).)

Figure 1.6 Galvanic corrosion of painted auto panel in contact with stainless steel wheel opening molding. (Reproduced by permission, ASM International (6).)

For example, it is not desirable to have a small anode connected to a large cathode as this favors accelerated localized anodic dissolution. Rivets of copper on a steel plate and steel rivets on a copper plate on immersion in seawater for a period of 15 months resulted in the steel plate covered with corrosion products while the steel rivets were corroded completely and disappeared. As copper is more noble than iron, it accelerated the hydrogen reduction reaction for the oxidation of the steel plate. In the case of the copper plate with steel rivets, the steel rivets corroded because of the relatively important cathodic surface of copper. The same reasoning applies to the corrosion of noncoated auto parts in contact with a large stainless steel surface (Fig. 1.6).

Galvanic corrosion occurs because of : (i) nonmetallic conductors and corrosion products, (ii) metallic coatings and sacrificial anodes, (iii) polarity inversion, (iv) deposition corrosion, (v) hydrogen cracking or damage, (vi) high temperature.

(i) *Nonmetallic Conductors and Corrosion Products.* Carbon brick in vessels, graphite in heat exchangers, carbon-filled polymers, oxides such as mill scale (magnetite Fe_3O_4), iron sulfides on steel, lead sulfate on lead can act as effective cathodes with an important area to that of anodes. Very often the pores of the conductive film are the preferred anodic sites that lead to pitting corrosion.

(ii) *Metallic Coatings and Sacrificial Anodes.* Some sacrificial metal coatings provide cathodic protection for base metals, such as galvanized steel or Alclad aluminum. If the metal coating is more noble than the base metal such as nickel on steel pitting of the base metal at pores, damage sites and edges can occur. It is important to note that the coatings should be kept free of pores, scratches, or any penetrating chemical attack or deterioration of the coat such as paint.
Sacrificial anodes, such as magnesium, zinc, and aluminum are used extensively for cathodic protection in some locations where impressed current systems are forbidden because of stray currents.

(iii) *Polarity Inversion.* The properties of the electrolyte such as pH, potential, temperature, fluid flow, concentration of different ions, dissolved gases, and conductivity can change with time and influence the polarization, the properties of the interface, and the galvanic potential of the components. The change in ion activity of one metal can reverse the polarity of iron–tin couple for iron-plated tin food cans. This shows the importance of media in galvanic corrosion where the canned food boxes are made of iron with a coated inside layer of tin. The tin may then react with food constituents forming a soluble complex. The concentration of tin is sufficient to the extent that it becomes anodic to that or iron and begins to corrode. Also, iron is protected by zinc coating as a sacrificial anode (galvanization) because of the formation of $Zn(OH)_2$ and white rust, and at temperatures above 60° a hard compact ZnO layer is formed, which is cathodic to iron, and this layer is able to reverse the polarity sign of the couple Fe–Zn. This phenomenon is observed with Cu–Al or Ag–Cu couples (8).

(iv) *Deposition Corrosion.* Dissimilar metallic corrosion can occur following the cementation of a more noble metal. In copper pipes carrying soft water containing carbonic acid into galvanized tank, any dissolved copper ions can be deposited according to the reaction

$$Cu^2 + Zn \rightarrow Cu + Zn^{2+}$$

which causes additional galvanic corrosion of zinc.

Severe corrosion may occur in active Al or Mg alloys in neutral solutions of heavy metal salts (salts of Cu, Fe, or Ni). This type of corrosion occurs when the heavy metal salts plate out to form active cathodes on the anodic magnesium surface. This type of galvanic corrosion can lead to localized pitting corrosion.

(v) *Hydrogen Cracking.* Self-tapping of martensitic stainless steel screws attached to aluminum roof in seacoast atmosphere failed by cracking. Hardened martensitic stainless steel propellers coupled to steel hull of a ship failed by cracking in service. Tantalum is embrittled by hydrogen on polarization or coupled to a more active metal in an electrolyte (8).

(vi) *High-Temperature Galvanic Corrosion.* High-temperature galvanic corrosion is involved in the reaction of silver with gaseous iodine at 174 °C in 1 atm oxygen, which is accelerated by contact of silver with tantalum/platinum/graphite (8).

1.2.1 Factors involved in Galvanic Corrosion

Electromotive force (emf) series and "practical nobility" of metals and metalloids are given in Table 1.2. The emf series, also known as the Nernst scale of solution potentials, are proportional to the free energy changes of the corresponding reversible half-cell reactions with respect to the standard hydrogen electrode. The "thermodynamic nobility" may differ from "practical nobility" because of the formation of passive layer and electrochemical kinetics.

TABLE 1.2   List of Some Systems Leading to SCC (9)

Alloy	Environment
Aluminum alloys	Aqueous chloride, cyanide, high-purity hot water
Carbon steels	Aqueous amines, anhydrous ammonia, aqueous carbonate, CO_2, aqueous hydroxides, nitrates
Copper alloys	Aqueous amines, aqueous ammonia, hydrofluoric acid, aqueous nitrates, aqueous nitrites, steam
Nickel alloys	Aqueous chlorides, concentrated chlorides, boiling chlorides, aqueous fluorides, concentrated hydroxides, polythionic acids, high-purity hot water
Austenitic stainless steels	Aqueous/concentrated chlorides, aqueous/concentrated hydroxides, polythionic acids sulfides plus chlorides, sulfurous acid
Duplex stainless steels	Aqueous/concentrated chlorides, aqueous/concentrate hydroxides, sulfides along with chlorides
Martensitic stainless steels	Aqueous/concentrate hydroxides, aqueous nitrates, sulfides plus chlorides
Titanium alloys	Dry hot chlorides, hydrochloric acid, methanol plus halides, fuming nitric acid, nitrogen dioxide
Zirconium alloys	Aqueous bromine aqueous chloride, chlorinated solvents, methanolic halides, concentrated nitric acid

Ignoring the kinetics and assuming that the passivating films are protective, the practical nobility depends on (i) immunity and passivation domains, and (ii) the stability domain of water. Practical nobility is greater when the immunity and passivation domains extend below and above the stability domain of water and the greater the overlap of these domains with the part of the diagram between pH 4 and 10. Table 1.3 shows the classification of 43 elements according to thermodynamic stability and practical nobility. This table of thermodynamic nobility and practical nobility must be regarded as a guide as the electrochemical equilibrium diagrams are themselves approximate in nature.

1.2.2   Galvanic Series and Corrosion

The practical change of the potential of the components of a galvanic couple as a function of time is important. When the potential difference between two metals is sufficient to form a sustained galvanic cell, the potential of every electrode can be varied because of the active–passive behavior, the properties of the passive or corrosion barriers, and the change in ion concentrations. The galvanic series is a list of corrosion potentials, each of which is formed by the polarization of two or more half-cell reactions to a common mixed potential, E_{corr} measured with respect to a reference electrode such as a calomel electrode. The galvanic series is a list of corrosion potentials in seawater as shown in Figure 1.7. The material with the most negative

TABLE 1.3   Comparative Features of HIC and SSC (10)

Phenomenon	Hydrogen-Induced Cracking	Sulfide Stress Cracking
Crack direction	Parallel to applied stress	Perpendicular to stress
Applied stress	No effect	Affects critically
Material strength	Primarily in low-strength steel	Primarily in high-strength steel
Location	Ingot core	Anywhere
Microstructure	Trivial effect	Critical effect Quenching and tempering enhances SSC resistance
Environment	Highly corrosive conditions, considerable hydrogen uptake	Can occur even in mildly corrosive media

potential has a tendency to corrode when it is in contact or connected to a metal with more positive or noble potential.

The galvanic series is useful in giving a qualitative indication of the possibility of galvanic corrosion in a given medium under some environmental conditions.

1.2.3   The Nature of the Metal/Solution Interface

Cast iron corrodes because of the exposure of graphite content of cast iron (graphitic corrosion), which is cathodic to both low alloy and mild steels. The trim of a valve must be cathodic to the valve body to avoid pitting attack. Thus in aggressive media valve bodies of steel are preferred to cast iron bodies. Steel bolts and nuts coupled to underground mild steel pipes or a weld rod used for steel plates on the hull of a ship should always be of a low nickel, low chromium steel, or from a similar composition to that of the steel pipe (8).

1.2.4   Polarization of the Galvanic Cell

The different phenomena of polarization of the anodic and cathodic reactions (activation, diffusion, convection) should be well known as a function of the evolution and change of the properties of the interface as a function of time. The polarization behavior of the cathode and anodic reactions on the two electrodes should be examined. In natural atmospheres, the cathodic reaction controls frequently the rate of attack. The diffusion of oxygen is an important parameter to avoid control and polarization of the corrosion by the rate of the cathodic reaction (8) (Fig. 1.8).

The resistance overpotential of the cell IR is mainly a function of the conductivity of the electrolyte solution and the distance between the electrodes as the electrolytic resistance is more important than the electric resistance of the metals. Thus if dissimilar pipes are butt-welded with the flow of electrolyte, the most severe corrosion will occur near the weld on the active metal. In soft water, the critical distance between

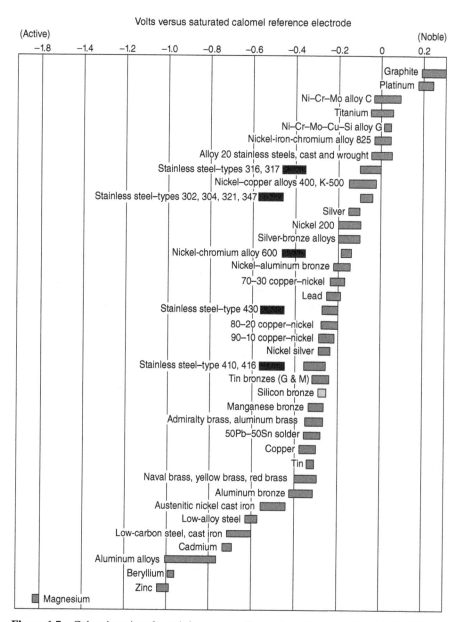

Figure 1.7 Galvanic series of metals in seawater. (Reproduced by permission, National Association of Corrosion Engineers International (11).)

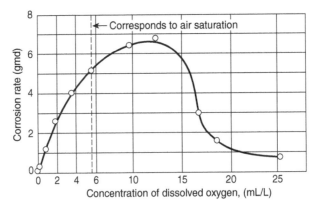

Figure 1.8   Effect of oxygen concentration on the corrosion of mild steel in slowly moving water (8).

copper and iron may be 5 mm and several decimeters in seawater. The critical distance is greater when the potential difference between the anode and cathode is larger. The geometry of the circuit affects galvanic corrosion as observed in stray current corrosion (8).

The relative area ratio of anodic to cathodic sites is critical for general and/or localized corrosive attack. The anode to cathode area ratio along with the conductivity of the electrolyte controls the corrosion rate. When the surface area of the anode is small as compared to the cathode and the solution is of low conductivity, the uniform corrosion can change to quasi-uniform corrosion, severe pitting, or other types of localized corrosion. When the diffusion of oxygen is rate-determining, large cathode/anode area ratios will result in severe galvanic corrosion (12).

In the case where diffusion of corrosive ions is a rate-controlling reaction, it has been found that the relationship

$$P = P_0 \left(1 + \frac{A_c}{A_a}\right)$$

is valid, where P is the penetration that is proportional to the corrosion rate, P_0 is the corrosion rate of less noble uncoupled metal, A_c and A_a are the areas of more noble and active metal, respectively (8). When galvanic cell cannot be avoided, a large anode and a limited size cathode are recommended for use. Stagnant condition and weak electrolytes may cause pitting corrosion in spite of the large area of the exposed active metal.

An example illustrative of the effect of relative areas of anodic and cathodic surfaces consists of steel reservoirs covered with phenolic paint with the bottom clad with stainless steel filled with corrosive solutions. After a few months the perforation of the steel wall at about 5 cm away from the weld of steel wall close to the stainless steel bottom was observed. Large cathodes and small anodes are not desirable as there is no perfect coating, and the paint on steel has some defects such as pores.

The pores act as anodes and the stainless steel bottom as the cathode. The corrosion current can be multiplied by a factor of 10–20 resulting in a quick penetration of the anodic sites. This failure shows (i) the cathodic control, which limits the corrosion current in several aqueous solutions is not operational; (ii) the distance factor and its effect on the failure is evident as the perforation occurred near the welding junction in low-conducting solutions. Galvanic corrosion can be prevented or reduced by (i) avoiding contact between metals with different potentials; (ii) protection by metallic, nonmetallic, nonorganic, organic (paints, lacquer) coatings is recommended; (iii) large cathodes and small anodic surfaces should be avoided; (iv) electrochemical testing and determination of polarization characteristics of all the components as recommended in Pourbaix diagrams of the system should be consulted; (v) use of corrosion inhibitors should be considered; (vi) prediction of the anodic and cathodic components of the galvanic cell and inversion of the polarity of the cell should be considered; (vii) cathodic protection is the only complete protection of the metallic surface.

1.2.5 Testing of Galvanic Corrosion

Corrosion testing for galvanic corrosion may be predicted by ASTM standards in the form of potential measurements. The driving force for galvanic corrosion is the potential difference between the anode and cathode. The galvanic currents between two dissimilar metals are measured using a zero resistance ammeter (ZRA) for a chosen length of time. The ratio of anode to cathode areas is 1:1.

1.3 STRAY CURRENT CORROSION

In the past, stray currents resulted from DC-powered trolled systems, which are now obsolete. An electric welding machine on board the ship with a grounded DC line located on shore will cause accelerated attack of the ship's hull as the stray currents at the welding electrodes pass out of the ship's hull through the water back to the shore. Houses in close proximity can dramatically corrode at the water line. The pipe in one house may be heavily corroded while the pipes in a neighboring house may be intact.

The majority of stray current problems occur in cathodic protection systems. The current from an impressed current cathodic protection system will pass through the metal of a neighboring pipeline at some distance before it returns to the protected surface. Increased anodic corrosion is frequently localized on the pipe at the zone where the current leaves the pipe back to the protected steel tank.

Stray current flowing along a pipeline very often will not cause damage inside the pipe, because of the high conductivity of the electric path compared with the electrolytic path. The damage occurs when the current reenters the electrolyte and will be localized on the outside surface of the metal. If the pipe has insulated joints and the stray current enters the internal fluid, localized corrosion on the internal side of the pipe will occur. The best solution to avoid this mode of corrosion is the electrical bonding of nearby structure and adding additional anodes and increasing rectifier capacity (13). Stray currents follow paths other than their intended circuit. They leave their intended path because of poor electrical connections or poor insulation around the intended conductive material. The escaped current then will pass through the soil, water, or any other suitable electrolyte to find a low-buried path such as metallic pipe. Stray currents cause accelerated corrosion when they leave the metal structure and enter the surrounding electrolyte. These corrosion sites can be several hundreds of meters away.

At the points where the current enters the structure, the site becomes cathodic in nature because of changes in potential, and the area where the current leaves becomes anodic. Electric railways, cathodic protection, electric welding machines, and grounded DC electrical sources are prone to stray current corrosion (14).

Although the damage because of stray current is localized in a part of the system, stray current may lead to uniform corrosion of this part of the system and hence considered as a general form of corrosion. The attack because of stray current is generally more localized, sometimes leading to a concentration of pits. Stray current corrosion can cause penetration along the boundaries or a selective attack of the ferrite within the matrix of gray cast iron. Aluminum and zinc (amphoteric metals) can show signs of corrosion at cathodic portion of the metals because of the localized alkalinity. Buried power lines can give rise to AC stray currents. In general, AC currents cause less damage. DC and the stray current corrosion decreases with increasing frequency. However, damage to passive alloys such as stainless steels and aluminum alloys is important because of the alternating reduction and oxidation of the passive or barrier layer on the surface, leading to porous and nonprotective passive layers (14).

The escaping current can be monitored by measuring the current before it enters the soil as around the electrolyte. Good electrical connections and insulation can stop the current leakage from the metallic structure to the ground. The stray current conductor is connected with the source ground via a separate conductor resulting in the elimination of the need for the current to leave the metal and enter the soil. Sacrificial anodes may be used to prevent stray current corrosion. The insulation should be used with care to reduce the current to a negligible value. Coatings are not useful as cracks, pinholes, or pores will promote localized corrosion. However, coatings on cathodically protected structures are useful and make the stray current less severe and more easily controlled.

1.4 LOCALIZED CORROSION

This is the most insidious form of corrosion as it is less predictable than general corrosion and can cause serious failures. All the forms of general corrosion that result in a nonuniform surface can be considered as localized corrosion. Figure 1.9 shows the various forms of localized corrosion.

The two major types of localized corrosion to be discussed are pitting corrosion and crevice corrosion, including filiform corrosion. Although the morphological appearance of these two types of corrosion is different, the electrochemical basis of

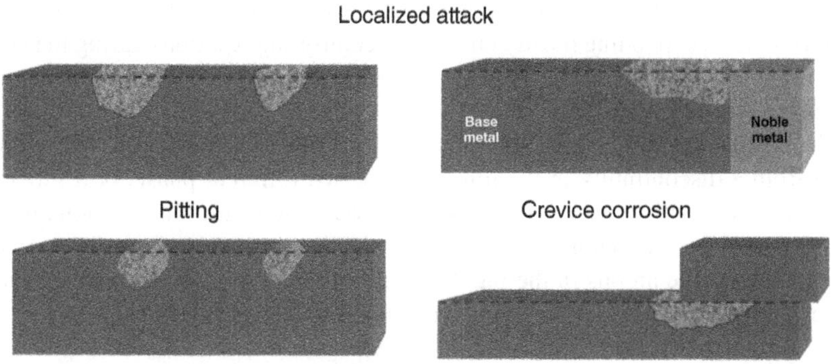

Localized attack

Pitting   Crevice corrosion

Figure 1.9 Various forms of localized corrosion. (Reproduced by permission, National Association of Corrosion Engineers International (3).)

these two forms of corrosion is nearly the same. The morphological differences may be because of the different causes in the initiation step involved in pitting or crevice corrosion (2).

1.4.1 Pitting Corrosion

Pitting corrosion may be caused by a number of factors such as mill scale, which is a very common form of pitting corrosion of steel. Areas where a brass valve is located in steel or galvanized pipeline form good examples. The junction of the two areas is usually pitted, and if the pipe is threaded, the thread in close contact with the brass valve undergoes pitting leading to a leak. This form of corrosion damage occurs frequently in industry, as well as in homes and farms. The deep pitting of tankers on the horizontal surfaces of cargo ballast tanks is a particularly aggravated pitting involving deep and frequent pits. The pits result from frequent changes of cargo and salt water that perpetrate the oxidation reduction corrosion cycle (15).

Pitting can also occur under atmospheric conditions. The corrosion can start at the break and continue to undercut the coating, forming a rather heavy tubercle of hard rust or scale with the pit underneath the original metal. The corrosion products help to isolate the aggressive medium in the pit. This type of corrosion is common in marine environments as well as in other industrial environments where strong corrosive conditions exist (15). Also, pits with open mouths (uncovered) exist and are responsible for loss of thickness and can also act as stress raisers.

Pitting corrosion is of concern in applications involving passivating metals and alloys in aggressive environments. Pitting corrosion can also occur in nonpassivating alloys with protective coatings or in certain heterogeneous corrosive media. Although pitting corrosion may appear harmless and less severe in appearance, the depth of the pit and the pit propagation rate are extremely dangerous and happen to be one of the most serious forms of corrosion. When the pits and crevices are active over an extended period of time, rapid dissolution of the metal occurs. The resulting pit and crevice geometries as well as the surface states within the pits vary markedly from open and polished hemispherical pits on free surfaces to etched crack-like shapes within crevices, depending mostly on the rate-controlling reactions during the growth stage (16). It is also possible that stress corrosion cracking (SCC) and fatigue corrosion may result from pitting.

The mechanism of pitting is self-initiating and self-propagating. Pit initiation can result from a discontinuity in the film, an impurity, different phase, or a scratch on the surface. The active metal immersed in aerated sodium chloride solution dissolves in the pit, and the oxygen moves toward the pit. The positively charged iron cations attract the chloride anions in the pit. The resulting iron chloride hydrolyses and the sequence of the reactions in the pit are as follows:

$$M^+ + Cl^- \rightarrow MCl$$

$$MCl + H_2O \rightarrow MOH + HCl$$

$$HCl \rightarrow H^+ + Cl^-$$

The hydrolysis of iron chloride regenerates HCl, which can further react with the iron metal leading to further iron metal dissolution. The pH changes during pitting corrosion arise from (i) reduction of dissolved oxygen and (ii) reduction of hydrogen. The precipitation of metal hydroxide at the mouth or the sides of the pit can strengthen the autocatalytic nature of the penetration of the pit. The rust tubercles on cast iron indicated pitting in progress, and the inner portion of the bubble was lower in pH and higher in chloride than outside the bubble.

The mechanism of iron corrosion is depicted in the literature (Fontana and Greene, (17)). It represents a section of the pit and a growing pit inside the metal. The pitting factor P/d consists of the deepest pit in comparison with uniform corrosion loss as shown in Figure 1.10. However, pitting corrosion is generally reported as the average of ten deepest pit depths, as recommended in ASTM G48.

There are two distinct processes before the occurrence of stable pit formation: (i) pit nucleation and growth of metastable pit and (ii) enabling the pitting potential – metastable pits cannot grow otherwise (18). There are many examples of pitting in practice as follows:

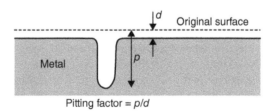

Pitting factor = p/d

Figure 1.10 Deepest pit in relation with penetration in metal and pitting factor. (Reproduced by permission, John Wiley and Sons (8).)

(i) Underground structures: the bottom of a metallic pipe buried in earth, with a relatively limited surface of the metal poorly aerated, tends to become anodic in relation to the large aerated surface of the rest of the metal.

(ii) Mill scale (rust) and pitting: three layers of iron oxide scales formed on steel during rolling results in mill scale, which, when placed in an electrolyte, undergoes corrosion. The defects in the mill scale act as anode with the remaining area acting as a large cathode. An electric current is easily produced between the steel and the mill scale leading to the corrosion of steel without affecting the mill scale. A galvanic cell with an emf of 200–300 mV similar to copper/steel couple is created.

Oxygen comes from the mill scale or diffuses from the air to the steel surface. The coating breaks down because of the movement of the electrolyte, oxygen, and moisture through the film to the pores. Water and gases passing through the mill scale film dissolve ionic material and cause osmotic pressure. Water diffusion and visual blistering occur. The osmotic pressure, thermal agitation, and vibration of the coating film molecules lead to electroendosmatic gradient between the corroding area and the protected areas in electrical contact.

1.4.2 Poultice Corrosion

This is a special case of localized corrosion because of differential aeration, which usually appears as pitting. This form of corrosion occurs when an absorptive material such as paper, wood, asbestos, sacking, cloth, is in contact with a metallic surface that periodically becomes wet. Because of drying periods, adjacent wet and dry regions develop. Near the edges of wet ones and because of limited amounts of dissolved oxygen differential aeration cell develops, and this leads to pitting corrosion. An example of this form of corrosion is the extensive damage of aluminum surface of fuel tanks in aircraft because of bacterial and fungal growth in jet fuel. Design to avoid contact of absorptive material with metallic surface or painting can prevent poultice corrosion (19).

1.4.3 Crevice Corrosion

This form of corrosion occurs in the presence of stagnant corrosive solution near a hole, under a deposit, or any geometric shape that can form a crevice. It is also known as cavernous corrosion or underdeposit corrosion. This form of corrosion results from a concentration cell formed between the electrolyte within the crevice, which is oxygen starved, and the electrolyte present outside the crevice where plenty of oxygen is present. The metal within the crevice acts as anode, and the metal outside the crevice functions as the cathode. The difference in aeration produces a different equilibrium potential, given by the Nernst equation applied to the reaction

$$\tfrac{1}{2}O_2 + 2H^+ + 2e^- \rightarrow H_2O$$

Using the activities of the dissolved species in water and assuming water activity to be 1, we can write

$$E = E^0 + \frac{0.0592}{2} \log{(O_2)^{1/2}(H^+)^2} /H_2O$$

$$= 1.23\text{--}0.0592 \text{ pH (volts at 25 }°\text{C)}$$

Thus localized corrosion ensues because of the difference in potential, which is in turn a result of the difference in oxygen concentration. The more aerated surfaces act as cathodes because of their more noble potential. Differences in metal ion concentrations can result in localized corrosion where the crevice rich in ions acts as the cathode.

Crevices result by design or accident. Crevices by design occur at gaskets, flanges, rubber O-rings, washers, bolt holes, rolled tube ends, threaded joints, riveted seams, overlapping screen wires, lap joints, underneath coatings (filiform corrosion), or insulation (poultice corrosion) and any point where close-fitting surfaces are present (9).

Oxygen differential cells can be established between oxygenated seawater outside or at the opening of the crevice surfaces such as inside crevice anodic areas. It is necessary that the crevice is large enough to allow the entry of the corrosive solution and narrow enough to form a stagnant state and hold the solution with the desired corrosive properties. The opening of the crevice is generally of the order of 50–200 µm. The narrow space present between two metals or a metal and a nonmetal is a favorable site for crevice corrosion. The hydrolysis reaction of iron and acid formation within the crevice produces change in pH and chloride concentration in the crevice environment. The space between the two materials is less aired, has a weak surface, and contains a concentrated salt solution. In some special cases such as in magnesium and magnesium alloys, crevice corrosion is initiated by hydrolysis reaction and acid formation, and oxygen does not play the conventional major role (14).

Crevice corrosion and pitting corrosion are similar in that both involve autocatalytic propagation. The two modes differ in causes of initiation; the morphology and the degree of penetration of pitting are different (17).

1.4.4   Filiform Corrosion

This is a special form of filamentary corrosion occurring on metallic surfaces and is related to crevice corrosion also known as underfilm corrosion. This form of corrosion is generally apparent under painted body of some used cars. It appears as a blister under the paint. The filament propagation underfilm may appear split or joined together, as they propagate in direct lines, some of them reflecting because of obstacles such as adhesive parts of the organic film to the substrate that become trapped in a very narrow place (17).

The filament occurs on metals covered by an organic film and because of a certain discontinuity in the film, air, and water penetrate through the coating and reach the underlying metal. This humid layer becomes saturated or rich in corrosive ions from soluble salts and forms a zone known as the active head of the filament. The dissolution of the metal decreases as the solubility of the oxygen increases. The metallic ions oxidize and form compounds or corrosion products. The zones are called tails.

Filiform corrosion of AZ91 magnesium alloy involves a corrosion mechanism different from the conventional mechanism. In this case, dissolved oxygen is not necessary, and the filiform corrosion propagation is fueled by hydrogen evolution at the filament head and is controlled by mass transfer because of the salt film on the tip of the filament (20, 21).

1.4.5 Breakdown of Passivation

Pitting and crevice corrosion are usually associated with the breakdown of passivity. During pitting corrosion of passive metals and alloys, local metal dissolution leads to cavities within passivated surface area. Pitting corrosion usually occurs in the presence of chlorides. Pitting may also occur in pure water as in the case of carbon steel at high temperatures or aluminum in nitrate solutions at high potentials. In all these forms of localized corrosion, both active and passive states are stable on the same metal surface over an extended period of time so that local pits can grow to a larger size (22).

It is necessary to exceed the critical anodic potential (23) E_{bd} for the electrochemical breakdown of passivation by pitting and consists of these factors: (i) presence of halides at the interface; (ii) induction time for the initiation of the breakdown process, leading to localized conditions that may increase the localized corrosion current density; (iii) development of favorable conditions inside the pits for propagation when the local sites become immobile and localized at certain sites. Electrochemical breakdown of some metal oxides is possible in the case of copper, lead, and tin cathodically to metal while ferric oxide is reduced to the ferrous ion in aqueous solutions. Zinc and aluminum oxides are not cathodically reducible and in these cases hydrogen is reduced. The vigorous evolution of hydrogen assisted by electron conducting zinc oxide can accelerate the breakdown of passivity.

Among metals there are differences in composition and stoichiometry of the oxide films. Halides such as chlorides play an important role in the growth and breakdown of passive films. Borates help stabilize the oxide film. Chloride ions cause severe localized corrosion such as pitting. Well-developed pits have high chloride ion concentration and low pH. Pitting can be random and amenable to stochastic (statistical) theory and very sensitive to experimental parameters such as induction time and electrochemical properties, which are difficult to reproduce. Electrochemical noise (EN) can clarify the initial conditions for pit initiation (24).

The tendency of halides to form metal halide complex is very important in understanding the stabilization of corrosion pit by prevention of the repassivation of a defect site within the passive layer. Among the halides, fluoride forms strong complexes with metals. The resistance of chromium to localized corrosion is because of slow dissolution kinetics of Cr(III) salts. Higher-valence oxides are the best passivators (films) because of their slow rates of dissolution.

Breakdown of passivity is the first stage in pitting corrosion. Corrosion passivation consists of pit growth and repassivation. Metals show different patterns of passivation. Aluminum and copper are not passive in strongly acidic solutions while Fe, Ni, and steels are passive even in strong acidic electrolytes in disagreement with the predictions of Pourbaix diagrams. Localized acidification by the hydrolysis of metal cations may be a stabilizing factor for pitting in some metals. Addition of alloying elements such as Cr and Mo in steel can produce a beneficial effect on the oxide film and result in effective passivation. For austenitic steels, the barrier oxide layers, the salt deposit layers, and the alloy surface layers play an important role in passivity and the breakdown of passivity. It is useful to note that modeling passivation can lead to insights into improved performance and development of new corrosion-resistant alloys (25, 26).

There are two types of pitting that follow the breakdown of a passive metal at surface and can be distinguished as pitting at low and high potential. Pitting at low potential is influenced by cathodic or self-activation leading to emerging etch pits that can eventually lead to general corrosion with etching. High-potential pitting results in hemispherical pits corresponding to anodic dissolution in the electrobrightening mode. This requires a random dissolution because of the presence of film breakdown factors but independent of the crystal structure of the metal. This occurs through a random defective solid film, which is a very good ion conductor. Pits of both types result in occluded corrosion (2). This is also in agreement with the level of potential of the passive metal, that is, noble or active potential (27).

Recently, statistical and stochastic approaches involving Gaussian and Poisson distribution of localized corrosion have been reviewed (28). Application of Poisson distribution to pit generation was successful. Different pit generation rates were observed as a function of time. The models consider either pit generation rate alone or pit generation and repassivation. The three models that are concerned with initiation process leading to passive film breakdown are (i) adsorption and adsorption-induced mechanisms where chloride adsorption is the main process; (ii) ion migration and penetration models, and (iii) mechanical film breakdown model (29).

1.4.6   Coatings and Localized Corrosion

This type of corrosion occurs when protective coatings are applied over metal and where there is a break in the coating so that the large coated area acts as a cathode and the small defective area as the anode (15).

1.4.7   Electrochemical Studies of Localized Corrosion

There are numerous electrochemical methods to determine the electrochemical conditions of pitting and crevice corrosion. Here we are concerned with pitting corrosion.

1.4.7.1   Cyclic Potentiodynamic Polarization Method   Electrochemical studies of pitting corrosion indicate that pitting occurs within or above a critical potential or potential range. The electrochemical methods involved in pitting corrosion studies are potentiodynamic or potentiostatic methods. The important parameters are (i) critical current density, i_{crit} characterizing the active–passive transition; (ii) pitting potential where stable pits grow, and (iii) the repassivation or protection potential (after reversal of potential scan direction), below which the already growing pits are repassivated and the growth stops (18, 29).

Cyclic potentiodynamic polarization used in determining pitting potential consists of scanning the potential to more anodic and protection potentials during the forward and return scans and compare the behavior at different potentials under identical conditions. The polarization curve of an alloy (with or without coating showing active–passive behavior may be obtained in a chosen medium as a function of chloride concentration). E_b, E_{bd}, E_{pit}, or E_p represent pitting potential or breakdown potential, while E_{prot}, E_{rep} refer to protection potential and repassivation potential, respectively. Scan rates of 0.05–0.2 mv/s may be used along with argon/nitrogen bubbling. The breakdown potential corresponding to considerable increase of anodic current at a certain scan rate corresponds to the condition for the initiation of localized attack. The more noble the breakdown potential, the greater is the resistance of the metal/alloy to pitting or crevice corrosion. The potential at which the hysteresis loop is completed on reverse polarization scan determines the potential below which there is no localized attack (30) (ASTM G5) (ASTM G61). The absolute values of pitting or breakdown potential and the protection potential depend on the scan rate and do not reflect the induction time for pitting. Allowing too much pitting propagation to occur along with changes in chemistry can influence the reversal in the scan rate (22).

Some experimental work suggests the convergence of pitting potential (E_{pit}) and protection potential (E_{prot}) to a unique pitting potential (31, 32). Later on, the concept of unique pitting potential corresponding to the most active value of E_p determined after a long incubation time and the most noble value of E_r measured following minimal pit growth have been advanced. It has also been suggested that the stationary pitting potential corresponds to a value between that of pitting and protection potentials. A critical pitting temperature has been defined below which a steel in chloride solution such as $FeCl_3$ would not pit irrespective of potential and exposure time (16). A good measure of pitting susceptibility is the difference between pitting and protection potentials. Alloys that are susceptible to pitting tend to exhibit a large hysteresis. This range of potentials can correspond to metastable pitting corresponding to a region where pits initiate and grow for a limited time before repassivation. The reasons for stop in the growth of large pits are different. Metastable pits are typically micrometer in size with a lifetime of seconds or less, but may continue to grow to form large pits under certain conditions (4, 31).

Pitting tendency increases with increasing temperature. At low temperatures, high pitting potentials are observed. Temperature dependence of pitting susceptibility of stainless steels has been used in the ranking of steels with respect to their pitting resistance as high pitting potentials at low temperatures and low pitting potentials at high temperatures are observed.

Methods used to study localized corrosion consist of galvanostatic methods and potentiostatic methods. At constant chosen currents the evolution of potential as a function of time is noted until the rate of change in potential approaches zero. This technique has been described for aluminum alloys in ASTM G1 as a method for Al alloys (22).

1.4.8 Potentiostatic Methods

When the breakdown potential is determined by cyclic potentiodynamic polarization methods, polarization of the sample at potentials below and above the breakdown potential gives information on the initiation and propagation of pits at different levels. Another method involves initiation of pits above the pitting or breakdown potential and then shifts to lower values above or below the protection potential. It is assumed that at imposed values below the protection potential the current is expected to decrease until complete repassivation.

The critical pitting potential (E_{cpp}) lies between the breakdown potential and protection potential and may be determined by scratch repassivation method. The scratch repassivation method for localized corrosion involves scratching the alloy surface at a fixed potential. Then the change in current is monitored as a function of time, which will show the effect of potential on the induction and repassivation times. An informed choice of the level of potential between the breakdown potential and the critical pitting potential leads to the critical potential for a particular value in the given conditions (22).

1.4.9 Prevention of Localized Corrosion

Knowledge and data on the various physical and chemical aspects of passivity such as composition, thickness, structure, growth, and properties of passive layers may be used in the studies of localized corrosion knowledge of the surface reactions in the formation and composition of passive films, passivation/repassivation, which is useful in the development of highly corrosion-resistant alloys (CRA). A good understanding of metallurgical factors and a rational use of alloying elements can be useful in the control of pitting corrosion. It is also useful to know the mechanism of failure or breakdown of passive films in pitting corrosion for safety purposes (2).

Design-to-prevent is a useful approach to avoid pitting and crevice corrosion. Some examples of surface treatments and coatings to mitigate pitting and crevice corrosion are the following:

(i) Sacrificial zinc coatings for steel.

(ii) Anodizing and sealing accompanied by painting.

(iii) Steel phosphating before painting or using corrosion inhibitors such as chromate, phosphate, molybdate, or zinc rich primers (70–75 μm thick).

(iv) Light barrier protective paint coating ∼100 μm thick or heavy 6–13 mm thick to acid proof brick lining.

(v) A thin organic coating for sacrificial zinc can extend life by a factor of 10. A comparable organic coating together with cathodic protection is probably the best solution.

(vi) It is useful to examine the strained portions of metal such as weld zones that tend to be anodic with respect to unstrained areas (cathodic areas) and the bottom of tanks in aggressive environments.

Some selected procedures/standards: measurement of weight loss along with visual comparison of the pitted surface may suffice to rank the alloys with respect to corrosion resistance. Both visual and microscopic examination of pits can determine the size, shape, and density of pits. Metallographic examination should show the correlation with the microstructure and to distinguish pits between pitting, intergranular corrosion (IGC), or dealloying (30). The test ASTM G 46 gives the standard rating chart for pits and the various methods of examination of pits and describes nondestructive examination of pitting that includes radiographic, electromagnetic, ultrasonic, and dye penetration methods. A brief outline of the statistical approach may be found in ASTM G 46 and the detailed approach (30) in ASTM G 16. It is customary to compare the resistance to pitting as a function of the chloride concentration, which causes pit initiation in different alloys. It is customary to determine the depths of 10 deepest pits and the deepest one along with pitting potentials. The pit depths may be determined with the aid of micrometer, a depth gage, or a microscope. The average of 30 deep pits and the deepest pit depth are usually determined. From these data, the pitting factor is the ratio of the deepest metal penetration to the average depth of penetration. Standard test procedures for pitting and crevice corrosion resistance are described in ASTM G48. Standard procedure for crevice corrosion is given in ASTM G78 (30).

1.4.10 Corrosion Tests

Five corrosion tests for iron- and nickel-based alloys have been identified of which two tests pertain to crevice corrosion. The method MTI-2 originating from ASTM G 48 involves the use of 6% ferric chloride for determining the relative resistance of alloys to crevice corrosion. The method MTI-4 uses high levels of chloride concentration in the range 0.1–1% NaCl to establish the minimum concentration of chloride that results in crevice corrosion (33, 34).

In one of the tests, the corrosion potential of the sample in 9 g/l sodium chloride is noted. The pH, initial, and final potential are noted. The current at +0.8 V versus SCE is noted. If localized corrosion is not simulated in the first 20 s, the polarizing currents will be small and decrease with time. If localized corrosion is not stimulated in 15 min, the test is terminated, and the sample is resistant to localized corrosion. Localized corrosion is indicated by increasing polarizing current with time ($>500 \, \mu A/cm^2$). The potential is then returned to corrosion potential to determine if the sample will repassivate or the localized corrosion will continue to occur. Evidence of pitting and crevice corrosion should be noted in ASTM F 746 (30).

1.4.11 Changes in Mechanical Properties

The change in mechanical property may be used to identify as well as quantify the extent of pitting corrosion in the case of intense pitting corrosion. It is also useful to note that replicates of exposed and blank unexposed samples are recommended. It is useful to note that factors such as edge effects, direction of rolling, and surface conditions must be considered (22, 35).

1.4.12 Electrochemical Techniques for the Study of Localized Corrosion

Electrochemical noise (EN) has been found to be useful in metastable pitting by recording the galvanic current between two nominally identical electrodes at the corrosion potential of a single electrode (18, 35). The spatial separation of the anodic processes in the pit and the cathodic processes in the surrounding surfaces necessitates the passage of current that gives rise to the noise signals. As the localized corrosion sites are small and of the order of 100 μm in diameter or less, current densities inside the cavities can amount to the order of 1 A/cm^2.

EN studies can be performed under open-circuit potential and very close to the natural conditions of pitting. EN measurements in chloride medium have been studied extensively. It has been suggested that the current transients observed below the pitting potential may be attributed to the formation of pit embryos (18). EN measurements are widely used in the studies of metastable pits. There are many factors involved in the interpretation of EN data (26). Some relevant factors are sample size, sampling time, and system noise. It appears that current noise gives a clearer picture than potential noise (18, 36).

EN gives instantaneous data that can show the relative corrosion resistance of different alloys in a particular medium over a long period. However, agitation or convection can mask the noise signals, making it difficult to simulate the operational conditions in certain situations. The ZRA can measure the current flowing in a circuit without introducing the additional voltage drop associated with the standard ammeter. As is the case with the potentiostat, the main functional component is an operational electronic amplifier that supplies the current necessary at its output to maintain zero potential difference between the two input potentials so that no current flows into or out of its input terminals (31). EN data obtained from microelectrochemical studies of stainless steels under potentiostatic conditions showed the current noise expressed as a standard deviation σ_i of the passive current increases linearly with the size of the exposed area whereas the pitting potential decreases (37). However, to complete the electrochemical studies and distinguish between repassivating superficial pits and penetrating pits, microscopic studies are highly desirable. The scanning reference electrode (SRET) should prove to be a suitable complementary technique (18).

1.4.13 Electrochemical Impedance and Localized Corrosion

Electrochemical impedance spectroscopy for studying localized corrosion involving pitting has been the subject of interest in recent times. The statistical variation of pit nucleation and the absence of steady states prevent long-term measurements in the low-frequency region. In addition, in the pitting region, a complicated Nyquist plot is obtained and difficult to interpret. However, Mansfield et al. (39) demonstrated that characteristic changes occur in the low-frequency region. It is useful to note that impedance spectra for pits in stainless steels and magnesium are different from those of aluminum (18, 29).

1.4.14 The SRET

This technique has enabled the measurement of localized corrosion current densities in the vicinity of pits in stainless steel in natural water. Novel potentiodynamic pitting scans were obtained for localized areas adjacent to accurately defined regions of the electrode surface. The SRET has been applied by Isaacs for studying pitting and IGC in stainless steels. Scanning vibrating electrode technique has also been proposed for studying localized corrosion (39). It has also been used in localized measurement of electrochemical impedance spectra (39). The electrochemistry background of the method is given in the literature (35).

A new microelectrochemical technique using microcapillaries as electrochemical cells has been developed. Small surface areas a few micrometers or even nanometers in diameter are exposed to the electrolyte. This leads to current resolution of the order of picoamperes. Microelectrochemical techniques combined with statistical approach of the data evaluation may result in greater understanding of the mechanism involved in these processes (29).

1.5 METALLURGICALLY INFLUENCED CORROSION

Very pure single crystals have defects that can effect corrosion, but impurities and alloying elements, grain boundaries, second phases, and inclusions often have serious effects. Welded structures invariably corrode first at the welds because of metallurgical heterogeneities that are present at the welds. It is an obvious fact that the most susceptible site or defect in a metal will be the first point of attack on exposure to a corrosive environment. Sometimes such attack can result in innocuous removal of susceptible material leaving a surface with improved corrosion resistance (4).

Metallurgically influenced corrosion is because of chemical composition (such as alloying elements, metalloids, and impurities), metallurgical properties (metallic phases, grain joints), and fabrication procedures (thermal treatments, lamination, and welding). Metallurgically influenced corrosion consisting of (i) weld decay; (ii) dealloying; (iii) exfoliation; and (iv) internal attack is shown in Figure 1.11.

The selective corrosion of cast iron known as graphitization, the preferential corrosion of the steel welding known as grooving corrosion, sensitization and knife-line attack of welded stainless steels are typical examples of corrosion influenced by metallurgical parameters.

1.5.1 The Influence of Metallurgical Properties in Aqueous Media

1.5.1.1 Chemical Composition and Microstructure When an alloy composed of various elements corrodes, usually one or more elements dissolve preferentially

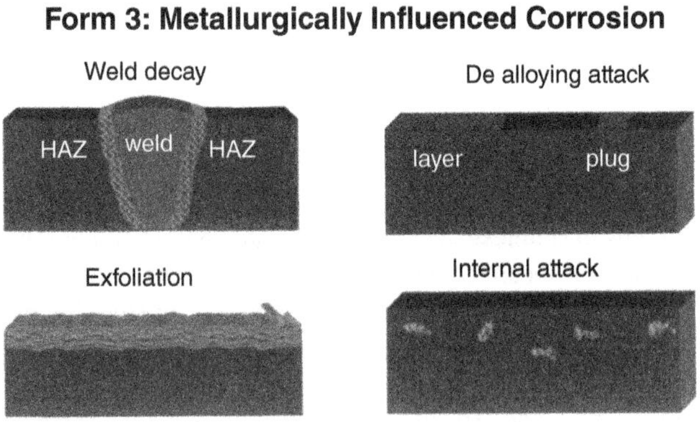

Form 3: Metallurgically Influenced Corrosion

Figure 1.11 Metallurgically influenced corrosion. (Reproduced with permission of NACE International from Reference 3.)

resulting in a surface enriched with other elements. This dealloying process depends strongly on the potential and environmental conditions. The dealloyed microstructure is considerably altered resulting in loss of strength and other properties (4). Glassy metals have been formed by very rapid cooling at the rate of ~106 k/s. This rapid cooling "freezes" the atoms, and the resultant material is chemically and structurally homogeneous and free from defects, secondary phases, and grain boundaries and hence differs in physical, chemical, and mechanical properties from those of the corresponding crystalline alloys. These glassy metal alloys can be more corrosion resistant than the corresponding crystalline alloys in large part because of the absence of multiple phases, grain boundaries, and other defects (4).

The various metallic phases encountered in crystalline alloys consist of pure elements, solid solutions of one element in another, and intermetallic compounds. In crystalline form, alloys have the same type of defects as pure metals. Crystalline alloys consist of a solid solution of one or more elements as a major component or may contain more than one phase. Adjacent grains may have a different composition, which can lead to different mechanical properties and chemical reactivities.

1.5.1.2 Grain Boundaries Usually, the spatial orientation of different grains as defined by intrinsic crystallographic planes is random with respect to each other, that is, the existence of a zone of transition over which the crystallographic orientation changes from one grain to another. This transitional zone is known as the grain boundary. This disorder at the grain boundary is energetically favorable for the accumulation impurities. Similarly, solute atoms and impurities tend to congregate at defects within the grains. Grain boundaries are often more resistant to mechanical deformation and have different chemical reactivities from the grains (40).

1.5.1.3 Point Defects These are zero dimensional consisting of atoms present in the spaces between the lattice positions, vacancies, and foreign atoms in lattice positions. Line defects are one dimensional consisting of edge dislocations and screw dislocations. An edge dislocation is a region of imperfection that lies along the internal edge of an incomplete plane of atoms within a crystal. In a screw dislocation, a portion of the crystal is displaced consisting of continuous ribbon-like structure. Plane defects are two-dimensional stacking faults and are imperfect regions of a crystal resulting from errors in the positioning of atomic layers. Stacking faults arise during crystal growth or as a result of plastic deformation, which can occur along the close-packed planes in a crystal (4).

1.5.1.4 Inclusions Inclusions are three-dimensional defects consisting of soluble particles of foreign material in the metal. Voids, a three-dimensional defect, are empty or gas-filled spaces within the metal. Metal oxides, sulfides, and silicates are common inclusions. For example, manganese sulfide in stainless steel provides a favorable site for pitting corrosion.

1.5.1.5 Passivation For a passivating metal, the corrosion rate is less dependent on potential and hence IGC driven by the free energy differences arising from the disorder in the metal are even less likely. However, the surfaces protected by passive film, grain boundaries, and defects can cause preferential attack by disrupting the formation of a protective layer (41). Texture can have an impact on the passivation characteristics as passivation and repassivation are predominant on densely packed crystallographic planes because of fewer steps and kinks involved. Thus texture can have an impact on passivation characteristics. The susceptibility to SCC can also be impacted by material texture, because the bias in grain orientation will favor alignment of stacking faults in different grains along preferred directions with consequent effects on slip processes and ultimately SCC, depending on the direction of tensile stresses (41).

1.5.1.6 Breakdown of Passivation and Pitting The local breakdown of passivity of metals such as stainless steels, nickel, or aluminum occurs preferentially at sites of local heterogeneities such as inclusions, second-phase precipitates, or dislocations. The size, shape, distribution as well as the chemical or electrochemical dissolution behavior of these heterogeneities in a given environment determine to a large extent whether pit initiation is followed either by repassivation (metastable pitting) or stable pit growth (29).

Localized corrosion of passivating metals initiates at local heterogeneities, such as inclusions, second-phase precipitates, grain boundaries, dislocations, flaws, or sites of mechanical damage. In the case of stainless steels, pit initiation occurs at sites of MnS inclusions. Rapid quenching or physical vapor deposition are useful in the exclusion of inclusions and precipitates with the resulting structure either amorphous or nanocrystalline. Sputter-deposited aluminum alloys containing Cr, Nb, Ta, W, Mo, or Ti show an increase of 0.2–1 V in pitting potential, and the increase in pitting resistance has been attributed to reduced pit initiation tendency as well as a more protective passive film favoring rapid repassivation (37).

With chromium contents of ≥ 20 wt% and 2–6 wt% of molybdenum the pitting potential of steels is increased. Molybdenum in superaustenitic stainless steels improved the repassivation behavior without any effect on pit initiation. Total dependence on metallurgical factors was noted in the corrosion resistance of aluminum alloys (42).

1.5.1.7 Internal or Subsurface Attack (Oxidation) It is usually identified by simple visual observation of the sample surface. However, subsurface phenomena within the matrix of the alloy sample, as well as obscured relations at the interface of the alloy with the surface films, formed in high temperatures may be noted as in Figure 1.11. Electrochemical corrosion at high temperatures at the interface also involves the diffusion of the aggressive gas phase to the vulnerable phase in the subsurface, leading to corrosion most of the time.

1.5.1.8 Dealloying or Selective Dissolution Dealloying is a corrosion process involving the selective dissolution of one or more elements, leaving behind a porous residue of the remaining elements. This phenomenon is also known as selective leaching or parting corrosion and is a corrosion process in which the more active metal is selectively removed from the alloy, leaving behind a porous weak deposit of the more noble metal. For example, selective leaching of zinc from brass is known as dezincification. In the case of gray iron, dealloying is known as graphitic corrosion (9). Dealloying can occur in nearly any system in which a large difference in equilibrium potential between the alloying components and the fraction of the less noble component(s) exists and is significantly high (4).

1.5.1.9 Dezincification Copper–zinc alloys containing more than 15% of zinc are susceptible to dezincification. The dezincification of brass involves selective removal of zinc leaving behind a porous and weak layer of copper and copper oxide. Corrosion of a similar form leads to replacement of sound brass by weak porous copper. The uniform dealloying in admiralty brass (dezincification) is shown in Figure 1.12 (8, 4, 43, 44).

Brass is only one strong phase of dissolved copper and zinc. Under certain conditions, preferential dissolution of brass occurs. This dezincification can be localized (plug dezincification) (Fig. 1.11) or more uniformly distributed (layer dezincification).

Dezincification of α-brass can be minimized by adding 1% tin and further inhibited by the addition of less than 0.1% of arsenic, antimony, or phosphorus. When dezincification is a problem, red brass, commercial bronze, inhibited admiralty metal, and inhibited brass can be successfully used.

1.5.1.10 Graphitic Corrosion This form of corrosion is observed on buried pipelines after many years of service. Gray cast iron consists of continuous graphite network in its microstructure that is cathodic to iron and remains behind as a weak porous mass as the iron is selectively removed from the alloy (4). This form of corrosion (graphitic corrosion) can be reduced by the use of ductile or alloyed iron in place of gray iron, changing the environment such as raising the pH to greater than 7, the use of corrosion inhibitors, avoiding stagnant conditions, and use of cathodic protection.

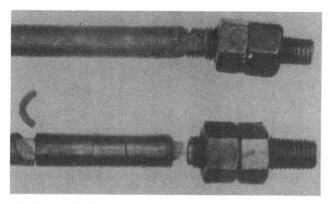

Figure 1.12 Uniform dealloying of admiralty brass. (Reproduced by permission, John Wiley and Sons (8).)

1.5.1.11 Dealuminification Recent studies have shown the importance of the dealloying of S-phase (Al_2CuMg) particles on the corrosion of aluminum aircraft alloys such as 2024-T3. In this alloy, the S-phase particles represent nearly 60% of the particle population. These particles are of the order of 1 mm in diameter with a separation of the order of 5 mm amounting to a surface area fraction of 3%. The selective removal of aluminum and magnesium from these alloy particles leaves behind porous copper particles that become the preferential site for oxygen reduction (45, 46).

Dealloying has also been observed in the case of Ag–Au, Cu–Au, Cu–Pt, Al–Pt, Al–Cu, Cu–Zn–Al, Cu–Ni, and Mn–Cu alloys (46).

Evidence for dealloying has been reported in austenitic stainless steel and iron–nickel alloys in acidified chloride solutions, reduction of titanium dioxide in molten calcium chloride, Cu–Zn–Al alloy in NaOH solutions giving rise to Raney metal particles (46).

1.5.1.12 Mechanism of Dealloying In the dealloying process, the mechanism involves alloy dissolution and replating of the cathodic element or selective dissolution of an anodic alloy constituent. In both types of mechanism, the residual product left behind is spongy and porous and loses most of its strength, hardness, and ductility (9). In the case of brass, the attack involves dissolution of both zinc and copper and subsequent deposition of copper. In the case of gold and silver, selective removal of silver is observed in 0.1 M $HClO_4$.

Two models have been advanced to explain the dissolution and rapid supply of more active or less noble metal. According to one model, the less noble metal dissolves. The remaining more noble element is now in a highly disordered state and begins to reorder by surface diffusion and nucleation of islands of almost pure metal. The coalescence of these islands continues to expose fresh alloy surface where further dissolution will occur, leading to the formation of tunnels and pits (4).

The second model is an extension of the surface diffusion model in the sense importance is given to the atomic placement of atoms in the randomly packed alloy. The model considers that a continuous connected cluster of the less noble atoms must exist to maintain the selective dissolution process for more than a few monolayers of the alloy. This percolating cluster of atoms provides a continuous pathway for the corrosion process as well as a pathway for the electrolyte to penetrate the solid. This mechanism is expected to depend on a sharp critical composition of the less noble element below which dealloying does not occur (44).

1.5.1.13 IGC and Exfoliation   IGC consists of preferential attack of either grain boundaries or areas adjacent to grain boundaries in a sample material exposed to a corrosive environment, but with little corrosion of grains themselves. This dissolution is caused by potential differences between the grain boundary region and any precipitates, intermetallic phases, or impurities that form at the grain boundaries. IGC susceptibility depends on the corrosive solution and on the extent of intergranular precipitation, which in turn is a function of alloy composition, fabrication, and heat treatment (41).

Steel phases have an influence on the corrosion rate. Ferrite has a weak resistance to pitting. Martensite can increase the fragilization of steel. Intermetallic phases such as Fe_2Mo in high Ni bearing alloys can influence the corrosion resistance. The $CuAl_2$ in series 2000 aluminum alloys is more noble than the matrix with corrosion around the precipitate. The majority of IGC processes occur in austenitic stainless steels and aluminum alloys and to a lesser extent in some ferritic stainless steels and nickel-based alloys (46).

Impurities that segregate at grain boundaries may promote galvanic action in a corrosive environment by acting either as anodic or cathodic sites. In 2000-series (2xxx) aluminum alloys, the copper-depleted (anodic) band on either side of the grain boundary is dissolved while the grain boundary is cathodic because of $CuAl_2$ precipitates. In the 5000-series (5xxx) aluminum alloys, intermetallic precipitates such as Mg_2Al_3 (anodic) are attacked when they form a continuous phase in the grain boundary. In chloride solutions the galvanic couples between the precipitates and the alloy matrix can lead to severe intergranular attack. The actual susceptibility to intergranular attack and the extent of corrosion depends on the corrosive environment and the extent of intergranular precipitation, which is a function of alloy composition, fabrication, and heat treatment parameters (46).

Precipitates that form on the exposure of metals to high temperature (during production, fabrication, and welding) often nucleate and grow preferentially at grain boundaries. If these precipitates are rich in alloying elements that are essential for corrosion resistance, the regions adjacent to the grain boundaries are depleted of these elements. The metal/alloy is thus sensitized and is susceptible to intergranular attack in a corrosive medium. An example is AISI type 304 steel corrosion because of precipitation of chromium (46).

At temperatures above 1035 °C, chromium carbides are completely soluble in austenitic stainless steels. However, on slow cooling of these steels or reheated in the range 425–815 °C, chromium carbides are precipitated at the grain boundaries. The precipitation of the chromium carbides depletes the matrix of chromium adjacent to the grain boundary, and the chromium depleted zone undergoes corrosion.

In most cases, intergranular attack arises from compositional dissimilarities rather than structural defects. One of the forms of corrosion in which structural defects play a role is SCC. When a large number of stacking faults are present, it is easier for grains to slip (facilitate shear displacement of one part of grain with respect to another). Thus the material has a tendency to creep rather than crack to relieve tensile stresses. In some materials, decreasing the free energy stored in stacking faults can change the nature of SCC from intergranular to transgranular process (41).

IGC at elevated temperatures is a serious problem in the sulfidation of nickel alloys. Deep penetration can occur rapidly through the thickness of the alloy. This type of IGC can be evaluated by (i) X-ray mapping during examination by a scanning electron microscope equipped with an energy-dispersive X-ray detector and transmission electron microscopy (4).

IGC of aluminum alloys is controlled by material selection and proper selection of thermal (tempering) treatments that can affect the amount, size, and distribution of second-phase intermetallic precipitates. Resistance to IGC is achieved by the use of heat treatments that cause general uniform precipitation throughout the grain structure. General guidelines for selecting suitable heat treatments for these alloys are available (45).

Exfoliation is a form of macroscopic IGC that affects aluminum alloys used in industrial or marine environments. Corrosion starts laterally from initiation sites on the surface and in general proceeds intergranularly along planes parallel to the surface. The corrosion products formed in grain boundaries force metal away from the underlying base material resulting in a layered or flake-like appearance. In certain materials, corrosion progressing laterally along planes parallel to rolled surfaces is known as exfoliation, and it generally occurs along grain boundaries – hence IGC. A layered appearance is a common manifestation of exfoliation (also known as layer corrosion) resulting from voluminous corrosion products prying open the material such as aluminum alloys (4).

Exfoliation corrosion is common among high-strength heat-treatable 2xxx and 7xxx alloys. Exfoliation corrosion of Al 6xxx in salt medium has been observed. Exfoliation corrosion in these alloys is usually confined to relatively thin sections of highly worked products. Exfoliation corrosion is observed in unalloyed magnesium above a critical chloride concentration, but this morphology was not seen in magnesium alloys, in which individual grains were preferentially attacked along certain crystallographic planes. The early stages of this form of attack caused swelling at points on the surface because of apparent delamination of the magnesium crystals with interspersed corrosion products. As the attack progressed, whole grains or parts of grains disintegrated and dropped out, leaving the equivalent of large irregularly shaped pits (47).

1.5.1.14 Testing of Intergranular Attack This necessitates metallographic examination (30). The Standard Practice ASTM G110 (1992) involves evaluation of IGC resistance of heat-treatable aluminum alloys by immersion in a sodium chloride and hydrogen peroxide solution. ASTM A262 test consists of six practices for detecting susceptibility to IGC in austenitic stainless steels using different oxidizing agents at different temperatures, and kinetics is examined by microscopic investigation of the etched structure for $Cr_{23}C_6$ sensitization and weight loss measurement. Similarly, testing for susceptibility to IGC has been described in ASTM 763 for ferritic stainless steels and ASTM G28 for wrought Ni-rich, Cr-bearing alloys.

A laboratory procedure for carrying out nondestructive electrochemical reactivation (EPR) test on types 304 and 304L stainless steels to quantify the degree of sensitization is given. The metallographically mounted and highly polished sample is potentiodynamically polarized from the normal passive condition in 0.5 M sulfuric acid and 0.01 M potassium thiocyanate solution at 30 °C to active potentials – a process known as reactivation. The amount of charge passed is related to the degree of IGC associated with $Cr_{23}C_6$ precipitation, which predominantly occurs at grain boundaries. After EPR test, the microstructure is examined. Certain media have been commonly used for the evaluation of the susceptibility of Mg, Cu, Pb, and zinc alloys to IGC.

1.5.1.15 Weldment Corrosion

The factors that can initiate or propagate different forms of corrosion of welded regions are numerous, interrelated, and difficult to define. However, some pertinent factors to take into account are: weldment design, fabrication technique, welding practice, welding sequence, moisture contamination, organic or inorganic chemical species, oxide film and scale, weld slag and spatter, incomplete weld penetration or fusion, porosity, cracks (crevices), high residual stresses, improper choices of filler metal, and final surface finish. Consequently, the corrosion resistance of welds may be inferior to that of the properly annealed base metal because of microsegregation, precipitation of secondary phases, formation of unmixed zones, recrystallization and grain growth in the weld heat-affected zone (HAZ), volatilization of alloying elements from the molten weld pool, and contamination of the solidifying weld pool (4).

Welded microstructures can be extremely complex and often change drastically over a very short distance. The fusion zone or weld metal is a dendritic structure that solidified from a molten state. Bordering the fusion zone are transition, unmixed and partially melted zones, and the HAZ. These zones can be reheated and altered by subsequent weld passes in multipass welding. For alloys with structures dependent on thermal history such as steels, the final microstructure can be very complex. As welded structures are often quite susceptible to corrosion, overalloyed filler metals are often used to increase the weld corrosion resistance. In the case of stainless steels with high levels of carbon content, sensitization in the HAZ is another problem (4).

1.5.1.16 Weld Corrosion of Carbon Steels

The corrosion behavior of carbon steel weldments obtained by fusion welding can be because of preferential corrosion of the HAZ or weld metal or associated with geometric aspects such as stress concentration at the weld toe, or creation of crevices because of joint design.

In addition, some environmental conditions such as temperature, conductivity of the corrosive fluid, or thickness of the corrosive liquid film can cause localized corrosion. In some cases, both metallurgical and geometric factors will influence behavior, such as in SCC. Preferential weldment corrosion of carbon steels has been studied since the 1950s, commencing with the problem on icebreakers, but the problem continues today in different applications (4).

Weldment corrosion has clear microstructural dependence, and studies on HAZ show corrosion to be more severe when the material composition and welding parameters are such that hardened structures are formed. It is a well-known fact that hardened steel corrodes more rapidly in acid solutions than fully tempered steel, as local microcathodes on the hardened surface stimulate the cathodic hydrogen evolution reaction (4).

1.5.1.17   Grooving Corrosion   This is a particular case of preferential weldment corrosion occurring in electric-resistance-welded/high-frequency-induction-welded pipe where attack of the seam weld HAZ/fusion line is affected in aqueous media. This grooving corrosion has been attributed to inclusions in the pipe material. This type of corrosion is probably because of the redistribution of sulfide inclusions along the weld. It has been suggested that MnS is concentrated by the movement of liquid metal during welding. At these high temperatures, MnS can dissociate into manganese and sulphur and form iron sulfide. A suitable heat treatment can prevent the formation of iron sulfide. And selection of cleaner alloyed steel is recommended.

Galvanic corrosion between the HAZ/fusion line and the parent material because of the unstable MnS inclusions produced during the welding cycle has been observed: corrosion of the weld metal because of electrochemical potential differences between the weld metal and base metal with the weld metal as anodic metal in the galvanic couple. Although the potential difference may only be 30–70 mV, the low surface area ratio of anode to cathode leads to corrosion rates of the order of 1–10 mm.

SCC failure may occur by both active path and hydrogen embrittlement (HE) mechanism, and in the latter case, failure is likely at low-input welds because of the greater susceptibility of the hardened structures formed. Most SCC studies of welds in carbon and carbon-manganese steels have evaluated resistance to hydrogen-induced-SCC under sour conditions prevalent in the oil and gas industry, and commonly referred to as sulfide stress cracking (SSC). The overriding influence of hardness leads to the conclusion that soft transformed microstructures around welds are preferable. Failures in oil refineries have shown cracks parallel or normal to welds, depending on the orientation of principal stresses. Both transgranular and intergranular cracks have been observed.

Equipment such as tanks, absorbers, carbon treater drums, skimming drums, and piping suffer from cracking. All welds of deaerator vessels made from carbon steel should be postweld stress relieved to minimize cracking and pitting.

Backing rings are sometimes used when welding pipe. It is necessary that the backing ring insert is consumed during the welding process to avoid a crevice.

It is important that the welding filler metal must at least match the contents of the base metal with respect to specific alloying elements such as chromium, nickel, and molybdenum. The thermal cycle of heating and cooling during the welding process affects the microstructure and surface composition of welds and adjacent base metal. Hence the corrosion resistance of welds made without the filler metals and welds with filler metals may be inferior to that of properly annealed base metal because of microsegregation, precipitation of secondary phases, formation of unmixed zones, recrystallization, and grain growth in the weld HAZ as well as volatilization of alloying elements from the molten pool and contamination of the solidifying weld pool (22, 48).

Unmixed zones result when welding stainless steels with a filler metal. An unmixed zone has the composition of base metal with the microstructure of an autogenous weld. The microsegregation and precipitation phenomena typical of autogenous weldments decrease the corrosion resistance of an unmixed zone relative to the parent metal. Unmixed zones bordering welds made from overalloyed filler metals can be preferentially attacked on exposure on the weldment surface (4, 48).

1.5.1.18   Sensitization   Welding is the common cause of sensitization of stainless steels to IGC. The main weld metal precipitates in austenitic stainless steels are δ-ferrite, σ-phase, and $M_{23}C_6$ with small amounts of M_6C carbide. Although the cooling rates in the weld and the adjacent base metal are high enough to avoid carbide precipitation, the weld thermal cycle brings part of the HAZ into the carbide precipitation temperature range. The carbides can precipitate and a zone somewhat removed from the weld becomes susceptible to intergranular corrosion. Reheating the alloy to above 1035 °C and cooling it rapidly removes the carbide precipitate, and this practice is termed solution annealing (4).

The well-known weld-related corrosion problem in stainless steels is weld decay (sensitization) caused by carbide precipitation in the weld HAZ. This sensitized microstructure is much less corrosion resistant as the chromium depleted layer and carbide precipitate are subject to preferential attack.

1.5.1.19   Pitting and SCC   Under moderately oxidizing conditions, as in pulp and paper bleach plants, weld metal austenite may undergo preferential pitting in alloy-depleted regions. This mode of attack is independent of any weld metal precipitation and results from microsegregation or coring in weld metal dendrites (48). In the 18-8 austenitic steels, the ruptures of SCC are intergranular when steel is subjected to nonsuitable thermal treatment such as in the zone of carbide precipitation, between 400 and 800 °C. When the steel is subjected to "hyperquenching," ruptures are generally transgranular.

Austenitic stainless steels that are susceptible to IGC are also prone to intergranular SCC. The problem of intergranular SCC of sensitized austenitic stainless steels in boiling high-purity aerated water has been well studied. Cracking of sensitized stainless steels in boiling water nuclear reactors has been observed. Cracking of sensitized stainless alloys in polythionic acid that is formed during the shutdown of desulphurization units in petroleum refineries has been observed. IGC of 430, 434 and 446 steels during welding has been observed (48). Weld decay (sensitization) in an austenitic steel weldment occurs. The grain joints become impoverished in chromium because of the formation of chromium carbides. Knife-line corrosion is a form of IGC encountered in stabilized austenitic stainless steels. During welding, the base metal adjacent to the fusion line is heated to high enough temperatures to dissolve the stabilizing carbides followed by sufficiently rapid cooling to prevent carbide precipitation. When weldments in stabilized grades are then heated in the sensitizing temperature range of 425–815 °C during stress-relieving operations, high-temperature service, or subsequent weld passes chromium carbide may precipitate. The precipitation of chromium carbide leaves the narrow band adjacent to the fusion line susceptible to IGC. Knife-line attack can be avoided by the proper choice of welding variables and by the use of stabilizing heat treatments (4).

1.5.1.20 Localized Biological Corrosion of Stainless Steels There are three sets of conditions under which localized biological corrosion of austenitic stainless steel occurs. These conditions should be examined for metals that show active–passive behavior. Microbiological corrosion in austenitic steel weldments has been studied and documented (4, 48).

1.5.1.21 Prevention of IGC In North America, susceptibility to IGC is achieved by the use of low carbon steels such as 316 L (0.03% carbon maximum) in place of type 316 (0.08% C maximum). In Europe, it is common to use steels with 0.05% C, which are fairly resistant to sensitization, in particular, when they contain molybdenum and nitrogen, which appear to raise the tolerable level of carbon and/or heat input (4, 48). This method is not effective for eliminating sensitization for long-term exposure at 425–815 °C.

Above 815 °C titanium and niobium form more stable carbides than chromium and are added to stainless steels to form stable carbides and prevent formation of chromium carbide.

Duplex alloys with a composition around Fe–26Cr–6.5 Ni–3Mo with low carbon and nitrogen have been found to be resistant to chloride SCC, pitting corrosion, and IGC in the as-welded condition (48).

Choosing the proper welding parameters, balancing alloy compositions to inhibit precipitation reactions, shielding molten and hot surfaces from reactive gases in the weld environment, removing chromium enriched oxides, and chromium depleted base metal from thermally discolored surfaces are the factors to note in the welding process (48).

1.5.1.22 Corrosion Resistance of Aluminum Alloys Both welded and unwelded aluminum alloys are corrosion resistant in uninhibited nitric acid up to 50 °C. Above this temperature, most of the aluminum alloys exhibit knife-line attack adjacent to the welds. No knife-line attack was observed for any commercial aluminum alloy or weld at 70 °C in inhibited fuming nitric acid containing 0.1% hydrofluoric acid.

Weldments of nonheat-treatable alloys are resistant to corrosion. In the case of heat-treatable alloys, corrosion is selective in the weld or in the HAZ. Welding can crack because of mercury-zinc amalgam.

SCC in weldments because of residual stresses introduced during welding is rare. Brazed joints in aluminum alloys have good resistance to corrosion. Soldered joints may be useful in milder environments but not in aggressive media (49).

1.5.1.23 Corrosion Resistance of Nickel-Based Alloys There are three types of nickel-based alloys: (i) CRA; (ii) heat-resistant alloys; and (iii) high-temperature alloys or super alloys. The corrosion performance can change because of the presence of a weld seam as a second phase.

Common failures are because of: oxidation, carburization, metal dusting, sulfidation, chlorination, and nitridation. The most common high-temperature degradation mode is oxidation, and the protection against oxidation is achieved by the formation of chromium oxide scale. Small amounts of aluminum or silicon may improve the resistance of chromia alloy. The attack by chlorine and sulfur depends strongly on the partial pressure of oxygen in the system.

Industrial environments can be oxidizing or reducing in nature in terms of electrode potential that the alloy experiences, which is controlled by the cathodic reaction. Uniform corrosion occurs under reducing conditions and localized corrosion such as pitting and crevice corrosion occur under oxidizing conditions. SCC can occur in any potential range (50).

Corrosion-resistant alloys of nickel are Ni–Cu, Ni–Mo, Ni–Cr–Mo, and Ni–Cr–Fe alloys. The cast nickel alloys do not have the same degree of corrosion resistance as the corresponding wrought alloys because of the higher carbon and silicon contents and the anisotropic structure of the cast alloys (50).

The performance of cast nickel-based alloy depends on the microstructure quality such as the amount of interdendritic segregation, secondary carbides, and intermetallic phases. The corrosion rate of a nickel alloy can vary by several orders of magnitude depending on the microstructure. The two most important metallurgical factors are second-phase precipitation because of thermal instability and the presence of cold work. Cold work is important where SCC is expected (50).

1.6   MICROBIOLOGICALLY INFLUENCED CORROSION (MIC)

1.6.1   Growth and Metabolism

Growth and metabolism depends on the availability of water as microorganisms take up the nutrients present in water and produce cell material (51). Under favorable conditions, some bacteria can double in number every 20 min or less. Thus a single bacterium can produce a mass of over a million microorganisms in <7 h. The bacteria can survive from −10 to >100 °C, pH ~0–10.5, dissolved oxygen 0 to saturation, pressure of vacuum to >31 MPa, and a salinity of ppb to 30%. Most of the bacteria encountered in corrosion grow best at 15–45 °C and pH of 6–8 (41).

A majority of these microorganisms can form extracellular polymeric materials known as polymer or slime. The slime helps glue the organisms to the surface, trap, concentrate nutrients as food for microbes and shields the organisms from biocides.

The slime can also act as a diffusion barrier and change the concentrations of different elements and pH at the electrochemical interface (41).

1.6.2 Environments

A great variety of microscopic organisms are present in all natural aqueous environments such as bays, estuaries, harbors, coastal, and open ocean seawaters, rivers, streams, lakes, ponds, aqueous industrial effluents, and waste waters. Barnacles and mussels are also present in many environments. In natural conditions, sulfate-reducing bacteria (SRB) grow in association with other microorganisms and use a range of carboxylic acids and fatty acids, which are common by-products of other microorganisms. Biological slimes are commonly found in water phases of industrial process plants. A wide range of common bacteria is thought to secrete large amounts of organic matter under aerobic and anaerobic conditions (41).

1.6.3 Biological Corrosion in Freshwater Environments

There exist bacteria and algae (yeasts and molds) in fresh water media. The organisms attach themselves to and grow on the surface of structural materials resulting in the formation of a biofilm. The film can range from a microbiological slime film on fresh water, heat-transfer surfaces to heavy encrustation of hard-shelled fouling organisms on structures in coastal seawater. The presence of a biofilm does not mean that it will always have a significant effect on corrosion (41).

Microbial films will affect the general corrosion only when the film is continuous. This is not the case in general and microorganisms form in discrete deposits or colonies and the resulting corrosion is localized.

A uniform slime film formation on the piping of potable water systems and on the heat-transfer surfaces of low-temperature heat exchangers is inconsequential unless it obstructs the flow leading to a health hazard by growth of the organisms or localized corrosion (41).

1.6.4 Biological Corrosion in Marine Environments

A heavy fouling of macroorganisms such as barnacles and mussels decreases the amount of dissolved oxygen at the interface and acts as a barrier on structural steel in the splash zone and shields the metal from damage caused by wave action. A continuous film of bacteria, algae, and slime (microorganisms) can have the same beneficial effect as the macroorganisms. In the majority of cases, the microbial films are not continuous and an oxygen preferential cell is created. These films are suspected to be capable of inducing pit initiation on stainless steels and copper alloys in marine environments. Natural seawater is more corrosive than artificial saline water because of the living organisms in seawater (41).

Incomplete coverage by barnacles is more likely to initiate pitting and crevice corrosion. The barnacles are attached to the periphery of a high-strength steel rudder, which is coated with an antifouling paint. During use, the paint around the edges has been removed by mechanical action, allowing the attachment of barnacles. The other alternative, a continuous film of bacteria, algae, and slime can occur and provide a barrier film and limiting corrosion. However, it should be noted that it is rare that a film of microorganisms in the marine environment is continuous over a large area of exposed surface (41).

1.6.5 Industries Affected

Many industries are affected by microbiological problems, which involve the material and the features of the medium. The industries affected by MIC are chemical processing, energy generation systems, pulp and paper, hydraulic systems, fire protection, water treatment, sewage handling and treatment, highway maintenance, buildings and stone works, aviation, underground pipeline and onshore, and offshore oil and gas equipment.

1.6.6 Role of Some Microbiological Species in Corrosion

Some bacteria are involved directly in the oxidation or reduction of metal ions such as iron and manganese. Some microbes produce organic acids such as formic and succinic acid or mineral acids such as sulfuric acid. Some bacteria can oxidize sulphur or sulfide to sulfate or reduce sulfates to hydrogen sulfide (H_2S) (41).

SRB are anaerobic bacteria that obtain the carbon from organic nutrients and their energy from the reduction of sulfate to sulfide. Sulfate is abundant in fresh water, seawater, and soils. Sulfide appears as dissolved or gaseous H_2S, bisulphide, or sulfide, metal sulfides, or a combination of these according to the conditions. Sulfides are highly corrosive. SRB are anaerobic and facilitate the cathodic reaction, which controls the corrosion rate in these media. In the case of buried steel in soil containing near neutral-pH solution, the presence of SRB accelerates the electrochemical reactions of corrosion as given below:

$$4Fe \rightarrow 4Fe^{2+} + 8e^- (anodic)$$

$$8H^+ + 8e^- \rightarrow 8H \text{ (cathodic)}$$

$$SO_4{}^{2-} + 8H^+ \xrightarrow{\text{SRB}} S^{2-} + 4H_2O$$

$$Fe^{2+} + S^{2-} \rightarrow FeS \text{ (corrosion product)}$$

$$Fe^{2+} + 2(OH)^- \rightarrow Fe(OH)_2 \text{(corrosion product)}$$

$$2H^+ + S^{2-} \rightarrow H_2S \text{ (possible gas product)}$$

The ferrous sulfide (FeS) is not continuous, and the base iron can corrode. H_2S may also be produced. The SRB have been found to contribute to corrosion of stainless steels, copper, and aluminum alloys.

Most sulfate-reducing bacteria are obligate anaerobes, yet they accelerate corrosion in aerated media. This is possible when aerobic organisms form a film or colony

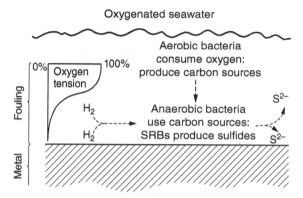

Figure 1.13   Association of anaerobic and aerobic bacteria. (Reproduced with permission of NACE International from Reference 53.)

and then through their metabolism create a microenvironment favorable for anaerobic bacteria.

Aerobic organisms near the outer surface of the film consume oxygen and create a habitat for the SRB at the metal surface (52) (Fig. 1.13). The accompanying flora delivers the nutrients SRB need such as acetic and butyric acids and consumes the oxygen that is toxic for SRB. Sulfate-reducing bacterial corrosion is encountered in the oil and gas industry (51).

1.6.7   Attack by Organisms Other than SRB

Ammonia and amines may be obtained by microbial decomposition of organic matter under aerobic and anaerobic conditions. These compounds are oxidized to nitrite by aerobic bacteria such as *Nitrobacter. Nitrobacter* destroys the corrosion inhibition properties of nitrite-based corrosion inhibitors by oxidation in the absence of a biocidal agent in the formulation. The release of ammonia at the surfaces of heat exchanger tubes has a detrimental effect (50).

Thiobacillus oxidizes inorganic sulfur compounds such as sulfides to sulfuric acid. Some *Thiobacilli* leach metal sulfide ores as follows:

$$4FeS_2 + 15O_2 + 2H_2O \rightarrow 2Fe_2(SO_4)_3 + 2H_2SO_4$$

It is useful to note that it is necessary to have oxygen and reduced sulfur for *Thiobacilli* to act on and produce sulfuric acid. The growth of anaerobic sulfate-reducing bacteria in sewage produces H_2S, which migrates to the air space at the top of the line, where it is oxidized to sulfuric acid in the water droplets at the crown of the pipe by *Thiobacilli*. The corrosion problem is because of the dissolution of alkaline mortar by the sulfuric acid produced by the bacterial action, followed by the corrosion of the ductile iron in the sewage pipe (50).

The influence of microbiological organisms can be the initiation of either general or localized corrosion. This influence is because of the effect of organisms to change variables such as pH, oxidizing power, flow velocity, and concentration of chemical species at the metal/solution interface.

Corrosion can be influenced by microorganisms in many ways such as: (i) production of differential aeration cell; (ii) production of biofilms; (iii) production of sulfides; (iv) production of organic and inorganic acids; (v) production of corrosive gases.

(i) Production of differential aeration cell: a scatter of barnacles on a stainless steel surface creates oxygen concentration cells. The formation of biofilm generates several critical conditions for corrosion initiation. Uncovered areas with access to oxygen act as cathodes, and the covered areas act as anodes. Underdeposit corrosion (crevice corrosion) or pitting corrosion can occur. The corrosion rate may be accelerated depending on the oxidizing capacity of bacteria and the chloride ion content.

Considering pit formation on the surface of iron, the anodic and cathodic reactions are

$$2Fe \rightarrow 2Fe^{2+} + 4e^- \text{(anodic)}$$

$$O_2 + 2H_2O + 4e^- \rightarrow 4OH^- \text{(cathodic)}$$

The insoluble $Fe(OH)_2$ can help bacterial film to control the diffusion of oxygen to the anodic sites in the pit thus forming a typical tubercle. In the presence of chlorides, the pH of the solution trapped in the tubercle can become highly acidic because of the autocatalytic propagation mechanism of localized corrosion because of the reaction involving the

$$FeCl_2 + 2H_2O \rightarrow Fe(OH)_2 + 2HCl$$

formation of iron hydroxide and hydrochloric acid.

1.6.8 Production of Biofilms

The bacteria implicated in corrosion may begin their lives on a metal surface as a scatter of individual cells. As the biofilm matures, the organisms are found as individuals or in colonies embedded in the matrix of a semicontinuous and highly heterogeneous biofilm. Microorganisms begin as individual bacteria on the surface of the metal and progress to thick semicontinuous films or colonies in the form of slime or polymer that can influence corrosion.

Depending on the flow velocity, the thickness may vary from 10 to 100 mm and may cover from less than 20% to more than 90% of the metal surface. Biofilms or macrofouling in seawater can cause redox reactions that initiate or accelerate corrosion. Biofilms accumulate manganese and iron in concentrations far above those in surrounding bulk water. Biofilms can also act as a diffusion barrier. Some bacteria can also take part in redox reaction of metal ions such as iron and manganese. These bacteria can shift the chemical equilibrium between Fe, Fe^{2+}, and Fe^{3+}, which influence the corrosion rate.

1.6.9 Production of Sulfides

This may involve the production of FeS, $Fe(OH)_2$, and so on and aggressive H_2S or acidity. Microorganisms may also interact with oxygen or nitrite inhibitor and thus consume chemical species that are important in corrosion reactions. Microorganisms may form a slime or poultice leading to the formation of a differential aeration cell attack or crevice corrosion. Microorganisms may also affect the desirable properties of lubricants or protective coatings (52).

1.6.10 Formation of Organic and Inorganic Acids

The sulphur oxidizing bacteria can produce up to nearly 10% sulfuric acid, which is highly corrosive to metals, coatings, ceramics, and concrete. Other bacteria can produce formic and succinic acids, which are also harmful especially to some organic coatings (52).

1.6.11 Gases from Organisms

Organisms of fermentative type metabolism produce CO_2 and H_2. Some bacteria can produce oxygen. Some bacteria can convert nitrates to nitrogen dioxide or ammonia. Some of these gases are corrosive (52).

1.6.12 MIC of Materials

Biodeterioration by bacteria and/or fungi of architectural building materials, stonework, fiber-reinforced composites, polymeric coatings, and concrete can occur (54). Biodeterioration results in staining, patina formation, pitting, etching, disaggregation, and exfoliation (4).

1.6.13 Wood and Polymers

Natural materials and materials from plant or animal origin such as wood, cotton, paper products, wool, and leather are fully biodegradable under aerobic conditions. Combined chemical, physical, and biological attack may cause polymers to be biodegradable by a combination of chemical, physical, and biological attack (51, 55).

1.6.14 Hydrocarbons

This term refers to crude oils, distillation, and cracking products (coal tar) and emulsions of these substances. All of these substances are microbially biodegradable. Hydrocarbons are conducive to microorganisms, and microbial growth causes damage to materials such as fuel tanks and pipelines (51).

Concrete can be damaged by acids, sulfates, ammonia, and other corrosives produced by microorganisms. Steel reinforcing bars in concrete corrode in the presence of microorganisms as a result of corrosive agents. Hydrogen sulfide generated by SRB can cause corrosion of the rebar in reinforced concrete structures. *Thiobacillus* bacteria are responsible for the deterioration of concrete. *Thiobacillus* converts sulfur and its compounds into sulfuric acid, which reacts with calcium hydroxide and calcium carbonate to form calcium sulfate. This is one of the modes of destruction of concrete sewage pipelines (51).

1.6.15 Types of Corrosion of Metals and Alloys

Microorganisms are more likely to cause localized corrosion than general corrosion because of the differential oxygen cell. In most cases the localized attack was observed beneath macrofouling layers. Corrosion of copper, steel, and aluminum anodes occur when connected to cathodes on which biofilms grow. Unexpectedly, rapid localized corrosion of steel bulkheads in marine harbor environments and of ship hull plating of several tankers has been observed (22).

Biofilms on passive alloy surfaces can increase cathodic kinetics by way of increasing the propagation rate of galvanic corrosion. Cathodic kinetics increased during biofilm formation on passive alloy surfaces. Crevice corrosion initiation times were reduced when natural biofilms were allowed to form on passive alloys S 30400 and S 31600.

Pitting corrosion of integral wing aluminum fuel tanks in aircraft, which use kerosene-based fuels, has been a problem for six decades. The fuel becomes contaminated with water by vapor condensation. The attack occurs under microbial deposits in the water phase and at the fuel–water interface. *Cladosporium resinae* is the organism involved, and it produces a variety of organic acids of pH 3–4 or lower and metabolizes fuel constituents. These organisms act in concert with slime-forming pseudomonas to produce oxygen concentration cells under the deposit. Active SRB have been detected under these deposits (56).

Hormoconis resinae poses a constant problem in fuel storage tanks and in aluminum integral fuel tanks of aircraft. Brown, slimy mats of *Hormoconis resinae* may cover large areas of aluminum alloy, causing pitting, exfoliation, and intergranular attack because of the organic acids produced by the microbes and the differential aeration cells. The problem of fungal growth in fuel tanks of jet aircraft has diminished as improved design to facilitate better drainage of condensed water and biocides such as organoboranes are gaining acceptance as fuel additives (41).

Organisms having a high tolerance for copper such as *Thiobacillus thiooxidans* can tolerate copper concentration as high as 2%. Localized corrosion of copper alloys by SRB in estuarine environments has been observed. Copper-nickel tubes from fan coolers in a nuclear power plant showed pitting corrosion under bacterial deposits. The slime-forming bacteria acting in concert with iron- and manganese-oxidizing bacteria were responsible for the deposits. Monel 400 tubing containing Ni, Cu Fe was pitted severely after exposure to marine and estuarine waters containing SRB (41).

It has been shown that welds provide unique environments for the colonization of SRB with subsequent production of sulfides that affect the weld seam surface of the HAZ. Exposure of sulfide-derived surfaces to fresh, aerated seawater resulted in rapid spalling on the downstream side of weld seams. The bared surfaces became anodic to the sulfide-coated weld root, initiating and accelerating localized corrosion (41).

1.6.16 Microbiological Impacts and Testing

Evidence of harmful impact: Microorganisms, including the corrosion-inducing microorganisms, are present in soils, freshwater, seawater, and air. In a majority of cases, these organisms influence corrosion. In a few cases, the organisms can reproduce the attack on introduction into a sterile system (57). Waters that may be untreated, fresh well waters, have been used in the hydrotesting of fabricated stainless steel structures. These waters may contain microorganisms such as *Gallionella*, which can cause corrosion.

1.6.17 Recognition of Microbiological Corrosion

This is done based on four types of evidence: (i) metallurgical; (ii) microbiological, (iii) chemical; and (iv) electrochemical evidence (41).

Metallography of the samples shows the types of MIC by the patterns of the corrosion products on the surface.

Microbiological evidence consists of gathering data on wet samples. It is also necessary to obtain photographs of the sample while the organism is live. This is followed by analysis of biological materials and corrosion products.

Chemical evidence consists of detailed chemical analysis of the corrosion products and any biological mounds present at or near the corrosion site. Evaluation of the chemistry of the liquid phase and its variability, both spatially and with time in relation to corrosive attack, is necessary. The important factors are color, texture, odor, and distribution of materials and organic and inorganic chemistries.

The color change of the corrosion product from black to brown indicates a sulfide corrosion product. Tests for sulfate, total organic carbon, pH, sulfide, and oxygen concentrations are good indicators of the potential for SRB growth (4, 41).

Electrochemical techniques such as variation of corrosion potential and corrosion rates as a function of time and EN may be useful in monitoring biological corrosion evolution (57).

An effective monitoring scheme for controlling both biofouling and biocorrosion should involve acquiring data on (i) sessile bacterial counts of the organism in the biofilm on the metal surface (57); (ii) direct observation of the community struc-ture of the biofilm; (iii) electrochemical corrosion measurements such as polarization resistance; (iv) water quality and redox potential measurements; (v) identification of microorganism in water and on metal; (vi) identification and analysis of corrosion products and biofilms; (vii) evaluation of morphology, form, and type of corrosion after removal of corrosion products.

Risk assessment of carbon steel pipelines, on the basis of the details of water chem-istry and operation parameters, has been documented.

Prevention of biocorrosion problems varies with the types of materials of con-struction, environment, economics, and duty cycle of equipment. The most common approaches to prevention of biocorrosion involve the use of sterilization, coatings, cathodic protection, and proper selection of materials. A general rule is to start with a clean system and keep it clean.

Sterilization of the system by gamma or UV irradiation for disinfection of mate-rials and environments may be useful to mitigate biocorrosion (51). Sterilization by chemical methods such as the use of biocides to control biofilm formation in closed systems such as heat exchangers, cooling towers, and storage tanks will mitigate bio-corrosion (51, 52).

1.7 MECHANICALLY ASSISTED CORROSION

Mechanically assisted degradation can consist of the following types of corrosion: erosion–corrosion, water drop impingement corrosion, cavitation erosion, erosive and corrosive wear, fretting corrosion, and corrosion fatigue (CF) (Fig. 1.14). Erosion–corrosion consists of the corrosion process enhanced by erosion or wear. Fretting corrosion consists of the wear process enhanced by corrosion. CF consists of the combined action of fluctuating or cyclic stress and a corrosive environment.

1.7.1 Corrosion and Wear

The progressive deterioration, because of corrosion and wear, of metallic surfaces leads to loss of plant efficiency, and in the worst case, to shutdown. For instance, both direct and indirect costs to the US economy have been estimated to be nearly $300 billion per year. The wear of materials has been estimated to cost $20 billion a year (in 1978 dollars) compared to $80 billion for corrosion during the same period (9, 58).

The combined action of wear or abrasion and corrosion results in more severe dam-age than with either mechanical or corrosive attack alone. The metal removed from the surface can be metal ions, or as particles of solid corrosion products or as elemen-tal metal. The erosion–corrosion process ranges from primarily erosive attack such as sandblasting, filing, or grinding of metal surface to essentially corrosion failures devoid of mechanical action.

Corrosion can occur in the absence of mechanical wear, but the opposite is rarely true. Corrosion often occurs in a wear process to a certain extent in all environments,

Mechanically Assisted Degradation

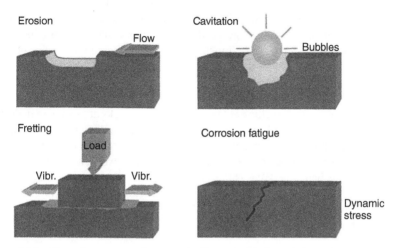

Figure 1.14 Corrosion types of mechanically assisted degradation. (Reproduced with permission of NACE International from Reference 3.)

except vacuum and inert atmospheres. Corrosion and wear can combine and cause damage in many industries, such as mining, mineral processing, chemical processing, pulp and paper production, and energy production. Wear debris and corrosion products formed during mineral processing operations can affect product quality and subsequent beneficiation by altering the chemical and electrochemical properties of the mineral (9).

Exposure to gaseous and humid environments affects mechanical properties, friction, and wear of polymers. Synergistic effects between abrasion, wear, and corrosion result in amplified damage (59, 60). The dominant and synergistic influence of some factors may be noted as follows.

1.7.2 Abrasion

This attack involves removal of protective oxidized metal and polarized coatings to expose unoxidized metal as well as removal of metal particles. This results in the increase of microscopic surface area exposed to corrosion. Strain-hardened surface layers are removed. Cracks are located on brittle metal constituents resulting in sites for impact hydraulic splitting. Plastic deformation by high-stress metal–mineral contact causes strain hardening and susceptibility to chemical attack.

1.7.3 Wear Impact

Plastic deformation makes some constituents more susceptible to corrosion. Wear can cause cracking of brittle constituents, tear apart ductile constituents to form sites for crevice corrosion and hydraulic splitting, supplies kinetic energy to drive abrasion mechanism, pressurizes mill water to cause splitting, cavitation, and jet erosion of metal and protective oxidized material, pressurizes mill water and gases to produce unknown temperatures, phase changes, and decomposition on reaction products from ore and water constituents, heats ball metal, and fluids to increase corrosive effects.

1.7.4 Corrosion Effects

Pits that can induce microcracking are produced. Microcracks at pits invite hydraulic splitting because of impact. Roughens surface and reduces energy required to abrade metal. May also produce hydrogen followed by absorption and cracking in steel. Grain boundaries may be selectively attacked followed by weakening adjacent steel. Wear damage mechanisms: Wear is the surface damage or removal of material from one or both of two solid surfaces in a sliding, rolling, or impact motion relative to one another. Wear damage precedes actual loss of material and it may also occur independently.

1.7.5 Wear Damage Mechanisms

Wear is the surface damage or removal of material from one or both of two solid surfaces in a sliding, rolling, or impact motion relative to one another.

Wear damage precedes actual loss of material and may also occur independently. Wear as in the context of friction is not an inherent material property. It depends on the operating conditions and surface conditions. Wear rate does not relate to friction. Wear occurs by mechanical and/or chemical means and is generally accelerated by frictional heating.

The principal wear mechanisms are: (i) adhesive; (ii) abrasive; (iii) fatigue; (iv) impact by erosion and percussion; (v) chemical; and (vi) electrical arc-induced.

There are other mechanisms such as fretting, fretting corrosion, and fretting CF, which is a combination of adhesive, corrosive, and abrasive forms of wear. Wear by all mechanisms except fatigue mechanism, occurs by gradual removal of material. One or more of these mechanisms may be operating in a particular machine. In many cases, wear may be initiated by one mechanism but may proceed by other mechanisms and thereby complicate failure analysis (60).

1.7.6 Adhesive Wear

Adhesive wear occurs because of adhesion at asperity contacts at the interface. These contacts are sheared by sliding, which may result in the detachment of a fragment from one surface to another surface. Some of the surfaces are fractured by a fatigue process during repeated loading and unloading resulting in the formation of loose particles. During sliding, surface asperities undergo plastic deformation and/or fracture. The subsurface up to several micrometers in thickness also undergoes plastic deformation and strain hardening with microhardness, by as much as a factor of two or higher than the bulk hardness.

1.7.7 Abrasive Wear

This type of wear occurs when asperities of a rough hard surface or hard particles slide on a softer surface and damage the interface by plastic deformation or fracture in the case of ductile and brittle materials, respectively. In most cases, the wear mechanism at the start is adhesive, which generates wear particles that get trapped at the surface, leading to three-body abrasive wear. In most of the abrasive wear situations, scratching is observed with a series of grooves parallel to the sliding direction (60).

1.7.8 Fatigue Wear

Subsurface and surface fatigue are observed during repeated rolling (negligible friction) and sliding with coefficient of ≥ 0.3, respectively. The repeated loading and unloading cycles to which the materials are exposed may induce the formation of subsurface and surface cracks, which, eventually after a critical number of cycles, lead to the breakup of the surface with the formation of large fragments leaving large pits on the surface. In this mode, negligible wear occurs, as wear does not require direct physical contact between two surfaces. Mating surfaces experience large stresses, transmitted through the lubricating film during the rolling motion, such as in the case of well-designed rolling element bearings. The failure time in wear fatigue is statistical in nature. Chemically induced crack growth such as in ceramics is referred to static fatigue. In the presence of tensile stresses and water vapor at the crack tip in ceramics, a chemically induced rupture of the crack tip bonds occurs rapidly, which increases the crack velocity. Chemically enhanced deformation and fracture result in increased wear of surface layers in static and dynamic conditions such as in rolling and sliding operations.

Prevention of wear fatigue corrosion involves the inherent physical properties of the alloy. For example, a gear must be tough and fatigue resistant and also have a wear-resistant surface. In the case of applications requiring a moderate degree of impact strength, fatigue resistance, and wear resistance, higher carbon through hardening steel may suffice. For more severe conditions, surface-hardened steel may be used (9).

1.7.9 Impact Wear

This wear consists of: (i) erosive wear and (ii) percussive wear. Erosion can occur by jets, liquid droplets, and implosion of bubbles formed in the fluid and streams of solid particles. Solid particle erosion occurs by the impingement when discrete solid particles strike the surface and the contact stress arises from the kinetic energy of the particles flowing in an air or liquid stream as it encounters the surface. Wear debris formed in erosion occurs as a result of repeated impacts (60). Neighboring particles may exert contact forces, and flowing fluid when present will cause drag. Under some conditions, gravitational force may be important.

The erosive wear mechanism, as in the case of abrasion, involves both plastic deformation and brittle fracture. The particle velocity and impact angle combined with the size of the abrasive particles gives a measure of the kinetic energy of the impinging particles, which is proportional to the square of the velocity. The ductile materials undergo wear by plastic deformation in which the material is removed by displacing or the cutting action of the eroded particle. Some brittle materials undergo wear predominantly either by flow or fracture, depending on the impact conditions. Wear rate dependence on the impact angle for ductile and brittle materials is different (60, 61).

Slug flow is the dominant flow regime in multiphase systems. Flow visualization has shown that bubbles distort and elongate in the vicinity of a pipe wall in a manner similar to collapsing bubbles. The corrosion rate increases because of a thinning of the mass transfer and corrosion product layers, as well as because of localized damage of the corrosion product film.

Multiphase environments are present in several industries. Changes in pressure and temperature in process equipment and the mixing of various streams can force the mixture into two or three phase environments. The corrosion because of the multiphase environment is of concern in nuclear and thermal power plants, chemical process industries, and in waste management systems. Slug flow occurs more readily in smaller diameter pipes, where the coalescence of bubbles can result in change of bubble flow into slug flow. Slug flow affects corrosion in two ways, namely: (i) a dramatic increase in turbulent intensity that can increase the mass transport of corrosive species by 1000 times; (ii) extensive damage of the corrosion product layer leading to extensive attack.

Percussion is a repetitive solid body impact, such as that encountered in print hammers in high-speed electromechanical applications and high asperities of the surfaces in a gas bearing. Repeated impacts lead to loss of solid material. Percussive wear occurs by hybrid wear mechanisms consisting of a combination of adhesive, abrasive surface fatigue, fracture, and tribochemical wear (60).

1.7.10 Chemical or Corrosive Wear

Corrosive wear occurs when sliding takes place in a corrosive environment. The most dominant corrosive agent in air is oxygen. Thus, the chemical wear in air is known as oxidative wear. In the absence of sliding, the oxides formed because of corrosion would slow down or even arrest the corrosion process. The sliding action can remove the oxide film and expose the bare metal to oxygen and formation of metal oxide. Thus, the chemical wear requires both sliding (rubbing) and oxygen. Corrosion occurs in a highly corrosive environment and in high-temperature and highly humid environments. The chemical wear requires a corrosive as well as a rubbing motion. Corrosive fluids provide a conductive medium for the electrochemical reaction to occur on sliding surfaces. In an aqueous medium, the corrosive agents are oxygen and carbon dioxide, which are responsible for corrosion (60).

Chemical wear is important in many industries, such as mining, mineral processing, chemical processing, and slurry handling. An example of a corroded roller subsequent to running in a bearing is shown in Figure 1.15. The corrosion caused a multitude of dark-bottomed pits, with surroundings polished by running. It is also

Figure 1.15   Quenched and tempered roller bearing after corrosive wear factor. (Reproduced by permission, John Wiley and Sons (60).)

found to create extensive, surface-originated spallings from a multitude of initiated points (60).

Friction modifies the kinetics of chemical reactions of sliding bodies with each other, and with gaseous or liquid environment, to the extent that reactions that occur at high temperatures occur at moderate, even ambient temperatures during sliding. Chemistry dealing with modification of chemical reactions by friction or mechanical energy is known as tribochemistry and the resulting wear as tribochemical wear (62). The mechanisms by which friction increases the rate of chemical reaction are frictional heat, removal of product scale to expose fresh surface, accelerated diffusion, and direct mechanochemical excitation of surface bonds. The tribochemical reactions result in oxidative wear of metals and tribochemical wear of ceramics (60).

1.7.11   Oxidative Wear

Interface temperatures produced at the asperity contacts during sliding of metallic pairs under nominally unlubricated conditions result in thermal oxidation, which produces oxide films several micrometers thick. This is beneficial as oxide film gives protection from further corrosion. The thick oxide film reduces the shear strength of the interface, which suppresses the wear as a result of plastic deformation. In many cases, tribological oxidation can reduce the wear rate of metallic pairs by as much as two orders of magnitude, compared with the same pair in an inert atmosphere. Tribological oxidation can also occur under conditions of boundary lubrication when the oil film thickness is less than the combined surface roughness of the interface. The oxidation can prevent severe wear. In oxidation wear, debris is generated from the oxide film (60).

At low ambient temperatures, oxidation occurs at asperity contacts because of frictional heating. At higher ambient temperatures, general oxidation of the entire surface occurs and affects wear. In the case of steel, the predominant oxide present in the debris depends on the sliding conditions. The predominant oxide at low speeds and ambient temperatures is α-Fe_2O_3, at intermediate conditions it is Fe_3O_4, and at high speeds and temperatures the oxide is FeO. Oxidation of iron and many metals follows a parabolic law with the oxide

$$h = Ct^{1/2}$$

film thickness increasing with the square root of time "t," h is the thickness, and C is the parabolic rate constant at elevated temperatures (60).

As diffusion is thermally activated, the growth rate in oxide film thickness during sliding as a function of temperature follows an Arrhenius equation

$$K = Ae^{(-Q/RT)}$$

where K is the parabolic rate constant for the growth of the oxide film, A is the Arrhenius constant (kg^2/m^4s), Q is the parabolic activation energy associated with the oxide (KJ/mol), R gas constant, and T the temperature. It has been observed that the Arrhenius constant for sliding is several orders of magnitude larger than that for static conditions. The oxidation rate during sliding may result from increased diffusion rates of ions through a growing oxide film, which has high defect density because of mechanical perturbations (60).

1.7.12 Electric-Arc-Induced Wear

When a high potential is present over a thin air film in a sliding process, a dielectric breakdown occurs, resulting in arcing. During arcing, a relatively high power density occurs over a very short period. The HAZ is usually shallow ($\sim$50 μm), and the heating results in melting and resolidification, corrosion, hardness changes, and other phase changes. Arcing might also cause ablation of material and craters. Any sliding or oscillation after an arc may either shear or fracture the lips, leading to three-body abrasion, corrosion, surface fatigue, and fretting (60, 63).

1.7.13 Erosion–Corrosion

All types of corrosive media such as aqueous solutions' organic media, gases, and liquid metals can cause erosion–corrosion. The corrodents can be a bulk fluid, a film, droplets, or an adsorbed substance. Hot gases may oxidize a metal at high velocity and abrade the protective scale. Slurries probably cause most damage in erosion–corrosion (9, 17). All types of equipment such as pipelines (curves, elbows, and T squares) floodgates, pumps, centrifugal fans, helixes, wheels of turbine, tubes of intersections of heat exchangers, and measuring devices are subject to this form of attack. Most metals and alloys are subject to erosion–corrosion, and the resulting failure can occur in a short time (17). The resistance of metals to erosion–corrosion damage depends on the physical and chemical properties of the products and/or the passive layers and their adhesion to the surface. Passivating metals such as stainless steels and titanium are relatively immune to erosion–corrosion in many oxidizing environments.

A stainless steel pump impeller with a projected life of 2 years failed in three weeks in a reducing solution. Soft metals such as copper and lead are readily damaged. Even noble metals such as silver, gold, and platinum are subject to erosion–corrosion (9, 17).

1.7.14　Impingement

This involves a liquid in turbulent flow, containing air bubbles and suspended particles that strike the metallic surface with a strong force and destroy the protective film on the surface. The shock of water bubbles or air against a metal surface results in wear of the surface. The wear takes the shape of the directional progress of the attack. The bottom part becomes anodic with respect to the adjacent outer surfaces. This mode of attack can occur even in the absence of air bubbles or suspended particles. Typical impingement is illustrated in Figure 1.16. Erosion–corrosion is characterized by the appearance of grooves, gullies, waves, rounded holes, and valleys and usually shows a directional pattern.

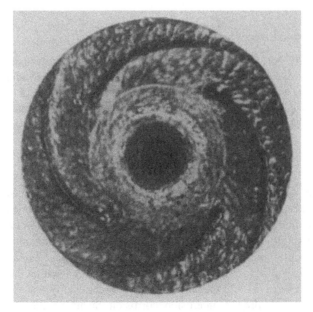

Figure 1.16　Erosion corrosion of a pump propeller. (Reproduced by permission, Elsevier Ltd., (2).)

Agitation or circulation of the electrolyte results in the increase in corrosion because of: (i) destruction or prevention of the formation of a protective film (50); (ii) increase in the rate of diffusion of aggressive ions because of the speed of the fluid thereby decreasing the cathodic polarization and increasing the current density as in the case of steels in the presence of oxygen, carbon dioxide, or bisulfate.

On the contrary, the speed of the fluid can decrease the corrosive attack by: (i) improving the inhibition efficiency through the quick transport of inhibitor to the metal–solution interface; (ii) by forming a passive layer for an active metal because of greater supply of oxygen to the surface; (iii) by sweeping away the corrosive agent (ionic species); (iv) by agitating and circulating at certain speeds of the medium and avoid pitting and crevice corrosion (17).

1.7.15 Effect of Turbulence

At very high flow rates of fluids, corrosive attack occurs by the combined action of erosion and corrosion. At high speeds, the abrasive action of the flow becomes severe. Thus the speed of flow strongly influences the mechanisms intervening in corrosion. The type of flow such as laminar or turbulent depends on the rate and quantity of fluid transported and on the geometry and design of equipment. Straight line flow is less damaging. Turbulent flow involves more intimate contact between the metal and the fluid. Other factors such as edges, cracks, deposits, abrupt changes of section, and other obstacles that disrupt laminar flow contribute to the turbulence and hence impingement attack (17).

1.7.16 Galvanic Effect

The galvanic cell between two different metals can have serious effects in a flowing system. In the case of 316 stainless steel and lead in 10% sulfuric acid under static conditions there was no galvanic cell. On increasing the flow rate to 11.89 m/s the rate of erosion–corrosion increased enormously because of the destruction of the passive film by the combined effect of galvanic corrosion and erosion–corrosion (17).

1.7.17 Water Droplet Impingement Erosion

In liquid impingement erosion, with small liquid drops stroking the surface of the solid at high speeds (as low as 300 m/s), very high pressures are experienced, exceeding the yield strength of most materials. Thus, plastic deformation or fracture can occur from a single impact, and repeated impacts lead to pitting and erosive wear (60). Water droplet impingement causes pitting and may cause cavitation damage. The damage may appear to be somewhat different from cavitation damage in ductile materials. The cavities in the surface show a directionality that is related to the angle of attack of the drop, as in erosion–corrosion. Steam turbines and water rotor blades are most susceptible to this form of attack. In turbines, condensation of steam produces droplets that are carried into rotor blades and causes surface damage. Rain drop erosion on helicopter blades can generate tensile stresses just below the surface and lead to cracking (52).

The following measures are useful in reducing or preventing erosion–corrosion: (i) a proper geometric design to get laminar flow with minimum turbulence, as in large diameter pipelines to avoid abrupt changes or streamline bends; (ii) the composition of the metal or alloy with inherent resistance to corrosion and the ease with which a protective film is formed, which is resistant to erosion–corrosion, plays an important role. Addition of 2% aluminum to brass, 1.2% iron to cupronickel results in a marked increase in resistance to impingement attack leading to corrosion. Condensation tubes made of brass-aluminum resist impingement because of the presence of iron in the protective film, which arises from corrosion products in water. It is also known that the addition of molybdenum to 18-8 stainless to form steel type 316 results in high resistance to erosion–corrosion; (iii) toughness can have an influence on the performance of materials subject to erosion–corrosion. Toughness is a good criterion for resistance to the mechanical erosion or abrasion, but this is not applicable to erosion–corrosion. Stellite (Co, Cr, W, Fe, C) alloy that has better toughness than 18-8 stainless steel showed better resistance to cavitation erosion on a water brake (2); (iv) There are several process procedures to harden alloys. Hardening can be done by the formation of a solid solution. Cold work can harden an alloy. For example, stainless steel resists cavitation erosion; (v) deaeration and addition of inhibitors are useful in reducing the aggressiveness of the environment. These methods are not economical. The suspended solid particles may be removed by filtration, and the temperature may be lowered; (vi) hard and tough coatings made of rubber and plastics may be used (9); (vii) cathodic protection may help reduce the electrochemical attack.

1.7.18 Cavitation

This form of corrosion is caused by repeated nucleation, growth, and violent collapse of vapor bubbles in a liquid against a metal surface. Cavitation erosion occurs when a solid and fluid are in relative motion, and bubbles formed in the fluid become unstable and implode against the surface of the solid. Cavitation erosion is similar to surface wear fatigue and the appearance of sample metal is similar to a pitted metal (9). Figure 1.17 shows the damage caused by cavitation erosion on a cylinder of a diesel motor. The cavitation is similar to pitting, except that surfaces in the pits are usually much rougher. On immediate observation, the surface affected region is free of deposits and accumulated corrosion products. Cavitation corrosion involves the destruction of the oxide layer formed by corrosion. When the mechanical effect damages the metal, it is known as cavitation erosion. This type of cavitation erosion damage is found in components such as ship propellers, runners of hydraulic turbines, centrifugal pumps, pump impellers, and on surfaces in contact with high-velocity liquids subject to variable pressure.

1.7.19 Cavitation Erosion

In certain conditions, a thin layer of liquid, nearly static at the metal–liquid interface, can prevent impingement of the surface by the turbulent flow of the liquid. However,

Figure 1.17   Cavitation erosion damage of a cylinder liner of a diesel engine. (Reproduced by permission, John Wiley and Sons (8).)

because of the water turbulence, bubbles of air or gas of a size larger than the thickness of the layer can interrupt, break the border layer, and cause a continuous rupture of the protective film. A film of semiconductor oxide on the surface such as Cu_2O, surrounding the damaged areas, produces a big cathode for reduction of dissolved oxygen on the surface and causes pitting.

1.7.20   Impacting Bubbles

When bubbles collapse that are in contact with or very close to a solid surface, they collapse asymmetrically. When a spherical bubble impacts a plane solid surface, the bubble becomes elongated with a tail and collapses. The jet from the bubbles can cause cavitation erosion on a solid wall (64). The implosion of a vapor bubble creates a microjet of liquid or microscopic "torpedo" of water that is ejected from the collapsing bubbles at velocities ranging from 100 m to 500 m/s. When the torpedo impacts on the metal surface, it dislodges the protective surface films and thereby deforms the metal locally.

The solid material absorbs the impact energy, leading to elastic or plastic deformation or even fracture. This may cause localized deformation and/or erosion of the solid surface. It has been suggested that the bubble grows and collapses, resulting in high pressure for a few milliseconds (65). The local pressure observed may be about 4000 atmospheres with a temperature increase up to 800 °C. These conditions of high pressure and temperature accelerate the corrosion rate. The cycle of exposure of fresh surfaces to corrosion, followed by reformation of protective films that leads to cavitation repeats itself (9, 60).

Single-phase flow has a definite pattern, whereas several flow patterns exist for multiphase flows. Multiphase flow may involve oil/water or oil/water/gas. Slug flow is the dominant flow regime in multiphase systems. It involves unique flow mechanisms with pulses of gas bubbles being released into a turbulent mixing zone because of a mixing vortex behind the slug front. These bubbles impact and collapse on the pipe wall, causing severe localized, cavitation-type corrosion, which can be up to 1000 times the values normally encountered in other flow regimes (64, 66).

1.7.21 Prevention

The following preventive measures may be used in addition to the measures listed under erosion–corrosion: (i) proper design should be used to minimize hydrodynamic pressure differences along with specifying a smooth finish on all critical metal surfaces. The pressure and temperature should be adjusted such that the formation of damaging steam bubbles is reduced. Proper operation of pumps and equipment is recommended. (ii) Cathodic protection may prove useful to avoid cavitation. Deaeration may be useful as dissolved air or gases cause nucleation of cavitating bubbles at low pressures; (iii) hard, tough metals, or elastomeric polymers may be useful in resisting cavitation erosion. Polymers and rubber have resilience, which is the capacity to dispose of the energy without absorption. The polymers and rubber are also resistant to abrasion (2).

1.7.22 Fretting Corrosion

This phenomenon is a combination of wear and corrosion in which the material is removed from the contacting surfaces when the motion of the surfaces consists of small amplitude oscillations with the relative movement ranging from fractional nanometers to fractional micrometers. Fretting occurs when low-amplitude oscillatory motion in the tangential direction takes place between two contacting surfaces, which are nominally at rest (67, 68). It is necessary that the load be sufficient to produce distortion of the surfaces. Fretting corrosion occurs in most machinery subject to vibration both in transit and in operation.

The most common factor in fretting is oxidation. In oxidizing systems, fine metal particles removed by adhesive wear are oxidized and trapped between the fretting surfaces. The oxides act like abrasives and increase the rate of material removal. The red material easily lost from between the contacting surfaces is an example of fretting corrosion in ferrous alloys. Fretting corrosion is encountered in shrink-fits, bolted parts, and splines. The contacts between hubs, shrink- and press-fits, and bearing housings on loaded rotating shafts or axles and many parts of vibrating machinery are prone to fretting corrosion. Fretting wear damage also occurs in flexible couplings and splines particularly where they form a connection between two shafts and are designed to accommodate some misalignment. Fretting corrosion is frequently observed between the crown of a ball bearing and its axle, or the head of a screw and the metallic surface, and in jewel bearings, elements of machines in movement, suspension springs, electric relay contacts, and kingpins of auto steering mechanisms (8, 69).

Fretting corrosion appears as discoloration and takes the form of local surface dislocations and deep pits. Fatigue cracks nucleate at these pits. The pits and cracks occur in regions where slight movements have occurred between mating and highly leaded surfaces (9). In the course of time, fretting corrosion can result in tarnished appearance of the metallic surface and variable piece sizes. Products of fretting corrosion may also cause blockages in machines in movement. Some examples (69) are: $NiO + Ni$; Cu_2O, CuO, Cu; Fe_2O_3, Fe, $Al_2O_3 + Al$.

1.7.23   Mechanism of Fretting Corrosion

Fretting is a form of adhesive or abrasive wear where the normal load causes adhesion between asperities, and oscillatory movement causes ruptures, resulting in wear debris. Fretting is generally associated with corrosion. In the case of steel particles, the freshly nascent surface oxidizes (corrodes) to Fe_2O_3 and the characteristic reddish-brown powder known as "cocoa" is produced. These oxide particles are abrasive. Because of the close fit of the surfaces and the oscillatory small amplitude motion (about a few tenths of micrometers) the surfaces are never out of contact and hence there is no opportunity for the products of action to escape. Oscillatory motion causes abrasive wear and oxidation. Thus the extent of wear per unit sliding distance because of fretting may be larger than that from adhesive and abrasive wear (60).

There are two approaches depending on the phenomenon that initiates and propagates the damage: wear-oxidation and oxidation wear. In wear-oxidation, two surfaces are in contact and the surfaces are imperfect. The surfaces are in contact through their asperities and the relative displacements of the two surfaces involve the wear of the crests (70). This phenomenon is similar to cold welding or fusion at the interface of metal surfaces under pressure. During displacements, the points of contact break and pieces of metal are produced. The small pieces oxidize following the heat generated because of friction. This process is repetitive and leads to accumulation of residue. Further oxidation of the damaged material has a secondary effect. Wear without debris occurs in the case of noble metals, mica, glass, and so on (17).

In oxidation wear, most of the metallic surfaces are initially protected from atmospheric oxidation by a thin adherent oxide film. When metals are in contact (under load) and subjected to repetitive weak movements, the oxide layer is broken at the level of asperities and it removes some of the oxide and the resulting metal is oxidized and the process repeats itself. Friction appears to be the driving force of oxidation wear (69).

A more plausible approach could be a combination of the two approaches cited under wear-oxidation and oxidation wear with relative importance of one or the other depending on the particular system and therefore a function of medium, surface finish and the nature of the materials in contact. It is useful to note that oxygen accelerates corrosion by fretting, in particular, in ferrous alloys (17).

Fretting is more severe in air than in an inert atmosphere (2). The damage in a humid atmosphere is less than in dry air as humidity has a lubricant action, and the hydrated oxides are less abrasive than dry oxides (2, 69, 71). Surfaces subjected to fretting wear have red–brown patches on ferrous metals and adjacent areas. There is no critical measurable amplitude below which fretting does not occur. When the deflection is elastic, fretting damage is not likely to occur. The wear rate increases with slip amplitude over a certain amplitude range. The fretting wear rate is directly proportional to the normal load for a given slip amplitude. The frequency of oscillation has little effect in total slip situation.

In a partial slip situation, the frequency of oscillation has little effect on the wear rate per unit distance in the low-frequency range, while the increase in strain rate at high frequencies leads to increased fatigue damage and increased corrosion because of an increase in temperature (60). The effect of temperature on fretting depends on the oxidation characteristics of the metals. An increase in temperature might result in the growth of a protective oxide layer that prevents metal–metal contact and hence a lower fretting rate.

Wear rate increases with slip amplitude over a range of amplitude. Wear debris can be plate-shaped, ribbon-shaped, spherical, and also irregularly shaped, on the basis of morphology.

Wear of material depends on the mating material, surface preparation, and operating conditions. Clean metals and alloys exhibit high adhesion and, as a result, high friction and wear. Any contamination prevents contact and any chemically produced films reduce friction and wear. In dry sliding, identical metals such as iron on iron exhibit high friction and wear and they must be avoided. Soft metals such as In, Pb, and Sn exhibit high friction and wear. Metals such as Co, Mg, Mo, and Cz exhibit low friction and wear. Lead-based white metals (babbits), brass, bronze, and gray cast iron exhibit low friction and wear and are used in dry and lubricated bearing and sea applications. In high-temperature applications, cobalt-based alloys that have good galling resistance are used (60).

1.7.24   Modeling Fretting Corrosion

An equation has been proposed for steel to evaluate loss of weight W caused by fretting corrosion on the basis of a model that combines the chemical and mechanical effect of corrosion by fretting. The chemical factor concerns the adsorption of oxygen resulting in oxidation of the metal to form the oxide, and the mechanical factor involves the loss of particles, at the asperities on the opposite surface

$$W = \left(K_0 L^{\frac{1}{2}} - K_1 L \right) \frac{C}{f} + K_2 \, ILC$$

Chemical factor   Mechanical factor

where L is the charge between surfaces, C is number of cycles, f is frequency of movement, I is slip, K_0, K_1, and K_2 are constants.

The chemical contribution decreases with increasing frequency of movement as there is little time for the chemical reaction. The mechanical factor is a function of the slip and the load. In the presence of nitrogen, the wear is a function of mechanical factor and is independent of the frequency (8).

1.7.25   Fretting CF

Fretting CF failures are encountered in aircraft engine parts, such as connecting rods, knuckle pins, splined shafts, clamped and bolted flanges, couplings, and other parts. Failures because of fretting CF also occur in railway axle shafts at the wheel seats and in automobile axle shafts, suspension springs, steering knuckles, and others (2, 69, 72). The oscillatory movement is usually the result of external vibration, but in many cases it is the consequence of one of the members of contact being subjected to cyclic stress (fatigue), which results in initiation of fatigue cracks leading to fretting fatigue. Fretting can also cause rupture of adhesive ties because of the strengths of oscillations that can generate fine cracks that can propagate to a major fracture of the sample.

It has been found that in some instances the quantity of metal lost by fretting is directly related to the reduction of the resistance to fatigue. The frequent oscillations cause formation of pits that initiate cracks of fatigue leading to increased suscepti-bility to fatigue fractures. Fretting corrosion causes loss of material in highly loaded bolted equipment. Bolts and studs are loosened and become more prone to fatigue failure. This is a serious problem when the bolt or stud is very short as in aircraft engine cylinder hold-down studs (72). The initiation of the fretting fatigue crack is located at the boundary of the fretting scar at the fretted zone. The fatigue crack then propagates into the surface at an angle to the surface.

1.7.26   Prevention of Fretting Wear

Various design changes minimize fretting wear. The machinery should be designed to reduce oscillatory movements, reduce stresses, and eliminate two-piece design. Some steps involved in the prevention of fretting wear are: (i) lubrication of fraying surfaces with low-viscosity, high-tenacity oils and grease to exclude direct contact with air (17). Phosphate conversion coatings in conjunction with lubricants, which are porous and provide oil reservoirs may be used (17); (ii) use of gaskets to absorb vibration and limit oxygen at bearing surfaces; (iii) restricting the degree of movement such as shot peening, which induces residual surface compression stress to retard corrosion and roughening the surface to increase friction (72); (iv) increasing the hardness of one or both of the contacting surfaces, surface hardening such as carburizing and nitriding or applying protective coatings by electrodeposition, plasma spraying, vapor deposition, or anodizing of aluminum alloys. Hard materials are more resistant than soft materials. A soft surface can yield by shearing instead of sliding (8) at the interface; (v) increasing the load to reduce the slip between mating surfaces (9, 17); (vi) using dissimilar metals that give wear resistance when coupled. Steels may be coupled with silicon bronze and stellite alloys. Surface treatment or adding a coating is useful in preventing wear (60).

1.7.27 Testing

The measurement of corrosion, wear, corrosion-wear interactions, and erosion–corrosion interactions involves a multistep process. Each aspect of the interaction must be measured separately. The resulting data must be combined to assess the synergistic effects and arrive at a total picture of the damage process. Measurement of the combined action of wear and corrosion is not easy. The ASTM, G119 (30) standard test applies to liquid systems or slurries and can be adapted to dry corrosion and wear combination as well (22).

1.7.28 Measurement of Wear and Corrosion

Jet and whirling arm tests are used to assess erosion (22). In the whirling arm test, the impact velocity is known, and the entire face of the sample is eroded resulting in a more uniform surface. The machining test is commonly used in high-temperature abrasive tests, as the machining process results in high temperatures. The high-temperature ring-on-disk test is commonly used for abrasive elevated-temperature tests (22).

1.7.29 Galling Stress

Wear galling is a good measure of wear resistance of a given material pair. Galling results in a groove or a score mark, leading to a mound of metal. Galling data show that identical metals do poorly than dissimilar metal couples with respect to galling wear resistance. Stainless steels when coupled together exhibit poor galling resistance compared to other steels by a factor of 2 or more (60).

1.7.30 CF

CF involves cracking in materials under the combined action of a fluctuating or cyclic stress and a corrosive medium. The damage because of CF is generally greater than the sum of the damage by corrosion and fatigue acting separately.

CF is involved in the case of the shaft of a ship propeller, slightly above the waterline, which can normally function until a leak occurs, resulting in the water impinging on the shaft in the area of maximum alternating stress. Failure of the shaft in a few days is an example of CF failure. Steel pumping tubes used in oil wells have a limited life span because of the CF experienced in oil well brines. Pipes carrying steam or hot fluids of variable temperature may fail because of thermal cycling (8).

1.7.31   Morphology of CF Ruptures

CF results in fine-to-broad cracks with little or no branching, unlike SCC that exhibits branching cracks. The cracks appear as families or parallel cracks and are filled with dense corrosion product. The sample may have pits, grooves, or some other forms of stress concentrator. Transgranular fracture paths, often ramified or branched are more common than intergranular fracture with the exception of lead and zinc (Fig. 1.18). Some systems show a combination of transgranular and intergranular forms of fracture (8, 9).

The surface of fatigue ruptured material on examination shows two zones; namely, (i) a smooth, silken zone not having undergone any plastic distortion and the second zone resulting from the propagation of fatigue crack through the metal. The cyclic constraint tends to smooth the surface of the rupture by wearing. A final rupture zone, rough and damaged because of plastic deformation, is noted. The crack propagates

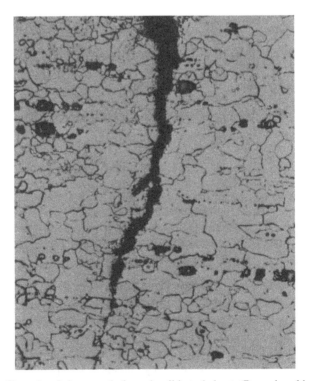

Figure 1.18   Corrosion fatigue crack through mild steel sheet. (Reproduced by permission, John Wiley and Sons (8).)

until the area of the transverse section of the metal is reduced sufficiently, such that the ultimate stress limit is reached and then a fragile and brutal rupture occurs.

1.7.32 Important Factors of CF

CF is not specific as most materials suffer degradation because of the fatigue properties in aqueous media. For example, steel undergoes CF in fresh waters, seawater, products of combustion condensates, and other chemical environments. In the case of materials, the fatigue strength (or fatigue life at a given maximum stress value) generally decreases in an aggressive environment. The effect varies widely depending on the particular metal-environment combination. The environment may affect the probability of fatigue crack initiation, fatigue crack growth rate, or both (80).

CF depends strongly on the combined interactions of the mechanical (loading), metallurgical, and environmental variables (9).

1.7.33 Stresses

The main mechanical properties of importance are (9): maximum stress or stress intensify factor σ_{max} or κ_{max}, cyclic stress or stress intensity range, $\Delta\sigma$ or $\Delta\kappa$, stress ratio R, cyclic loading frequency, cyclic load waveform (constant-amplitude loading), load interactions in variable-amplitude loading, state of stress, residual stress, crack size and shape, and their relation to component size geometry (9).

The greater the applied stress at each cycle, the shorter the time to failure. Mechanical damage is more likely when load and frequency are high and corrosion damage is likely at intermediate load and frequency. In these conditions, cracking may be transgranular or intergranular, and the morphology may be similar to SCC morphology. Cyclic stress has negligible influence on the resistance to fatigue while it has a distinct influence on CF. The corrosion influence in fatigue corrosion is influenced significantly by the frequency of the cyclic stress. The corrosion influence in fatigue corrosion is more pronounced at lower frequencies as the contact between the metal and corrosive agent is of longer duration.

Ultra-high-strength steels are very sensitive to the environment, such as distilled water, and are characterized by high growth rates, which depend on the stress intensity range $\Delta\kappa$ to a reduced power. Time-dependent CF crack growth occurs mainly above the threshold stress intensity for static load cracking and modeled by linear superposition of SCC and inert environment fatigue rates. The CF behavior may be described by taking into account the level of mechanical loading, the frequency, and the shape of the cycles. It is preferable to express CF as a function of crack growth rate rather than frequently used cycle-dependent crack growth rate (73).

Cyclic load frequency is the most important factor that influences CF in most material environment and stress intensity conditions. The dominance of frequency is related directly to the time dependence mass transport and chemical reaction steps involved for brittle cracking.

1.7.34 Stress Ratio

Rates of CF crack propagation generally are enhanced by increased stress ratio R, which is the ratio of minimum stress to maximum stress.

Some other factors, namely, the metallurgical condition of the material such as composition and heat treatment, and the loading mode, such as uniaxial, affect fatigue crack propagation (4).

Environmental factors to consider are: type of environments (gaseous, liquid, liquid metal) temperature, partial pressures of corrosive species in gaseous environments, concentration of corrosive species in aqueous or liquid media, corrosion potential, pH, conductivity, halogen or sulfide ion content, viscosity of the medium, oxygen content, solution composition, inhibitors, coatings (4), and so on. Corrosion products on or within fracture surfaces are identified. CF cracking of high-strength steel exposed to water vapor leading to hydrogen damage may be difficult to distinguish from other forms of hydrogen damage. At high frequencies, the fracture surface features because of CF crack initiation and propagation do not differ significantly from those produced by fatigue in nonaggressive environments (24).

The usual test of fatigue of metals in air is affected by oxygen or humidity and represents a measure of CF. Some tests showed the CF endurance limit for copper is 14% higher in vacuum than in air. Oxygen has very little influence on initiation but considerable effect on crack propagation (8). Brackish waters have a greater effect on CF of steel than that of copper. Controlled changes in the potential of a sample can lead to either the complete elimination or the dramatic increase in brittle fatigue cracking (4).

1.7.35 Material Factors

The most important metallurgical properties are: alloy composition, distribution of alloying elements and impurities, microstructure and crystal structure, heat treatment, mechanical working, preferred orientation of grain and grain boundaries (texture), mechanical properties such as strength, fracture toughness, and so on. (9).

A fatigue test involves subjecting a metal sample to alternate cyclic stresses, compression–tension of different values, and measuring the time (number of cycles, N) before rupture of the sample. A short characteristic of the fatigue test is known as the C–N curve, giving the number of cycles N to rupture. The value of maximum stress for which an infinite number of cycles can be supported without rupture is known as endurance limit or fatigue limit. The fatigue limit exists for steels, but not necessarily for other metals and generally equals half the tensile strength. Nonferrous metals such as Al, Mg, Cu alloys do not have a fatigue limit and are assigned fatigue resistance to an arbitrary number of cycles, such as 10^8 cycles (69).

Crack growth rates are influenced by metallurgical variables such as compositional impurities, microstructure, and the cyclic deformation mode. In carbon steels, cracks often originate at hemispherical corrosion pits and often contain significant amounts of corrosion products. The cracks are often transgranular with possible branching. Surface pitting and a transgranular fracture path are not prerequisites for CF cracking of carbon steels. CF cracks can occur in the absence of pits and follow boundaries or prior-austenite grain boundaries (50).

Aluminum alloys exposed to chloride solutions result in CF cracks originating at pitting sites or sites of IGC. Initial crack propagation is normal to the axis of principal stress. This is contrary to the behavior of fatigue cracks initiated in dry air, where initial growth follows crystallographic planes. Initial CF cracking normal to the axis of principal stress also occurs in aluminum alloys exposed to humid air, but pitting is not a prerequisite for crack initiation (50).

CF cracks in copper and its alloys initiate and propagate intergranularly. Cu–Zn and Cu–Al alloys show marked reduction in fatigue resistance in aqueous chloride solutions. This type of behavior is difficult to distinguish from SCC except that it may occur in environments that do not cause failures under static stress and in sodium chloride or sodium sulfate solutions (50).

The CF of low alloy steels in high-temperature waters is an example where crack tip chemistry has been identified as responsible for environmentally assisted cracking. In some cases, hot water may increase the fatigue crack propagation rate in low alloy steels. This effect is observed only when the sulphur content of the wrought steel exceeds 100–150 ppm (74). The cause of environmentally assisted cracking (EAC) has been attributed to the accumulation of sulfide ions in the crack tip environment, originating from the dissolution of MnS inclusions bared by the crack advance (75, 76). In aerated conditions the potential gradient inside the crack tends to inhibit outward diffusion of sulfide, which leads to environmentally assisted corrosion over larger loading conditions, particularly at lower frequencies (77, 78).

1.7.36 Mechanism of CF

The mechanism of fatigue in air proceeds by localized slip within grains of the metal caused by stress. The air adsorbed on the fresh surface exposed at slip steps prevents rewelding on the reverse stress cycle. Continued slip produces displaced clusters of slip bands that protrude above the metal surface and corresponding cracks (intrusions) from elsewhere. The corrosion process may remove barriers to plastic deformation such as dislocations, induce plastic deformation by reducing surface energy, and favor slip formation by injecting dislocations along slip planes (Fig. 1.19) (8). After or during initiation of microcracks, the propagation follows in part because of the adsorption of oxygen, water, or other ionic species along the partitions of the crack. The adsorption of oxygen or ionic species reduces the energy of the surface and prevents the welding of the metallic surface during the inverse constraint cycle. The formation of differential aeration cells because of different concentrations of oxygen in the localized sites can play a role in the dissolution of the metal at the bottom of the crack (anode) and hence contribute to the propagation of the crack. A corrosive medium eliminates the fatigue limit or shortens the life above the fatigue limit.

According to Wang (79) the four stages of fatigue are:

1. Precrack cyclic deformation that includes the formation of persistent slip bands (PSBs), extrusions, and intrusions.

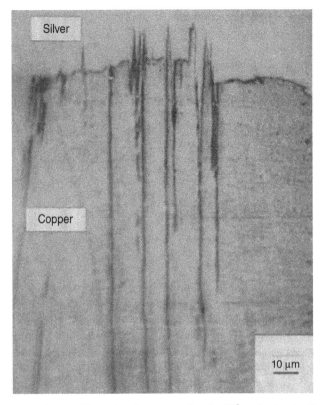

Figure 1.19 Extrusions and intrusions in copper after 6×10^5 cycles in air. (Reproduced by permission, John Wiley and Sons (8).)

2. Crack initiation and stage one growth that deepens the intrusions within the plane of high shear stress.
3. Stage two crack propagation of well-defined cracks on the planes of high tensile stress in the direction normal to the maximum tensile stress.
4. Ductile fracture propagation.

1.7.37   Crack Initiation

CF cracks are always initiated at the surface, unless there are near-surface defects that act as stress concentration sites and facilitate subsurface crack initiation. Crack initiation takes place independently of fatigue limit in air as it can be decreased or eliminated through the increase of dissolution rates at anodic sites. Localized corrosion such as pitting favors fatigue crack initiation through stress concentration and a local acidic environment. The two main mechanisms of CF are anodic slip dissolution and HE (80).

Cracks grow by slip dissolution because of diffusion of active water molecules, halide ions to the crack tip, followed by rupture of the protective oxide film because of strain concentration and fretting contact between the crack faces. This is followed by dissolution of the fresh surface and growth of oxide on the bare surface. In the alternate mechanism of HE, the steps involved are (i) diffusion of water molecules or hydrogen ions to the crack tip; surface diffusion of adsorbed atoms to preferential surface locations, absorption and diffusion to critical locations in the microstructure such as grain boundaries, regions of high triaxiality ahead of crack tip or void. Under cyclic loading, fretting contact between the mating crack faces, pumping of aqueous medium to the crack tip by crack walls, and continuous blunting and resharpening of the crack tip by reverse loading influence the rate of dissolution (79).

Fatigue crack initiation in commercial alloys occurs on the surface or subsurface and is usually associated with surface defects or discontinuities such as nonmetallic inclusions, notches, and pits. For low-stress, high-cycle fatigue, crack initiation spans a large part of the total lifetime. For high-strength steels in sodium chloride, the sulfide inclusions served as sites for corrosion pits and subsequent fatigue crack initiation. Corrosion was formed by selective dissolution of MnS inclusions. Cathodic polarization suppresses the dissolution rate and prevents formation, but hydrogen effects can increase the crack growth rates of well-defined cracks (79).

The materials and corrosive environments have been classified (81) into the following three groups on the basis of surface corrosion conditions:

1. Active dissolution conditions.
2. Electrochemically passive conditions.
3. Bulk surface films, such as three-dimensional oxides.

In the active dissolution group, emerging PSBs are preferentially dissolved. This dissolution attack results in mechanical instability of the free surface and generation of new and larger PSBs, followed by localized corrosive attack, resulting in crack initiation. Under passive conditions, the relative rates of periodic rupture and reformation of the passive film control the extent to which corrosion reduces the fatigue resistance. When bulk oxide films are present on the surface of the metal sample, rupture of the film by PSBs leads to preferential dissolution of the fresh metal that is produced (79).

1.7.38 Crack Propagation

The environment can affect crack propagation in CF, leading to an increase in crack growth rate. Three types of crack growth behavior have been documented (73, 82).

Figure 16.4 on page 199 of *Uhlig's Corrosion Handbook* (79) shows the sigmoidal variation of the fatigue crack growth as a function of stress intensity factor range on a log–log scale under purely mechanical loading conditions. The typical variation of the crack velocity, da/dt as a function of applied stress factor K is shown on a log–log scale in the figure for growth of cracks in metallic materials under sustained loading in the presence of an environment. It is clear from this figure that the environment has no effect on fracture behavior below a static intensity factor K_{ISCC} the threshold intensity stress factor for the growth of stress corrosion cracks in tensile operating mode. Above K_{ISCC}, the crack velocity increases with increasing stress intensity factor K (region I). In region II, da/dt is independent of K. In region III, there is a steep increase in crack velocity as the maximum intensity factor approaches the fracture toughness of the material (K_{IC} region III).

The figure illustrates type A true CF growth pattern in which the synergistic interaction between cyclic plastic deformation and environment produces cycle- and time-dependent crack growth rates. True CF influences cyclic fracture, even when the maximum stress intensity factor K_{max} in fatigue is less than K_{ISCC}.

The cyclic load form is important. The crack growth rate increases, and the fatigue threshold decreases as compared with that in air. The crack growth rate obeys a Paris Law with increase in crack growth rate and a decreased fatigue threshold compared with the behavior in air (73).

The figure shows the stress CF process, type B, purely time-dependent CF propagation, which is essentially a simple superposition of mechanical fatigue and SCC. Stress CF occurs only when $K_{max} > K_{ISCC}$. In this case, the cyclic character of loading is not important. The combination of true CF and stress CF results in type C, the most general form of CF crack propagation behavior (79).

This behavior is generally characterized by a plateau region, which prevails over a definite threshold K_{th}. It is usually referred to as stress CF as SCC systems usually exhibit this behavior, and the most common theory assumes that the crack growth rate is because of the addition of SCC and pure fatigue crack advance. This type of synergistic effect is observed in systems not sensitive to SCC such as ferritic stainless in seawater under cathode polarization. It is often associated with HE. It is possible that the plateau behavior is because of the control of crack growth rate by nonmechanical processes such as transport processes (73).

The figure shows the cyclic time-dependent acceleration in da/dN below K_{ISCC} combined with time-dependent cracking (SCC) above the threshold (79). This is a combination of environmental effects.

Examples of CF types and a combination of two types (A and C) have been observed in similar steels under different potential. Type C behavior is clearly associated with cathodic potential and thus with HE (73). Thus this is a mix of true fatigue and stress CF where one can dominate the other in its influence on crack growth, depending on the properties of the interface (83).

1.7.39   Prevention of CF

One or a combination of the following procedures is recommended to prevent CF:

1. Redesigning to reduce or eliminate both temporary and permanent cyclic stresses. It is recommended to reduce the magnitude of stress fluctuation (79).
2. Selecting a material or heat treatment with higher CF strengths.
3. Use of inhibitors, reduction of oxidizers, or pH increase, depending on the system and the environment, can delay the initiation of CF cracks. For example,

deaeration of saline solution with mild steel brings back the CF limit to the value in air (84). Addition of 200 ppm of sodium dichromate to the city water reduces the CF of normalized 0.35% carbon steel to a lower level than the value in air.

4. Cathodic protection, provided the material is not prone to embrittlement. Sacrificial zinc (galvanized) coatings may be used.

5. Surface treatments such as hot peening, nitriding of steels, sandblasting of the surface of the metal and other treatments that produce constraints of compression are beneficial.

6. Organic coatings, which can impede CF. The coatings must contain inhibitory pigments in the primary layer. Local defects in the coatings reduce the CF strength of carbon steel (9).

7. Noble metal coatings, which are useful as long as they remain unbroken and are of high density and thickness. Electrolytic deposits of tin, lead, copper, silver, or steel are useful as protectors of CF (60).

1.8   ENVIRONMENTALLY INDUCED CRACKING (EIC)

In some environments and under certain conditions, a microscopically brittle fracture of materials can occur at levels of mechanical stress that may be far below the level required for general yielding or those that cause significant damage in the absence of an environment. The susceptibility also depends on the chemical composition and microstructure of the alloy. This form of corrosion requires an interaction between the electrochemical dissolution of the metal, hydrogen absorption, and the mechanical loading conditions (stress, strain, and stain rate) (73). The nature of these fracture modes varies from one class of material to another. However, all fracture modes are largely similar to one another.

1.8.1   Testing of CF

Some of the factors that must be considered in CF testing are the following:

1. Stress intensity range, load frequency, and stress ratio.
2. Electrode potential in aqueous media and intended environment composition.
3. Metallurgical factors such as alloy composition microstructure and yield strength (4).

The scientific basis for the reliable estimate of fatigue life for intended conditions and situations remains elusive in spite of expenditures of the order of billions of dollars in combating fatigue. The fatigue test consists of subjecting a sample to a certain frequency of alternate cyclic stresses (compression-tension) of different values and plotting a $C–N$ curve. The presence of rust or other corrosion products do not indicate a decrease in fatigue. It is necessary to carry out fatigue tests and determine the decrease in fatigue resistance (17).

1.8.2   Types of Tests

(i) cycles-to-failure (complete fracture) or (ii) crack propagation (crack growth) test. In the cycles-to-failure test, samples are subjected to a number of stress cycles to initiate and propagate cracks until complete fracture occurs. This test involves using smooth or notched samples. This test cannot distinguish between corrosion fatigue crack (CFC) initiation and CFC propagation.

1.8.3   Sampling in CF Tests

A sample has generally three regions; namely, a test section and two grip ends. The design and type of sample depends on the fatigue testing machine and the objective of the study. The test section of the sample is of reduced cross section to prevent failure at the grip ends. Rounded samples used in axial fatigue may be threaded, buttonhead or constant-diameter types for clamping in V-wedge pressure grips. In rotating-beam machines, short, tapered grip ends with internal threads are used, and the sample pulled into the grip by a draw bar. Torsional fatigue samples are generally cylindrical. Flat samples for axial or bending fatigue tests may be generally reduced in width and thickness in the rest of the section (4).

The fracture mechanics approach in CF provides the basics for many fatigue crack growth studies. The relationship is

$$\Delta\kappa = \sqrt[\alpha]{\pi\alpha(\sigma_{max} - \sigma_{min})}$$

where $\Delta K = \kappa_{max} - K_{min}$ is the stress intensity range, and κ is the magnitude of the mathematically ideal crack tip stress field in a homogeneous linear-elastic body and is a function of applied load and crack geometry, σ is the stress, and α is a function of the geometry of the rupture and test sample.

The growth or extension of a fatigue crack under cyclic loading is mainly controlled by the maximum load and minimum/maximum stress ratio. However, as in crack initiation, there are other factors that have an effect, especially in the presence of an aggressive environment.

The established standard test method for CF crack growth rate is ASTM E647 (30). In this constant load test method, the crack length is measured visually as a function of elapsed cycles followed by numerical analysis of the data to obtain the crack growth rate. Crack growth rates are then expressed as a function of crack tip intensity range $\Delta\kappa$, which is based on linear-elastic stress analysis.

The crack growth rate da/dN where "a" is the e crack length and "N" the number of cycles as a function of $\Delta\kappa$ gives results that are independent of sample geometry, and this permits the comparison of data obtained from a variety of sample configurations and loading conditions (4).

Various studies on crack growth rates for many metallic structural materials have shown that da/dN versus $\Delta\kappa$ plots have three distinct regions of behavior. In an inert environment, the crack growth rate depends strongly on κ at κ levels approaching K_{1c} (plane-strain fracture toughness) at the high end (region III) and at levels approaching an apparent threshold $\Delta\kappa_{th}$ at the lower end (region I) with an intermediate region II that depends on some power of κ or $\Delta\kappa$ of the order of 2–10. The following power-law relationship

$$\frac{da}{dN} = c(\Delta K)^n$$

where c and N are constants for a particular material and stress ratio have been advanced. In an aggressive environment, the CFC curve can be different from the pure fatigue curve, depending on the sensitivity of the material to the given environment and the occurrence of various static stress fracture pathways. The environmental effects are quite strong above the threshold for sec (K_{ISCC}) and negligible below this level (K_{ISCC} is the stress intensity threshold for plane-strain environment assisted cracking). Other loading factors, such as frequency stress ratio and stress waveform can markedly affect crack growth rates in aggressive media.

CF tests may be carried out in an apparatus designed by the Continental Oil Company. The apparatus consists of a Monel tank with four samples subjected to cyclic bending. The first step consists of determining the displacement caused by the applied load. The exact stresses are determined by strain gauges. The electrolyte is deaerated with 3% sodium chloride. Polished or sand blasted samples are used, and the behavior of the alloy in CF may be studied at the free corrosion potential under different percentages of stress amplitude of the elastic limit. From potentiokinetic curves, $I = f(E)$: the protection or pitting potential applied during the stress test. Each test can have four samples and the difference between results (85) for similar tests does not exceed 15%.

Environmentally assisted cracking (EAC) occurs not only in metals but also in glasses (Plexiglas), ceramics, and polymers. Structural failures because of EAC can be sudden and unpredictable, occurring after a few hours of exposure or after months or years of satisfactory service (86, 87). The two types of environmentally assisted cracking (EAC) are mechanically assisted cracking and EIC.

The total cost of material fracture is about 4% of gross domestic product in the United States and Europe (88, 89). Fracture modes included in the cost estimates were stress-induced failures (tension, compression, flexure, and shear), overload, deformation, and time-dependent modes, such as fatigue, creep, SCC, and embrittlement. The environmentally assisted corrosion problem is very much involved in the maintenance of the safety and reliability of potentially dangerous engineering systems, such as nuclear power plants, fossil fuel power plants, oil and gas pipelines, oil production platforms, aircraft and aerospace technologies, chemical plants, and so on. Losses because of environmentally assisted cracking (EAC) of materials amount to many billions of dollars annually and is on the increase globally (87).

1.8.4 SCC

The necessary conditions for the occurrence of SCC are: (i) a crack-promoting environment; (ii) the susceptibility of the material to SCC; (iii) exceeding threshold value with regard to tensile stresses. One of the distinguishing features of SCC is that stress corrosion faces suffer very low corrosion even in solutions that damage free surfaces.

An example is the SCC of stainless steel at 200 °C in a caustic solution or in aerated chloride solution where no traces of dissolution are visible on the crack face. The three conditions, namely, tensile stress, susceptible sample material, and a corrosive environment are the conditions necessary for stress corrosion to take place (73, 90). For instance, SCC of metals has been by far the most prevalent cause of failure of steam generator components in pressurized water reactors (PWRs) to an extent of 69% of all cases, piping in boiling water reactors (59.7%) and PWRs (23.7%). More than 60% of inspected steam turbines in nuclear power plants have disks with stress corrosion cracks (91).

The two classic examples of SCC are the seasonal cracking of brass and the caustic embrittlement of steel. Seasonal cracking refers to SCC of brass cartridge cases. Cracks were observed during the period of heavy rainfall along with hot weather in the tropics. This intergranular SCC was attributed to the internal stresses in ammonia solution that resulted from the decomposition of organic matter in the presence of oxygen and humidity. Many explosions of riveted boilers occurred in the early steam-driven boilers at the tubes of riveted furnaces because of the fact that some areas were subjected to cold working during riveting operations. Carbon steel subjected to a stress close to the elastic limit and exposed to hot concentrated alkali solutions or nitrate solutions are susceptible to SCC. SCC was observed in rivets used in water boilers although the furnace water was treated with alkalis to minimize corrosion. Crevices between rivets and the boiler plate of the furnace allowed boiler water to concentrate until the alkali content was sufficient to reach the pH required to cause cracking (17).

1.8.5 Morphology

Failed samples appear macroscopically brittle and exhibit highly branched cracks that propagate transgranularly and/or intergranularly, depending on the metal/environment combination. Transgranular stress corrosion crack propagation is often discontinuous on the microscopic scale and occurs by periodic jumps of the order of a micrometer. Intergranular cracks propagate continuously or discontinuously, depending on the system (17, 73).

Intergranular and transgranular cracking often occur simultaneously in the same alloy. Such transitions in crack modes are observed in alloys with large amounts of nickel, iron chromium, and brasses. In corrosion under tension, ruptures are fragile and are sometimes characterized by the presence of cleavages, in particular, in the case of HE (17).

Cleavage is a brittle fracture that occurs along specific crystallographic planes. Cleavage has a well-defined crystallographic orientation and it is easy to recognize its occurrence by optical microscopy as it exhibits brilliant and flat fracture facets that are related to the dimension of the grain size of the material of interest. Examination with a scanning electron microscope shows flat fracture facets showing cleavage steps and river patterns caused by the crack moving through the crystal along a number of parallel planes that form a series of plateaus and connecting ledges.

1.8.6   Some Key Factors of SCC

The stress applied on a metal is nominally static or slowly increasing tensile stress. The stresses can be applied externally, but residual stresses often cause SCC failures. Internal stresses in a metal are because of cold work or heat treatment. In general all manufacturing processes create some internal stresses. Stresses caused by cold work arise from processes such as lamination, bending, machining, rectification, drawing, drift, and riveting. Stresses caused by thermal treatments are because of the dilation and contraction of the metal or indirectly by the modification of the microstructure of the metal or alloy. Welded steels have residual stresses near the yield point. Corrosion products can also function as a source of stress and can cause a wedging action.

The manner in which the logarithm of crack growth rate varies as a function of the crack tip stress intensity magnitude factor is normal. No cracking is observed below some threshold stress intensity. The stress corrosion cracks initiate when the stress exceeds a threshold value σ and propagate when the stress intensity factor exceeds the threshold value K_{ISCC} (17). The threshold stress intensity level, K_{ISCC}, is determined by the alloy composition, microstructure, and the environment (composition and temperature). The threshold stress intensity K_{ISCC} lies between 10 and 25 MPa/m^2 and σ_1 is usually in the order of 60–100% of the yield stress, but much lower values can be observed, such as for 304 stainless steel in boiling magnesium chloride at 154 °C. In some cases, stresses as low as 10% of the elastic limit caused the SCC. Such low levels must be viewed with prudence as the environmental conditions of a system can change at the metal/solution interface during service and accidental pit, or a slash can increase stresses locally and reach the level, K_{ISCC} (17).

At intermediate stress intensity levels (stage 2) the crack propagation rate levels off (plateau) $V_{plateau}$, which is independent of mechanical stress but depends on the alloy/environment interface and the rate-limiting environmental processes such as the mass transport of the aggressive species to the crack tip. The plateau in a quenched and tempered low alloy steel of 1700 MPa yield strength in deaerated water at 100 °C can be higher than that in a similar steel of 760 MPa yield strength by seven orders of magnitude (91, 92). Stage 3 corresponds to the critical intensity level for mechanical fracture in an inert environment.

The crack growth rate depends on the strength of the metal in almost all aggressive environments. Doubling the yield strength σ_{ys} of martensitic steel (from 800 to 1600 MPa) is accompanied by a tenfold (from 70 to 7 MPa $\sqrt{m}$) decrease of the threshold stress intensity K_{ISCC} corresponding to the onset of stress corrosion crack growth in alloys in chloride solutions at ambient temperatures (86, 93). Stress sources likely to promote cracking are weldments and inserts. Welded structures of these alloys require stress-relief annealing.

In some media, SCC can occur above a certain temperature. Increase in temperature generally lowers the threshold for cracking (σ_1 and K_{ISCC}) and increases the growth rate of propagation. An example is the SSC of stainless steel in neutral pH solution above 40° and 80° as opposed to SCC at pH of 1.0 and room temperature (73). Hydrogen absorption can favor local plasticity very near the crack tip region, because of enhanced dislocation velocities with hydrogen. Hydrogen penetration can

be accelerated very near the crack tip region by stress-assisted diffusion and dislocation transport (94).

1.8.7 Material Properties in SCC

The susceptibility to SCC is affected by: (i) the chemical composition; (ii) preferential orientation of the grains; (iii) the composition and distribution of intergranular
precipitates; (iv) the interaction of dislocations; (v) the progression of the phase transformation; and (vi) cold work (17).

The carbon content and its distribution in the steel matrix are the most important
factors controlling SCC. The threshold stress for cracking was found to depend on the
carbon content of the steel. The carbon in the steel affects the mechanical properties
of the steel in a favorable way, but the influence of carbon on the microstructure
is important and depends on the distribution of the different phases. For example,
carbon particles at ferrite grain boundary regions have been observed in the case of
intergranular cracking of carbon steels with >0.1% C. The other alloying elements
have either harmful or beneficial influence on SCC resistance, depending on their
effect on the segregation of the carbon particles (cementite) at grain joints. These
elements tend to segregate at the grain joints, but their influence is much weaker than
carbon because of their low concentrations.

Increasing zinc content of brasses increases the cracking rate in ammonia solutions
while low amounts of tin, lead, or arsenic improve the resistance (95). The addition
of molybdenum to austenitic steels increases their resistance to cracking in chloride
solutions and decreases it in caustic media. This susceptibility depends on the nature
of the environment. For example, a network of intergranular, coherent, chromium-rich
carbide precipitates increases the resistance to intergranular cracking in hot caustic
solutions, but the related chromium depletion of grain boundaries (sensitization) promotes intergranular cracking in hot, aerated "pure" water or in polythionates (73).

1.8.8 Potential–pH Diagram and SCC

Critical potentials for SCC of a metal/solution system can be related to its E–pH
diagram as these diagrams describe the conditions at which film formation and metal
oxidation occur. The potential–pH (E–pH) diagram of carbon steel shows that SCC is
associated with potentials and pH values at which phosphate, carbonate, or magnetite
films are stable, while Fe^{2+} and $HFeO_2^-$ are metastable. A comparable diagram exists
for a 70 Cu–30 Zn brass in a variety of solutions (96).

The effect of factors such as pH, oxygen concentration, and temperature can be
related to their effect on the E–pH diagram. Change in pH and/or potential corresponds to a region of stability because of the oxide film to a region of active general
corrosion or a zone of severe cracking susceptibility for a specific ion. An increase in
oxygen concentration shifts the potential to more noble or positive potentials. Hydrogen evolution or reduction of iron in water may be a dashed line. This shows that
hydrogen evolution becomes less endothermic with a shift to lower pH or more acidic

values, which plays an important role in SCC-hydrogen-induced subcritical crack growth mechanisms.

The E–pH diagram of the cracking metal/solution interface is a useful tool in the evaluation and understanding of the mechanism of SCC although it is difficult to evaluate in a precise manner. The E–pH behavior of the crack tip solution interface is substantially different from that in bulk solution (97–99).

1.8.9   Active–Passive Behavior and Susceptible Zone of Potentials

An example of the behavior is shown by stainless steel in 1.0 M sulfuric acid solution. Transgranular SCC can occur in two ranges of potentials. Intergranular SCC can occur in a wider potential range. The potential zones 1 and 2 correspond to the active–passive and passive–active state transitions. The crack tip corresponds to the crack tip and the passive state, or film formation corresponds to zone 2. Zone 2 is frequently above the pitting potential, indicating the possible pit initiation and propagation.

The potentials that indicate the susceptibility to SCC can be determined by the scanning of potential-current curves at different scan rates. An example for carbon steel is shown in Figure 1.20. Potentiodynamic polarization curves involve the recording of the values of current with changing potentials (scan rate 1 V/min). This simulates the state of crack tip where there is very thin film or no film at all. To simulate the state of the walls of the crack, a slow sweep rate of 10 mV/min is needed such that the slow scan rate permits the formation of the passive oxide film. The intermediate anodic region between the two curves is the region where SCC is likely to occur. This electrochemical technique anticipates correctly the SCC of carbon steel in many different media. The polarization curves also show the active zone of pitting and the stable passive zone before and after the expected zone of SCC susceptibility, respectively.

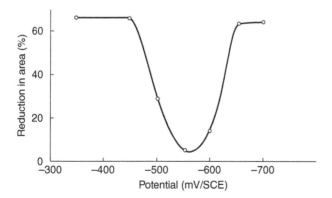

Figure 1.20   Effect of the potential cracking of mild steel. (Reproduced by permission, Elsevier Ltd., (2).)

1.8.10   Electrode Potential and its Effect on Crack Growth

In SCC systems, variation of material composition or microstructure, or modification of environmental composition or redox potential may change the sensitivity to cracking. The crack growth rate is frequently determined by the electrode potential whereby an insignificant change with all other conditions (specimen, pH, solution, temperature, frequency, stress ratio) being constant can lead to the acceleration of fatigue crack growth by a dozen times. This can cause the rapid growth of the crack that was "static" at open-circuit potential ($K < K_{ISCC}$), retard, or stop the crack growth at $K > K_{ISCC}$ (98, 100).

1.8.10.1   Inhibitors   The SCC of an alloy/solution system is generally associated with the presence of a specific corrosive environment that, mostly, would attack the alloy superficially in the absence of stress. Nitrates and hydroxides that generally act as anodic inhibitors for carbon steels also cause cracking of mild steels. Failure of storage steel reservoirs containing anhydrous ammonia generally occurs in the presence of air containing CO_2 that forms the inhibitor $(NH_4)_2CO_3$, which causes cracking. Current density differences slow in polarization curves at various potentials between fast and slow sweep rates for mild steel immersed in hydroxide, carbonate–bicarbonate, and nitrate solutions have been studied. The results obtained from controlled potential slow-strain-rate tests show that cracking is at a maximum between -500 and $-600\,mV$.

A differential aeration galvanic cell consists of differences in oxygen concentration, leading to pitting and cracking in distilled water (96).

1.8.10.2   Alloy/Liquid Interface   The following table 1.2 lists some alloy–environment combinations that result in SCC. This table serves as a general guide (9).

SCC is often encountered in hot gaseous atmospheres on materials under creep or fatigue conditions. SCC susceptibility of titanium alloys in moist chlorine, dry HCl, and dry hydrogen has been observed. The titanium alloy (Ti–8Al–1Mo–1V) exhibits cracking in moist chlorine gas at 288 °C. Binary titanium alloys also exhibit cracking in moist chlorine at 427 °C (96).

Hot salt SCC and hot gaseous SCC of titanium alloys have been reported to occur. Irradiation-assisted SCC has been reported to occur in nuclear reactors (96).

1.8.10.3   HE and Hydrogen Stress Cracking   The interaction between hydrogen and metals results in solid solutions of hydrogen in metals, gaseous products formed by the reactions between hydrogen and elements present in the alloy and metal hydrides. Depending on the type of metal/hydrogen interaction, hydrogen damage occurs as follows (9, 91).

HE is the loss of ductility of materials containing hydrogen, which occurs in high-strength steels, primarily quenched and tempered and precipitation-hardened steels, with tensile strengths greater than about 1034 MPa. There are two types of HE (9, 91).

(i) Hydrogen environment embrittlement occurs during the plastic deformation of alloys in contact with hydrogen-bearing gases or a corrosion reaction and is hence strain rate dependent. Some examples of this are the degradation of the mechanical properties of ferritic steels, nickel-based alloys, titanium alloys, and metastable austenitic stainless steel when there is low strain rate and pressure of pure hydrogen is high.

(ii) Hydrogen stress cracking or hydrogen-induced cracking (HIC) is typified by a brittle fracture under sustained load in the presence of hydrogen. This cracking mechanism depends on: (i) hydrogen fugacity; (ii) strength of the material; (iii) heat treatment/microstructure; (iv) applied stress; and (v) temperature. For many steels, a threshold stress exists below which hydrogen stress cracking does not occur, but this is not material property as it depends on the strength of the steel and the specific hydrogen environment. This is sometimes known as stepwise cracking (SWC).

Hydrogen stress cracking usually produces sharp, singular cracks in contrast to the extensive branching in SCC. Experimental evidence supporting a HE mechanism is that immersion in a cracking solution before stress application produces a fracture similar to a SCC fracture. The effect because of preimmersion in a cracking solution is reversed by vacuum annealing. Testing in gaseous hydrogen produces results similar to the crack characteristics in solutions. SCC occurs at crack velocities at which the adsorbed hydrogen is present at the crack tip.

A critical minimum stress exists below which delayed cracking will not occur. The value of critical stress decreases with increasing hydrogen concentration. These effects are seen in SAE steel (0.4% C) charged with hydrogen by cathodic polarization in sulfuric acid, followed by cadmium plating to retain hydrogen and finally subjected to a static stress (8, 22).

1.8.10.4 Formation of Metallic Hydrides

The precipitation of brittle metal hydride at the crack tip results in considerable loss in strength, ductility, and toughness of metals such as Mg, Nb, Ta, V, Th, U, Zr, Ti, and their alloys in hydrogen environments. Alloys that form hydrides fail by ductile fracture.

Nickel and aluminum alloys may form unstable hydrides leading to hydrogen damage. Some alloys may fail in hydrogen by other mechanisms.

1.8.10.5 Acceleration by Ions

It has been reported that hydrogen damage is accelerated by species such as hydrogen sulfide (H_2S), carbon dioxide (CO_2), chloride (Cl^-), cyanide (CN^-), and ammonium ion (NH_4^+). Some of these ions produce severe hydrogen charging of steel equipment leading to HIC and stress-oriented hydrogen-induced cracking (SOHIC), which can cause a failure. It is helpful to know the cracking severity of the environments so that either the environment can be modified or more crack-resistant materials are chosen. Cracking requires the presence of nascent hydrogen atoms at the steel surface as in the presence of H_2S-containing solution

$$Fe^{2+} + H_2S \rightarrow FeS + 2H$$

The hydrogen atoms combine to form innocuous molecular hydrogen. But this recombination of hydrogen atoms is prevented by cyanide and sulfide, and hydrogen atoms cause HE followed by a failure of the structure leading to general loss in ductility. Internal blisters may occur when large amounts of hydrogen are present in localized areas. Small amounts of dissolved hydrogen may also react with microstructural features of the alloys leading to failures at applied stress much below the yield strength.

HIC may propagate in either a straight or stepwise manner. Straight growth occurs in steels with pearlite structures and high levels of Mn and P segregation or the presence of martensitic or bainitic transformation structures. The Mn level around lean cracks may be twice the level in the matrix while the P level may be as high as a factor of 10.

1.8.10.6 SSC SSC is an important cracking phenomenon in hydrogen sulfide medium and a special case of HIC. Natural aqueous environments contaminated with hydrogen sulfide are very corrosive. The hydrogen sulfide present in salt water in sour oil wells places an upper limit of 620 MPa on the yield strength of steel that can be tolerated.

Sulfide stress cracking is considered to be a form of HE that occurs in high-strength steels and in localized hard zones in weldments of susceptible materials. In the HAZs adjacent to welds, there exist very narrow hard zones combined with regions of high tensile stress that may become embrittled to such an extent by dissolved atomic hydrogen that they crack. Sulfide stress cracking is directly related to the amount of atomic hydrogen dissolved in the metal lattice and usually occurs below 90 °C. Sulfide stress cracking also depends on the composition, microstructure, strength, and the sum of residual stress and applied stress levels of the steel. Sulfide stress cracking was first encountered in oil industry failures of tubular steels and wellhead equipment made from steels with hardness values greater than HRC 22. This type of failure can be eliminated by heat treatment of steels to hardness less than 22. The base metal of steel pipe has hardness well below HRC 22; service failures occurred in regions of high hardness in the weld HAZ. Thus HRC 22 limit is applied to the weld 1-HAZ areas in pipeline steels. There are a number of test procedures (10, 101) for SSC of which the most widely used is the NACE test.

Comparative features of HIC and SSC are given in Table 1.3.

1.8.10.7 Hydrogen-Induced Blistering and Precipitation of Internal Hydrogen
Hydrogen-induced blistering is a cracking process because of absorbed hydrogen atoms. The phenomenon is also referred to as HIC and occurs in lower strength (unhardened) steels with tensile strengths less than about 550 MPa (80 ksi). Pipeline steels exposed to sour gas are prone to HIC.

Penetration of hydrogen into the metal can result in the formation of blisters. Blister formation is because of a deterioration of mechanical property.

Consider a steel reservoir containing an acidic electrolyte and the outside of the reservoir is exposed to the atmosphere. Hydrogen will be evolved inside the walls of the vessel because of corrosion. A part of the atomic hydrogen will diffuse through the steel and a major portion will evolve on the external surface. But some of the atomic hydrogen may diffuse into a void and combine to form molecular hydrogen. Molecular hydrogen builds up pressure in the voids. The pressure of molecular hydrogen in contact with steel is several hundred thousand atmospheres, which can cause rupture of any known engineering material. Filiform corrosion can cause blisters, and hydrogen evolution under paint can result in swelling, but the mechanisms of formation of these similar defects in appearance are totally different from that of hydrogen blistering.

1.8.10.8 Hydrogen Blistering   Hydrogen blistering occurs because of hydrogen atoms diffusing through the steel and accumulation at hydrogen traps such as voids around inclusions. Hydrogen atoms combine and form molecular hydrogen in the trap. The hydrogen pressure builds up in the trap resulting in HIC and formation of blisters. Blisters occur in low-strength steels of yield strength <80 ksi or 535 MPa yield strength. The blisters are generally formed along elongated nonmetallic inclusions or laminations in the steel pipe (102, 103).

Two types of HIC cracks, namely, centerline cracks and blister cracks are shown in Figure 1.21. Blister cracks are related to the type and distribution of nonmetallic inclusions in the steel. MnS inclusions as well as planar arrays of other inclusions are generally initiation sites for cracking. The cracks propagate along a longitudinal direction, which is also the direction of alignment of the inclusions.

The steels that have high concentrations of manganese and sulfur lead to the formation of MnS inclusions. Rolling of steel leads to elongation of the inclusions followed by an increase in surface area of hydrogen traps. Low sulfur bearing steels are not necessarily HIC resistant as alloying for inclusion shape control, reduced centerline segregation, and reduction of nitrides and oxides are also necessary. High sulfur bearing steels are not necessarily susceptible to HIC as microsegregation and inclusion shape are more important factors than bulk sulfur content of the steel (105).

1.8.10.9 Hydrogen Sulfide Damage   The most extensively studied failures in practice is the pipeline steel used in the petroleum industry. Cracking failure in pipeline steel exposed to sour gas in an oil medium is commonly encountered.

1.8.10.10 SOHIC   This form of corrosion is caused by atomic hydrogen dissolved in steel combining to form molecular hydrogen. The molecular hydrogen collects at defect sites in the metal lattice similarly to HIC. Because of applied or residual stresses the trapped molecular hydrogen produces microfissures that align and interconnect in the through wall direction. Although SOHIC can propagate from blisters caused by HIC and SSC, and from prior weld defects, neither HIC nor SSC is a precondition for SOHIC (106, 107).

Just as in the case of HIC, the primary cause of SOHIC is probably atomic hydrogen produced at the steel surface by wet acid gas corrosion (108).

SOHIC tends to occur in the base metal adjacent to hard weldments in pipe and plate steels where cracks may initiate by sulfide stress cracking. SOHIC is characterized by interlinking microscopic cracks oriented both in the direction

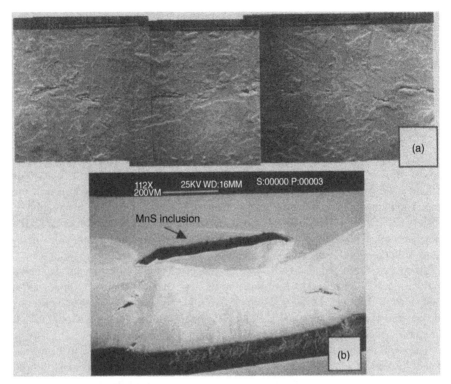

Figure 1.21   Hydrogen-induced cracking: (a) centerline cracks and (b) blister crack. (Figure originally published in Reference 104. Reproduced with permission of the Canadian Institute of Mining, Metallurgy and Petroleum. www.cim.org.)

perpendicular to the stress and in the plane defined by nonmetallic inclusions as depicted in Figure 1.22. SOHIC is a process by which SSC can propagate in relatively low-strength steels (hardness <22HRC), which would otherwise be considered resistant to SSC.

It is useful to note that there may be more than one cracking mode that can cause a failure. The following four main environmental cracking modes can result when steels are exposed to wet H_2S environments (108, 109).

1. Hydrogen blistering: blister formation, originating at nonmetallic inclusions.
2. HIC also known as SWC.
3. SOHIC.
4. SSC: cracking in high strength or hardness steels.

1.8.10.11   Precipitation of Internal Hydrogen   During cooling of the melt, hydrogen precipitates in voids and discontinuities. Cracking because of precipitation of internal hydrogen may be considered as a special case of blistering and occurs when

7J2 (A)   50×

Figure 1.22 SOHIC in plate steel exposed to sour gas. (Figure originally published in Reference 104. Reproduced with permission of the Canadian Institute of Mining, Metallurgy and Petroleum. www.cim.org.)

excess hydrogen is picked up during melting or welding. Some examples of SSC are: shatter cracks, flakes, and fisheyes found in steel forgings, weldments, and castings.

High-temperature high-pressure exposure to hydrogen is encountered usually in steels used in petrochemical plants that handle hydrogen and hydrogen–hydrocarbon streams at high pressure (21 Mpa or 3 ksi) and temperatures as high as 540 °C. This type of hydrogen damage is irreversible while HE is often reversible and occurs below 200 °C. The absorbed hydrogen reacts with carbides present in steel and forms methane bubbles along the grain boundaries. The methane bubbles grow and cause fissures. Fissures occur at grain boundaries and are accompanied by decarburization. Copper alloys containing small amounts of cuprous oxide are susceptible to hydrogen attack, leading to significant void formation (4).

$$Cu_2O + H_2 \rightarrow 2Cu + H_2O$$

Hydrogen gas exposure may also cause titanium embrittlement (96).

1.8.10.12   Other Types of Hydrogen Attack   Hydrogen trapping, microperforation, and degradation in flow properties have been observed (4). Liquid metal-induced embrittlement (LMIE) is the catastrophic brittle failure of a normally ductile metal when coated with a thin film of liquid metal and then stressed in tension (110). Solid-metal-induced embrittlement occurs below the melting temperature of the solid in certain liquid metal environments (LMIE couples). The severity of the embrittlement increases with temperature with sharp increase in severity at the melting point of the embrittler (4).

1.8.10.13   Mechanisms of EIC   . EIC extends over a wide range from brittle fracture to electrochemical process, and the underlying mechanisms are complex.

1.8.10.14 The Overlapping of Cracking Phenomena There are many corrosion-based causes of premature fracture of structured materials. The most common causes of premature fracture of components are given in the literature. CF cracking occurs only under cyclic or fluctuating operating loads, while SCC and HE occur under static or slowly rising loads.

It is possible that CF, SCC, and HE could occur at the same time in some service conditions. The simultaneous operation of SCC and HE can occur in some systems. The interrelationship among stress corrosion, CF, and HE are discussed in the literature. The cross-hatched regions represent the most serious practical situations involving ductile alloy/environment systems. These regions indicate the combination of any two failure mechanisms. In the center, all three phenomena interact, which is probably realistic in ductile alloy/aqueous environment systems (4).

For example, in CF and SCC, a surface pit constitutes a stress raisor, in particular, in the initiation process. CF can assist initiation of the fracture while SCC and/or HE can assist more or less intensively the crack propagation. In certain conditions, either the crack initiation or the crack propagation can dominate the morphology and mechanism of crack propagation. Under fatigue loading, modification of the number of PSBs in fatigue and of the slip offset height in PSBs had been observed in comparison with air in nickel single crystals in 0.5N H_2SO_4 and copper crystals according to the applied potential. Cyclic softening effects under anodic condition can occur in austenitic and ferritic stainless steels in NaCl solutions (73, 81). Both the decrease of K_{ISCC} and the acceleration of crack growth rates because of increase of metal strength become apparent under cyclic loading.

$K_{ISCC} \gg \Delta K_{th}$ the threshold intensity range at the corrosion crack growth rate $\leq 10^{-10}$ m/cycle. The rate da/dN in aqueous solutions is much higher (by a dozen times) than in ambient air. However, by suitable choice of solution composition, the CFC growth can be reduced in titanium and magnesium alloys (86, 100).

Cathodic protection generally mitigates CF and SCC but may increase the HE corrosion of susceptible materials.

The electrode pH and potential at an active crack tip surface may be significantly different from those on boldly exposed surfaces of a material. Low pH conditions can lead to local dissolution of the metal and crack tip blunting that reduces the stress concentration effects. In contrast, low pH conditions favor hydrogen generation and consequently increase risk of HE corrosion. The reduction in ductility associated with HE corrosion may produce sharp crack tips, which in turn may increase stress concentration effects for any synergistic SCC or CF (4).

In many systems, SCC occurs in a limited range of electrochemical conditions as shown for potential/pH domains for cracking of mild steel in different environments. Three main types of SCC as a function of potential have been identified (111):

1. Low-potential regions where significant amounts of hydrogen are produced that can be absorbed by the material leading to HE.
2. Passive regions close to active–passive transitions such as for Ni–Cr–Fe alloy 600 or steel in caustic solutions (112, 113).

3. Regions close to critical potential for localized corrosion, which is observed in the case of austenitic stainless steels in near-neutral solutions (80, 114).

Thus it is obvious from the foregoing discussion that complex relationships exist between corrosion fatigue cracking (CFC), SCC and hydrogen embrittlement cracking (HEC), and the distinction between CFC, SCC, and HEC is difficult as it is likely that hydrogen–metal interaction near the crack tip is the controlling process of SCC or CF crack propagation (73).

1.8.10.15   Mechanism of SCC   The crack initiation of environmentally assisted cracking (EAC) is complex and not well understood until now. The majority of SCC systems exhibit short initiation times ranging from minutes to weeks, and cracking often occurs because of a change in environment rather than long initiation time. The stress corrosion growth rates are in the range 10^{-11} to 10^{-6} m/s. In the case of stainless steels in chloride solutions, localized corrosion may create local conditions prone to crack development, but it is difficult to rationalize initiation of crack development in the absence of localized corrosion in environmental conditions different from that of propagation. It is useful to note that dealloyed surface layers such as copper alloys in ammonia solutions tend to undergo SCC (44).

Surface films appear to play a major role in the initiation of SCC and may also contribute to HE effects. The main role of the surface film is to localize the damage inflicted on the material by the environment. The damage can be caused by the mechanical breakdown of the passive film by slip step or electrochemical breakdown of the passive film (73). SCC may be related to the nature of the surface film. The SCC of carbon steels is related to the presence of magnetite in environments at 90 °C except when pitting is involved in the crack initiation process, as in nitrate medium or in high-temperature water (115, 116).

1.8.10.16   Mechanism of HE   The hydrogen from the environment is adsorbed at the crack tip resulting in reduction in the effective bond strength and lowered surface energy leading to the diffusion of hydrogen atoms into the metal (Fig. 1.23a, b) with decohesion of atoms by hydrogen influx to the dilated lattice (Fig. 1.23c). Some interactions may occur in advance of the crack tip where the stress and/or strain conditions are favorable for the nucleation of a crack. The nucleation of a crack may be followed by the formation of a brittle phase such as metal hydride (Fig. 1.23d). It is also possible that the hydrogen atoms can combine to form molecular hydrogen, which can cause pressure inside the metallic network and can cause inflation, and this is the mechanism of formation of blisters.

Hydrogen-induced crack growth as the dominant SCC mechanism has been suggested for ferritic steels, nickel alloys, titanium alloys, and aluminum alloys. The effects of factors such as yield strength, impurity segregation, and the temperature on the crack growth behavior of ferritic materials in aqueous solutions follow the trends of HE. Tin and antimony, known as hydrogen recombinant poisons, segregated to the grain boundaries of nickel alloy, increase the uptake of atomic hydrogen (117).

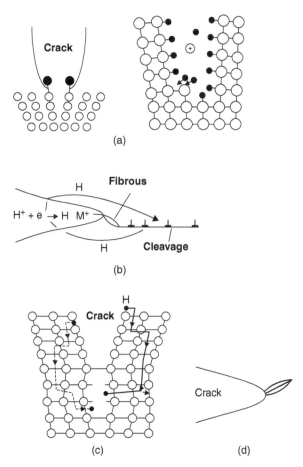

Figure 1.23 Possible mechanisms of hydrogen embrittlement: (a) chemical adsorption of hydrogen, (b) adsorption of atomic hydrogen, (c) decohesion of atoms, and (d) possible brittle hydride particle at the tip. (Reproduced by permission, Elsevier Ltd., (2).)

Temperature plays an important role in HE of ferrous alloys. Embrittlement is severe at room temperature and less severe or nonexistent at higher or lower temperatures. The diffusivity of hydrogen is low to fill sufficient traps, but hydrogen mobility is enhanced at higher temperatures. At a high strain rate, fracture proceeds without the assistance of hydrogen. For instance, the crack growth rate of 3% Ni steel as a function of temperature and that of 4340 steel in gaseous hydrogen have been studied. The maximum crack growth rate occurs at about 20 °C for steel with similar decreases at higher and lower temperatures for nickel. However, anodic stress corrosion processes become active above 100 °C for nickel tested in water suggesting SCC mechanism has assumed control (4).

Threshold stress intensity for crack growth in iron-based alloys generally decreases with increasing yield strength regardless of environments and very high-strength steels cannot be used in hydrogen environments. In this regard, H_2S is one of the most corrosive environments. At lower yield strengths, the mechanisms for hydrogen-assisted failures apparently changes, and blistering becomes the more common failure mode. The threshold stress intensities for high-strength steels subjected to hydrogen environments are significantly less than those measured under benign conditions (K_{ISCC} or K_{HI}).

In low-strength steels (700 MPa or 100 ksi yield strength or less) hydrogen damage occurs by loss in tensile ductility or blistering. In the case of loss in tensile ductility, hydrogen promotes the formation and/or growth of voids by promoting the decohesion of the matrix at carbide particles and inclusion interfaces. At higher hydrogen fugacities and often in the absence of stress, blistering, or a form of cracking also associated with inclusions termed as SWC, blister cracking or HIC is observed. This mode of cracking is not a function of steel strength but depends on steel composition, processing, and hydrogen environment. In a hydrogen sulfide environment, a cracking morphology has been observed in low-strength steels that combine features of hydrogen stress cracking and HIC, which has been termed as SOHIC (4).

Localized corrosion and stress corrosion may often be observed. Stress corrosion cracks usually initiate at pits in many systems. The role of pitting is to disrupt films that otherwise prevent the ingress of hydrogen (118, 119). Electrochemical polarization technique may be used to distinguish between SCC and HE mechanisms in high-strength steels in sodium chloride solutions (120).

1.8.10.17 Propagation Models   Generally, SCC is considered to be brittle, meaning that it occurs at stresses below yield stress and propagates as essentially an elastic body even though local plasticity may be required for the cracking process. Thus linear-elastic fracture mechanics is used in studying SCC. Propagation of the crack is usually because of periodic microruptures although anodic dissolution is an important controlling process. Local attack results in localized hydrogen absorption, which can cause local or bulk embrittlement and even plasticity. Thus anodic dissolution and hydrogen can play a role in the system. The local environment leading to crack propagation is not well defined particularly in SCC phenomena. The transport of water to the crack tip may be an important controlling process. With the exception of the slip dissolution model, environmentally assisted cracking models are notable for providing quantitative prediction of the crack propagation rate (73).

Major models of SCC mechanisms are:

Stress Corrosion Cracking Mechanisms			
Dissolution Models		Mechanical Fracture Models	
Film rupture	Active path process	Ductile fracture	Brittle fracture
		Corrosion tunneling	Plasticity	Tarnish rupture	Film-induced cleavage

1.8.10.18 Dissolution Models The model considers that the crack propagates because of the preferential dissolution at the crack tip giving rise to the formation of active paths in the material, stresses at the crack tip, and chemical–mechanical interactions.

1.8.10.19 Film Rupture or Slip Dissolution Model This model assumes the presence of an active protective film and that the stress opens the crack and ruptures the film. Localized plastic deformation at the crack tip ruptures the passivating film, exposing bare metal at the crack tip, and the freshly exposed surface dissolves rapidly. It is also suggested that the crack tip remains bare as the rate of repassivation is slower than the rate of film rupture during crack propagation. The low-stress limiting case of this mechanism is akin to the IGC mechanism, and this model is applicable to intergranular SCC and not transgranular SCC (4).

The limiting velocity for the crack growth according to the dissolution model may be written as

$$\frac{da}{dt} = \frac{i_a M}{ZFP}$$

where i_a is the anodic current, M the atomic weight, Z the valence, F the Faraday constant, and P the density of the material. In general, the crack growth rate depends on the rate at which the film is ruptured and reformed (121).

1.8.10.20 Ductile Mechanical Models Stress concentrations at the base of corrosion pits (slots) increase to the level of ductile formation or fracture.

In the *corrosion tunneling model*, a fine array of small corrosion tunnels are thought to form at emerging slip steps. The tunnels grow both in diameter and length until the stress in remaining ligaments result in ductile formation and fracture. It is suggested that the application of tensile stress results in a change in the morphology of the corrosion damage from tunnels to thin slots.

1.8.10.21 Adsorption-Enhanced Plasticity Models According to fractographic studies the cleavage fracture is not an anatomically brittle process, but occurs by alternate slip at the crack tip in conjunction with the formation of very small voids ahead of the crack. It is also thought that the chemisorption of environmental species facilitates the nucleation of dislocations at the crack tip, promoting the shear process responsible for brittle-like fracture (4).

1.8.10.22 Corrosion-Enhanced Plasticity Models These models are based on a localizing effect at the very crack tip because of corrosion. Fracture occurs because of the enhancement in plasticity along one slip system, inducing the formation of pile-up and a local decrease in cohesion energy because of the presence of hydrogen. The role of dissolution is to create vacancies that can enhance the plasticity at the crack tip and create defects in the passivated metals at the slip band emergence needed for hydrogen absorption. Vacancies may also act as traps for hydrogen and thus increasing the crack tip concentration. This model has also been applied to intergranular SCC on the basis of the same corrosion–deformation interactions in the vicinity of grain boundaries (122). The enhancement of creep by anodic dissolution is known in the case of copper in acetic acid (123) and austenitic stainless steels and nickel-based alloys in PWR environments. The initial vacancy injection from the surface is followed by vacancy attraction to the dislocations inside, which promotes easier glide, climb, and crossing of microstructural barriers. This mechanism illustrates the corrosion-enhanced plasticity approach (73).

1.8.10.23   Tarnish Rupture Model   In this model, fracture of the film exposes the fresh bare surface that reacts with the environment to reform the surface film. The crack propagates by alternating film growth and rupture. This model was first proposed to explain transgranular SCC and then applied in a modified form to explain intergranular SCC by assuming that the oxide film penetrates along the grain boundary ahead of the crack tip, which may not be the case in all systems (124).

1.8.10.24   Film-Induced Cleavage Models   Dealloying and/or vacancy injection could cause brittle fracture. In this model, the brittle crack initiates in a surface film or layer and this crosses the film/matrix without loss of speed. The brittle crack continues in the ductile matrix until it eventually blunts and stops. This model needs better verification and understanding of surface films and brittle fracture.

1.8.10.25   Adsorption-Induced Brittle Fracture   This model is based on the concept that adsorption of environmental species lowers the interatomic bond strength and the stress required for cleavage (4). This model can satisfactorily explain the susceptibility of certain alloys for bond cleavage when the alloys are bonded to certain ions. One of the important factors in support of this mechanism is the existence of critical potential below which SCC does not occur in some systems. This model of cracking underlines the relationship between the potential value and the capacity of adsorption of the aggressive ion. This also explains the prevention of SCC by cathodic protection. This model is also useful in explaining the rupture of plastic materials or glass. This model is referred to as the stress-sorption model and similar mechanisms have been proposed for HE and liquid metal embrittlement. In this model, the crack propagates in a continuous way at a rate dictated by the arrival of embrittling species at the crack tip. This model does not explain as to how the crack maintains a sharp tip in a normally ductile material (125).

1.8.10.26   Decohesion Models   Interactions between a localized dislocation array and the crack tip under an applied stress produce a maximum stress ahead of the crack tip to which hydrogen is driven under the stress fields from behind the tip. When the hydrogen concentration reaches a critical value, a microcrack is nucleated because either the local cohesive strength is reduced, dislocation motion is blocked in the hydrogen enriched zone, or both. The microcrack arrests about 1 mm ahead of the original location of the tip, and these processes repeat, ultimately leading to discontinuous microcracking (73). This phenomenon is sometimes referred to as hydrogen-enhanced decohesion.

1.8.10.27 Hydrogen-Induced Plasticity Two models have been proposed on the basis of the fact that hydrogen increases the local plasticity. Two models have been advanced on the basis of adsorption and absorption of hydrogen.

The adsorbed hydrogen-induced model is based on the fact that the adsorbed atoms weaken interatomic bonds at crack tips and thereby facilitate the injection of dislocations (alternate slip) at crack tips. Crack growth occurs by alternate slip the crack tips, which promotes the coalescence of cracks with small voids nucleated ahead of the cracks.

The second absorbed hydrogen-enhanced local plasticity mechanism is based on the fact that the local decrease of the flow stress by hydrogen leads to highly localized failure by ductile processes, while the local macroscopic deformation remains small. Shear localization results from local hydrogen absorption, resulting in a macroscopically "brittle" fracture related to microscopic localized deformation.

1.8.10.28 Prevention of SCC Although several factors are the causes of SCC, the measures to prevent or reduce the risk of SCC can be divided as a function of stresses, environmental factors, metallurgical properties, and surface treatments.

(i) *Stresses.* The sum of stresses in service and residual stresses, including fabrication stresses, should be below the threshold level, which, in the absence of reliable data, should be evaluated by testing as a percentage of the tensile yield strength. This can be achieved by a design that avoids initial concentration of stresses, reducing operating stresses, relieving fabrication stresses by heat treatment, and by choosing appropriate dimensions of the loaded alloy.

(ii) During use, avoiding stress concentrators by increasing susceptibility to different forms of degradation in service, especially localized corrosion, galvanic corrosion, and erosion–corrosion will be helpful. Accidental or nondesigned cyclic loading also should be avoided.

(iii) Bolted or riveted joints can produce high local stresses that can cause SCC, so that proper joint design and construction can be done. Some examples are the use of performed parts; avoid overtorqueing of bolts, provision of adequate spacing, and edge margins for rivets.

Environmental Considerations:

(i) Eliminating potential aggressive ions that cause SCC such as dissolved gases, heavy metal ions, and impurities by chemical or physical methods such as degasification, demineralization will prevent the risk of SCC.

(ii) Addition of inhibitors to form a strong and adhesive film will also help. Phosphates and other inorganic and organic inhibitors decrease the effects of SCC. A minimal amount of oxidizing inhibitor such as sodium nitrite is absolutely necessary to avoid pitting.

(iii) Mixture of inhibitors may be useful for corrosion control. Capillary condensation can fill crevices and result in corrosion that can be avoided by filling the crevices with corrosion-inhibited putty.

(iv) Cathodic polarization may reduce or prevent SCC but may only be used with care for alloys that resist all types of critical hydrogen damage.

(v) Avoiding temperature range that causes SCC, for example, SCC of steel in caustic solutions at $>105\,°C$ will also help.

Metallurgical Considerations:

(i) Selection of a resistant alloy with desirable microstructure and suitable chemical composition and microstructure is preferrable.

(ii) Shot peening and other mechanical processes that create compressive residual stresses at the surface are desirable.

(iii) Stress-relief at low temperature is used to lower corrosion susceptibility. Tensile residual stresses from welding are dangerous and a low-temperature thermal stress-relief treatment is desirable for welded assemblies.

(iv) Alloy composition and/or alloy microstructure may be changed as in changing stainless steel grade 304 by nickel base superalloy.

Surface Treatments:

(i) Surface modification by oxidation, phosphating ,and anodizing is helpful for certain alloys when these treatments are followed by sealing with suitable inhibitors or coatings.

(ii) Coatings can extend life but do not totally prevent SCC as defects in the coating lower the protection.

(iii) Cladding a susceptible alloy with a resistant alloy is also an effective method.

(iv) Carrying out nondestructive testing, inspection, and maintenance programs to avoid SCC precursors such as concentration of stresses by localized corrosion will help. In the case of coatings, routine inspection is essential as scratches could be favorable sites for the initiation of SCC.

1.8.11   Prevention of Hydrogen Damage

Metallurgical Considerations:

(i) Modifying the design to lower stresses as well as choice of materials resistant to HIC such as nickel-containing and nickel-based alloys because of their low hydrogen diffusion rates will be an advantage. Hydrogen damage can often be prevented by using more resistant material such as alloying high-strength steels with nickel or molybdenum that reduces susceptibility to HE.

(ii) Susceptibility of pipeline steels to HIC depends on nonmetallic inclusions and anomalous structures because of P and Mn segregation in steel. Lowering the sulphur content or modifying the morphology of segregation by the addition of Ca is useful in avoiding HIC. Tempering is effective in the elimination of low-temperature anomalous structure. Addition of Cu ($>0.2\%$) has been found to be effective. Calcium and rare earth metals inhibit HIC by modifying the morphology of inclusions (126). Calcium, La, and Ce spheroidize nonmetallic inclusions and thus raising the Cth and hence preventing the HIC. Cobalt, Ca, W, and Ni are also effective in the prevention of HIC. By reducing the sulfur content from 0.002 to 0.0005%, the inclusions such as MnS can be reduced and HIC can be prevented (127).

(iii) Hydrogen blisters may be avoided by using "clean" steel without voids such as killed steels instead of rimmed steels. Manufacturing processes and treatments affect MnS morphology and influence sensitivity; for example, rimmed and Si-killed steels have low susceptibility. If the MnS inclusion content is low, adverse effects of low-temperature rolling are minimized. Both quenching and tempering reduce the susceptibility. Tempering eliminates Mn and P segregation when the Mn level is more than 1%; tempering reduces hardness around inclusions and hence HIC susceptibility.

(iv) HE can be minimized by using dry conditions as well as low-hydrogen welding rods.

(v) Inhibitors may be used to reduce corrosion and hence hydrogen. Baking after processing reduces HE. Addition of inhibitors during pickling to avoid hydrogen pick up is recommended. The pH can be raised to reduce HIC.

(vi) Changing the environment can be very efficient. Blistering rarely occurs in pure acid corrosives without hydrogen evolution poisons such as sulfides, arsenic compounds, cyanides.

(vii) Coatings should be impervious to hydrogen penetration and resistant to a corrosive medium. Metallic, inorganic, and organic coatings are often used to prevent HIC and hydrogen blistering.

1.8.11.1 Corrosion Testing Objectives of SCC testing are: (i) determination of the risk of SCC for a given application and comparison of alloys; (ii) examination of the effects of chemical composition, metallurgical processing, fabrication practices for structural components; (iii) evaluation of protective systems and prediction of service life; (iv) development of new alloys, which are less expensive, and offer a longer, safer, and efficient performance for chosen environments; (v) evaluation of claims for SCC performance of improved mill products.

Prediction of corrosion performance can be done from published data and testing. Accelerated testing should involve the same mode of failure and reflect a known order of resistance of some alloys in service media (128). The common test objectives of SCC are high stresses, slow continuous straining, precracked specimens, higher concentration of corrosive agent than in service medium, higher temperature, and electrochemical stimulation (129). For electrochemical corrosion, the properties of the medium at the interface should be noted in accelerated tests.

1.8.11.2 Media Considerations SCC tests are conducted in: (i) natural atmospheres; (ii) seawater immersion, and (iii) laboratory or other fabrication conditions. Atmospheric exposure tests take a long exposure time, but are reliable as they represent the true conditions of projected use. The standard practice is to use 3.5% NaCl, pH 6.5, and the standard tests are AST B-117, 2003, and the conditions of the test AST G44 (130).

1.8.11.3   Test Samples   Commonly used specimens under elastic stress range are bend-beam specimens, C-ring samples, O-ring specimens, tension specimens, and tuning fork specimens. Plastic strain specimens and residual strain specimens have also been used. Static loading of precracked specimens as well as slow-strain-rate testing are also used. Stressed O-rings as well are used for protective treatments for SCC prevention. The samples are subjected to various loading conditions such as constant load, constant strain, or monotonically increasing strain to total failure in some of the slow-strain-rate tests. Other tests are cyclic loading as well as slow straining over a limited stress range (131).

1.8.11.4   Stressors   Corrosion accelerators for testing alloys is achieved through the use of "stressors," such as cold work of the material, high concentration of aggressive ion, low pH, high temperature, higher stress. Externally applied stresses are known and easy to control, but residual stresses are the ones that are normally responsible for stress-corrosion failures under service conditions. Although laboratory tests are useful in encouraging conservative design of the equipment, results of long-term atmospheric tests of tensile loaded specimens are considered to be more reliable.

Constant load SCC tests are more severe than constant deflection tests.

1.8.11.5   Slow strain test   The strain rate chosen frequently for tests indicates a value of about $2 \times 10^{-6}\,\mathrm{s}^{-1}$ for cracking of steels, aluminum, and magnesium alloys. This value for cracking refers to open-circuit conditions, and the stain rate sensitivity of cracking depends on the potential and the solution composition. Where necessary, the potentiostat may be used to control the potential of the samples during slow-strain-rate tensile testing (132). The reduction in area is a simple and reliable way to quantify the susceptibility to SCC. Both AC and DC potential drop measuring methods are proven techniques for obtaining information on the onset of stable crack growth.

REFERENCES

1. LL Shreir, RA Jarman, GT Burstein (eds.), *Principles of Corrosion and Oxidation*, Butterworth Heinemann, 1994.

2. LL Shreir, *Corrosion, Metal/Environment Reactions*, Vol. **1**, 2nd ed., Butterworths, London, 1976, 1:169–174; 8:1–129.

3. CP Dillon, *Introduction in Forms of Corrosion*, NACE Handbook, Vol. **1**, NACE International, Houston, TX, 1982, pp. 1–4; ED Verink, pp 5–18.

4. ASM Handbook Vol. 13A: Corrosion: Fundamentals, Testing and Protection, Frankel, pp. 236–241, 257; Kolman, pp. 381; Phull p. 568; Jones pp. 346–366; Craig pp. 367–380, American Society of Metals, 2003.

5. RJ Landrum, *Fundamentals of Designing for Corrosion Control*, NACE International, Houston, TX, pp. 1–24, 1992.

6. *ASM Corrosion*, Vol. **13A**, Baboian (ed.), ASM International, pp. 210, 2003.

7. DL Piron, *The Electrochemistry of Metals*, NACE, Houston TX, 1991.

8. HH Uhlig, RW Revie, *Corrosion and Corrosion Control*, 3rd ed., Wiley, NY, 1985, pp. 327–340; 28–30; 90–164; 198–199; 320; 386; 6–15; 123–164; 60–89; 405–414.

9. *ASM Surface Engineering for Corrosion and Wear Resistance*, JR Davis (ed.), ASM International, Ohio, pp. 16–27, 1–81, 2001.

10. M Elboujdaini, CANMET, MTL Report, Ottawa, Canada, pp. 1–13, 2005.

11. M Pourbaix, *Atlas of Electrochemical Equilibria in Aqueous Solutions*, NACE International, Houston, TX, 1974.

12. JC Scully, *The Fundamentals of Corrosion*, Pergamon Press, pp. 90–92, 1966.

13. *ASM Stress Corrosion Cracking*, RH Jones, RE Ricker (eds.), ASM Intl., Ohio, 1992, pp. 1–40, Anderson, pp. 181–210; Miller, pp. 251–263; Schutz, pp. 256–297; Sprowls, pp. 336–415.

14. *ASM Metals Handbook, Corrosion*, Vol. **13**, 9th ed., ASM International, Metals Park, OH, 1987, Craig and Pohlman, pp. 77–189; Craig pp. 367–380; Sprowls pp. 231–3.

15. WF Bogaerts, KS Agena, *Active Library on Corrosion*, Elsevier, Amsterdam, 1996, in conjunction with NACE, Houston, TX.

16. H Böhni, *Uhlig's Corrosion Handbook*, 2nd ed., RW Revie (ed.), Wiley, pp. 173–190, 2000.

17. MG Fontana, ND Greene, *Corrosion Engineering*, McGraw-Hill, NY, pp. 1–115, 1978, Fig. 3-24.

18. Z Szklarska-Smialowska, *Pitting and Crevice Corrosion*, NACE International, Houston, TX, pp. 5–43, 2005.

19. HP Godard, "Localized Corrosion" in *NACE Basic Corrosion Course*, NACE, Houston, TX, pp. 8:1–15, 1970.

20. O Lunder, JE Lein, SM Hesjevik, TK Aune, K Nisaneioglu, *Werkstoffe und Korrosion* **45**:331–340 (1994).

21. A Yamamoto et al., Proc. of 2nd Intl. Conf. on Environment Sensitive Cracking and Corrosion Damage,Nishika Printing, Hiroshima, Japan, pp. 160–167, 2001.

22. *ASM Metals Handbook, Corrosion*, Vol. **13**, 9th ed., LJ Korb, DL Olson (eds.), ASM International, Ohio, 1987, Scully pp. 212–220; Sprowls pp. 231–233; Adler pp. 333–344.

23. TP Hoar, *Corrosion Science* **7**:355 (1967).

24. B MacDougall, MJ Graham, *Corrosion Mechanisms in Theory and Practice*, P Marcus, J Oudar (eds.), Marcel Dekker, NY, pp. 143–173, 1995.

25. CR Clayton, I Defjord, Corrosion mechanisms, *Theory and Practice*, P Marcus, J Oudar (eds.), Marcel-Dekker, New York, pp. 175–199, 1995.

26. PC Pistorius, *Corrosion* **53**:273 (1997).

27. J Kruger, Uhlig's Corrosion Handbook, 2nd ed., RW Revie (ed.), Wiley, pp. 165–170, 2000.

28. T Shibata, Corrosion **52**:813 (1996).

29. H Böhni, Localized corrosion of passive metals, *Uhlig's Corrosion Handbook*, 2nd ed., RW Revie (ed.), Wiley, pp. 173–190, 2000.

30. *Annual Book of ASTM Standards*, ASTM International, Vol. 01.03; 03.01; 03.02; 13.01, XXX.

31. NG Thompson, JH Payer, *Corrosion Testing Made Easy*, BC Syrett (ed.), NACE International, Houston, TX, pp. 72–77, 1998.

32. NG Thomson, BC Syrett, *Corrosion* **48**:649 (1992).

33. RM Kain, *Corrosion* **40**:313 (1984).

34. RS Treseder, EA Kacick, *ASTM Special Technical Publication* **866**:373–399 (1985).

35. DA Eden, Uhlig's Corrosion Handbook, RW Revie (ed.), Wiley, pp. 1227–1238, 2000.

36. PC Pistorius, GT Burstein, *Philosophical Transactions of the Royal Society*, London, **A341**:531 (1992).

37. T Suter, H. Bohni, *Eletrochim Acta* **43**:2843 (1998).

38. F Mansfield, Y Wang, H Shih, Critical Factors in Localized Corrosion, GS Frankel, RC Newman (eds.), Electrochemical Society, Proceedings, Pennington, NJ, p. 497, 1999.

39. HS Isaacs, MW Kendig, *Corrosion* **36**:269 (1980).

40. LE Samuels, *Metals Engineering: a Technical Guide*, ASM International, pp. 5–6, 238, 1988.

41. ASM Corrosion, Vol. 13A, Noël, pp. 258–265; ASM Committee 16-27, 1-81 (2001); ASM Intl. Corrosion, Vol. 13A, p. 568–616, 2003.

42. L Muller, JR Galvele, *Corrosion Science* **17**:1201 (1977).

43. H Moore, S Beckinsale CE Mallinson, *Journal of Institute of Metals* **25**:35–153 (1921).

44. K Sieradzki, JS Kim, AT Cole, RC Newman, *Journal of Electrochemical Society* **134**:1635 (1987).

45. N Dimitrov, JA Mann, M Yukirovic, K Sieradzki, *Journal of Electrochemical Society* **147**:3282–3285 (2000).

46. *ASM Corr of Al and Al Alloys*, JR Davis (ed.), ASM International, p. 25, 1999; pp. 16–27, 1–81, 2001.

47. V Mitrovic-Scepanovic, RJ Brigham, *Corrosion* **48**:780–784 (1992).

48. A Wahid, DL Olson, DK Matlock, C Cross, *ASM Metals Handbook*, Vol. **6**, ASM International, Ohio, pp. 1065–1069, 1993.

49. E Ghali, Uhlig's Corrosion Handbook, 2nd ed., RW Revie (ed.), Wiley, NY, pp. 677–716, 2000.

50. ASM Corrosion, Vol. **13A**, 2003: Rebak, pp. 279–286; Stott, pp. 644–649; Dexter, pp 398–416, American Society of Metals, 2003.

51. W Sand, *Corrosion and Environmental Degradation*, M Schutze (ed.), Wiley, Weinheim, pp. 172–202, 2000.

52. *ASM Metals Handbook, Corrosion*, Vol. **13**, 9th ed., LJ Korb, DL Olson (eds.), ASM International, Ohio, 1987, Dexter pp. 87–88; Stott pp. 644–649.

53. WA Hamilton, S Maxwell, Proceedings of Biologically Induced Corrosion, NACE International, p.131, 1986.

54. TD Perry, M Branker, R Mitchell, Proc. of Corrosion 2002 Research Topical Symposium: Microbiologically Influenced Corrosion, NACE International, pp. 113–121, 2002.

55. PE Holmes, *Journal of Applied Environmental Microbiology* **52**:1391 (1986).

56. JJ Elpjick, *Microbial Aspects of Metallurgy*, JDA Miller (ed.), American Elsevier, pp. 157–172, 1970.

57. ASM Corrosion, Vol. **13A**, 2003, Little, pp. 478–486; Stott, pp. 644–649.

58. Battelle Columbus Laboratories, Effects of Metallic Corrosion in USA, 1978, 1995.

59. R Schubert, ASME, *Journal of Lubrication Technology* **93**:216–223 (1971).

60. B Bhushan, Introduction to Tribology, Wiley, pp. 331–420, 2002.

61. IM Hutchings, *Tribology: Friction and Wear Engineering*, CRC Press, Boca Raton, FL, pp. 171–197, 1992.

62. TE Fisher, MP Anderson, S Jahanmir, R Sather, *Wear* **124**:133–148 (1988).

63. B Bhushan, RE Davis, *Thin Solid Film* **108**:135–156 (1983).

64. M VanDyke, *An Album of Fluid Motion*, Parabolic Press, p. 107, 1982.

65. JF Douglas, JM Gasiorek, JA Swaffield, *Fluid Mechanics*, Pitman, p. 648, 1979.

66. M Gopal, WP Jepson, *Corrosion and Environmental Degradation*, M Schutze (ed.), Vol. **1**, Wiley, pp. 265–284, 2000.

67. PL Hurries, *Wear* **15**:389–409 (1970).

68. RB Waterhouse, Proceedings of International Conference on Wear, ASME, NY, pp. 1–22, 1981.

69. HH Uhlig, Corrosion et Protection, Dunod, Paris, France, pp. 98–108, 136–143, CF pp. 148–157, 1970.

70. JM Dorlot, JP Baillon, Des matériaux, 2nd ed., pp. 201–239, 1986.

71. G Wranglen, *An Introduction to Corrosion and Protection of Metals*, Chapman and Hall, London, England, pp. 118–119, 1984.

72. JO Almen, *Corrosion Handbook*, HH Uhlig (ed.), Wiley, pp. 590–597, 1998.

73. T Magnin, P Combrade, *Materials Science & Technology, Corrosion and Environmental Degradation*, Vol. **1**, M Schutz (ed.), Wiley, pp. 207–263, 216–318, 2000.

74. W Bamford, *Institute of Mechanical Engineers Conference Publication* **4**:51 (1977).

75. J Atkinson, D Tice, PM Scott, *Proceedings of 2nd IAEA Specialists' Meeting on Subcritical Crack Growth*, Sendai, WH Cullen (ed.), Vol. **2**, p. 201, 1985.

76. P Combrade, M Foucault, G Siama, ibid. Vol. **2**, pp. 201, 1985.

77. WA Van Der Sluys, Proceedings of 4th Intl. Symposium on Environmental Degradation of Materials in Nuclear Power Systems, Travers City, MI, p. 277, 1988.

78. LM Young, PL Andersen, Proceedings of 7th Intl. Symp. On Environmental Degradation in Nuclear Power Systems, Water Reactors, NACE, Vol. **2**, pp. 1193, 1995.

79. YZ Wang, Corrosion Fatigue in Uhlig's Corrosion Handbook, RW Revie (ed.), pp. 195–202, 2000.

80. Suresh , *Fatigue of Materials*, Cambridge Solid State Science Series, Cambridge University Press, pp. 363–368, 1991.

81. DJ Duquette, Environment-Induced Cracking of Metals, RP Gangloff, MB Ives (eds.) NACE 10, Houston, TX, pp. 45, 1990.

82. AJ McEvily, RP Wei, *Fracture Mechanics and Corrosion Fatigue*, NACE, Houston, TX, pp. 381–395, 1973.

83. O Vosikovski, *Transactions of the ASME*, **97**(4):298 (1975).

84. DJ Duquette, HH Uhlig, *Transactions of the ASME*, **61**:449 (1968).

85. M Elboujdaini, MT Shehata, E Ghali, Microstructural Science, DE Alman, JA Hawk, JW Simmons (eds.), IMS and ASM International, Ohio, Vol. **25**, pp. 41–49, 1997.

86. SA Shipilov, *Technology, Law and Insurance*, **1**(3):131–142 (1996).

87. SA Shipilov, Teaching and Education in Fracture Fatigue, HP Rossmanith (ed.), E&FN Spon, London, pp. 293–299, 1996.

88. RP Reed, JH Smith, BW Christ, *Economic Effects of Fracture in USA*, NBS Special Publication, 647-1; 19, 1983.

89. L Faria, The Economic Effects of Fracture in Europe, Final Report, study contract no. 320105, pp. 1–57, 1991.

90. DA Jones, Principles and Prevention of Corrosion, Prentice-Hall, NJ, pp. 235–291; 343–356, 1996.

91. SA Shipilov, Intl. Symp. on Environmental Degradation of Materials and Corrosion Control in Metals, M Elboujdaini, E Ghali (eds.), COM, Metsoc., pp. 225–242, 1999.

92. HH Strehblow, Corrosion Mechanisms in Theory and Practice, P Marcus, J Ouder (eds.), Marcel Dekker, NY, pp. 201–237, 1995.

93. MH Peterson, BF Brown, RL Newbigin, RE Groover, *Corrosion* **23**:142–148 (1967).

94. HP Van Leeuwen, *Engineering Fracture Mechanics* **6**:141 (1974).

95. EH Mazille, HH Uhlig, *Corrosion* **28**:427 (1972).

96. *ASM in Stress Corrosion Cracking*, RH Jones, RE Ricker (eds.), Jones and Ricker, Metals Park, Ohio, pp. 1–40, Schutze 256, 1992.

97. A Turnbull, Embrittlement by Local Crack Environment, RP Gangloff (ed.), Metallurgical Society, p. 3, 1984.

98. O Vosikowsky, *Journal of Engineering Materials and Technology* **97**(H) (4):298–304 (1975).

99. WT Tsai, A Moccaril, Z Szlarska-Smialowska, DD MacDonald, *Corrosion* **40**:573–584 (1984).

100. MJ Danielson, C Oster, RH Jones, *Corrosion Science* **32**:1 (1991).

101. NACE Standard MR175-90, *Standard Materials Requirements-Sulfide Stress Cracking Resistant Metallic Materials for Oilfield Equipment*, NACE International, Houston, TX, 1990.

102. M Elboujdaini, MT Shehata, W Revie, RR Ramsingh, Corrosion 98, Proc. of 53rd Annual Conference, San Diego, CA, NACE, Houston, TX, Paper no. 748, 1998.

103. MH Bartz, CE Rawlings, *Corrosion* **4**:187 (1948).

104. M Hay, COM, Shell Canada, 42nd Annual Conf. of Metallurgists of CIM, Aug. 24-27, Vancouver, BC, 2003.

105. EM Moore, DR McIntyre, *Journal of Material Engineering and Performance* **37**:77 (1998).

106. J Gutzeit, *Mat Perf*, **29**:54 (1990).

107. GM Buccheim, *Oil and Gas Journal* **92** (1990).

108. RD Merrick, Corrosion 87, NACE, Houston, TX, Paper no. 190, 1987.

109. RD Merrick, ML Bullen, NACE, Houston, Paper no. 269, 1989.

110. SP Lynch, P Trevena, *Corrosion* **44**:113–124 (1988).

111. FP Ford, *Corrosion Processes*, RN Parkins (ed.), Applied Science, pp. 271, 1982.

112. N Pessall, *Corrosion Science* **20**:225 (1980).

113. HH Lee, E Ghali, *Journal of Applied Electrochemistry* **19**:368 (1989).

114. G Cragnolino, LF Lin, Z Szlarska-Smialowska, Corrosion **37**:312–320 (1981).

115. RN Parkins, Prevention of Environment Sensitive Fracture by Inhibition, RP Gangloff (ed.), Metallurgical Society, pp. 385, 1984.

116. RN Parkins, *Mat Perf* **24**:9–20 (1985).

117. RM Latinision, Metallurgical Transactions **5**:483 (1974).

118. Z Szlarska-Smialowska, J Gust, *Corrosion Science* **19**:753 (1979).

119. Z Szlarska-Smialowska, *Hydrogen Embrittlement and Stress Corrosion Cracking*, R Gibala, RF Hehemann (eds.), ASM International, Vol. **99**, pp. 207–230, 1995.

120. HJ Bhatt, EH Phelps, Corrosion **17**:430t–434t (1961).

121. ASM, Stress Corrosion Cracking, RH Jones, RE Ricker (eds.), ASM International, Metals Park, pp. 1–40, 1982; Andersen 181–210; Miller 251–256; Shutz 256–297; Sprowls 336–415.

122. T Magnin, A Chambreuil, B Bayk, *Acta Metallurgica* **44**:1457 (1996).

123. RW Revie, HH Uhlig, *Acta Metallurgica* **38**:1313 (1990).

124. JA Beavers, IC Rosenberg, EN Pugh, Proc. 1972 Tri-Service Conference on Corrosion, MCIC 73-19, Metals and Ceramic Information Centre, pp. 57, 1972.

125. HH Uhlig, *Physical Metallurgy of Stress Corrosion Fracture*, TN Rhodin (ed.), Interscience, p. 1, 1959.

126. TE Perez, H Quintanilla, E Rey, Proceedings of NACE, Houston, TX, Paper no. 121, 1998.

127. RD Cayne, MS Cayard, Proceedings of Materials Resource Recovery and Transport, L Collins (ed.), Met. Soc. Of CIM, Calgary, pp. 3–49, Aug 1998.

128. BW Lifka, *Corrosion Testing and Standards*, R Baboian (ed.), ASTM, West Conshohocken, PA, XXX.

129. JE Hillis, Magnesium, in Corrosion Testing and Standards: Applications and Interpretation, R Baboian (ed.), ASTM, Philadelphia, PA, pp. 438–446, 1995.

130. Annual Book of ASTM Standards, Vol. **0302**, G.44–99˜R05; XXX

131. RN Parkis, Y Suzuki, *Corrosion Science* **23**:577 (1983).

132. K Ebtehaj, D Hardie, RN Parkins, *Corrosion Science* **28**:811–829 (1988).

2

CORROSION COSTS

2.1 INTRODUCTION

Studies on the cost of corrosion have drawn the attention of several countries, namely, the United States of America, the United Kingdom, Japan, Australia, Kuwait, Germany, Finland, Sweden, India, and China. The studies have ranged from formal and extensive efforts to informal and modest efforts. The common denominator of all these studies was that the annual corrosion costs ranged from 1% to 5% of the gross national product (GNP) of each country. Corrosion costs may be further divided into avoidable costs by using better corrosion control practices and the unavoidable costs that require new and advanced technology. The estimates of avoidable costs varied widely with a range of 10–40% of the total cost. Most of the studies allocated the corrosion costs to industrial sectors or to corrosion control categories such as products and services. All the studies addressed direct corrosion costs. A common conclusive feature was that the indirect costs because of corrosion damage are often significantly greater than the direct costs. It was also noted that the indirect costs were more difficult to estimate.

Potential savings and steps to achieve the savings have been discussed in most of the reports as formal results or as informal directions and discussion. The two most important and common findings are:

1. Better dissemination of the available existing knowledge and information through education and training, technical advisory and consulting services, and research and development activities.

2. The opportunity for significant savings through more cost-effective use of currently available means to reduce corrosion.

Challenges in Corrosion: Costs, Causes, Consequences, and Control, First Edition. V. S. Sastri.
© 2015 John Wiley & Sons, Inc. Published 2015 by John Wiley & Sons, Inc.

The review of prior studies on the costs of corrosion has provided useful background and direction for further study. The studies addressed only the extent and magnitude of possible savings, not the means of realizing such savings.

The review of the prior studies on the costs of corrosion is useful in the current and future studies. Both technical content and methods have been reviewed. Some specific areas where the previous studies have proven useful are:

1. Development of comprehensive lists of corrosion cost elements to be used in the analysis of total costs and costs in individual sectors.
2. Identification of categories in which to divide the total economy into two sectors, namely: (i) a set of industrial sectors, and (ii) a list of corrosion control methods.
3. Gathering background and reference information on the costs of corrosion and corrosion control methods.
4. Identification of preventive strategies and ascertaining potential savings.

2.2 DATA COLLECTION AND ECONOMIC ANALYSIS

The costs of corrosion have been a subject of interest to many researchers and the methods used for data collection and economic analysis are reviewed in chronological order.

2.2.1 The Uhlig Report (United States of America, 1949)

This study, concerned with "the cost of corrosion in the United States" and led by H.H. Uhlig (1), was the earliest effort to estimate corrosion costs. The annual cost of corrosion to the United States was estimated to be $5.5 billion or 2.1% of the 1949 GNP. This study arrived at the total costs by summing up the cost for both the owner/operator (direct cost) and for the users (indirect cost) of corroding components. The cost for the owners/operators was estimated by summing up cost estimates for corrosion prevention products and services used in the entire U.S. economy. They counted the total amount of corrosion prevention products and services through the entire economy (e.g., coatings, inhibitors, corrosion-resistant metals/alloys, and cathodic protection) and multiplied it by their prices. The costs for private consumers/users were evaluated as costs because of domestic water heater replacement, automobile repairs, and replacement of automobile mufflers. This method has the advantage in that the costs are readily available for well-defined products and services. The direct and indirect costs of corrosion are given in Table 2.1.

2.2.2 The Hoar Report (United Kingdom, 1970)

In March 1966, the UK Committee on Corrosion Protection was established by the UK minister of technology under the chairmanship of T.D. Hoar. In 1970, the

TABLE 2.1   Direct and Indirect Costs of Corrosion

	Item	Cost ($ million)	% of Total Cost
Direct costs	Paint	2000	36
	Metallic coatings and electroplate	472	9
	Corrosion-resistant metals	852	15
	Boiler and other water treatment	66	1
	Underground pipe maintenance and replacement	600	11
Indirect costs	Domestic water heater replacement	225	4
	Auto engine repairs	1030	19
	Auto muffler replacement	66	1

committee issued its report entitled, "Report of the Committee on Corrosion and Protection" (2).

The committee summarized the findings as follows: "We conservatively estimate the cost of corrosion in the United Kingdom as £1365 million per annum which amounts to 3.5% of the GNP in the year 1970. We believe that a saving of approximately £310 million per annum could be achieved with better use of current knowledge and techniques." The estimated savings represent approximately 20–25% of the total national corrosion costs.

The three most important findings of the Hoar report were the following:

1. The need for better dissemination of information on corrosion protection.
2. The need for more education in corrosion and protection.
3. The need for an increased awareness of the hazards of corrosion.

In order to achieve savings, a number of improvements have to be made on a national scale, particularly in the field of education and information dissemination.

The Hoar report determined the cost of corrosion for the industry sectors of the economy (2). The cost of corrosion for each industry sector was added together to arrive at the cost of corrosion for the entire UK economy. The report identified the sources for the cost of corrosion by sectors of the economy. It evaluated and summarized the direct expenditures (costs to owner/operator) in each economic sector. Indirect costs (costs for user) were not included in the studies.

Information was gathered from corrosion experts on site and surveys on expenditures for corrosion protection practices. Corrosion experts estimated the corrosion costs and the potential savings on the basis of their experiences with major economic sectors. Technical judgments and estimates of industry experts were used.

Information on education and research in corrosion was obtained by answers to a questionnaire distributed to academic institutions. The suitable questionnaire was also distributed to research associations, development associations, and government departments. Trade associations and professional bodies were used to gather information. The data gathered for a specific industry were used to estimate costs in other similar industries.

TABLE 2.2   UK National Cost of Corrosion

Industrial Sector	Estimated Corrosion Costs	
	£ x million	%
Building and construction	250	18
Food	40	3
General engineering	110	8
Government departments	55	4
Marine	280	21
Metal refining and semifabrication	15	1
Oil and chemicals	180	13
Power	60	4
Transport	350	26
Water	25	2
Total	£1365	100

The UK national costs of corrosion in major industrial areas are given in Table 2.2. These costs include direct costs of the industry and, in certain cases, those costs sustained by the users of the product because of maintenance or replacement. Costs from interactions among sectors were not included.

The studies reveal that the corrosion costs are substantial, but not higher than expected on the basis of consideration of annual expenditures for corrosion protection technologies. The annual expenditures in United Kingdom on protective coatings, including the cost of application, were estimated to be £772 million. Further, nearly £620 million were estimated for annual expenditures on corrosion-resistant austenitic stainless steels and nonferrous alloys. It was noted that these costs were not incurred solely for the purpose of corrosion resistance.

Avoidable corrosion costs: The Hoar report estimated that nearly 20–25% of the total corrosion costs could be saved by better use of current knowledge of corrosion control. For each industry, the percentage savings ranged from approximately 10% to 40% of the industry's corrosion costs.

The estimated potential savings by industry are given in Table 2.3.

Factors bearing on costs: The UK Committee and industrial organization listed 16 factors that could lower the cost of corrosion. The factors prioritized by combined judgment of experts in the field are as follows:

1. Better dissemination of existing corrosion control information.
2. Improved protective treatments.
3. Closer control over the application of existing protective treatments.
4. Improved design with the existing materials.
5. Greater awareness of corrosion hazards by the users.
6. Use of new materials.
7. Cost-effectiveness analysis of materials and protective treatments leading to procurement on the basis of total lifecycle costs.

TABLE 2.3   Estimated Potential Savings of the UK National Costs by Industry

Sector Names	Estimated Corrosion Costs £ × million	Estimated Potential Savings £ × million	Savings as Percent of Industry Corrosion Costs	Changes Required to Achieve Savings
Building and construction	50	250	20	More awareness in the selection, specification, control of corrosion protection
Food	4	40	10	More awareness in the selection of equipment and protection methods.
General engineering	35	110	32	Greater awareness of corrosion hazards in the design stage and throughout manufacture.
Government agencies and departments	20	55	36	More awareness mainly on defense items by better design and procedures
Marine	55	280	20	Improved design, awareness, and application
Metal refining and semifabrication	2	15	13	Improved awareness in plant and product protection
Oil and chemical	15	180	8	Improved effectiveness in the selection of materials and protection.
Power	25	60	42	Greater use of protection and improved awareness in design stage.
Transport	100	350	29	Change of exhaust system material and improved awareness in design stage.
Water	4	25	16	Improved awareness of corrosion protection

8. Previous feedback on service performance.
9. Improved specifications for protective treatments.
10. More basic research on corrosion mechanisms.
11. Improved communication between government departments.
12. Improved storage facilities.
13. Information on corrosion sensitivity of equipment.
14. Better nondestructive testing techniques.
15. Standardization of components.
16. More frequent or longer duration maintenance periods.

The single most important factor considered to be of great importance in the reduction of corrosion costs in the United Kingdom involves better dissemination of existing information on corrosion control. The effect of taxation in the United Kingdom on the costs of corrosion may play an important role. The UK taxation system encouraged a low capital investment and a high maintenance cost approach in some industrial sectors. Maintenance costs qualified for tax exemption since these costs could be expensed in the year in which they were incurred. Thus a company may choose to deliberately select inferior material for plant construction, resulting in a reduced capital investment but increased maintenance costs. From the foregoing discussion it is clear that the nature of the tax system results in an increase in corrosion costs.

2.2.3 Report of the Committee on Corrosion and Protection (Japan, 1977)

Japan conducted a survey of the cost of corrosion to its economy in 1977 through the Committee on Corrosion and Protection (3). The committee was chaired by G. Okamoto and was organized by the Japan Society of Corrosion Engineering and the Japan Association of Corrosion Control. The studies were supported by the Ministry of International Trade and Industry.

2.2.3.1 Total Costs The survey determined that the annual cost of corrosion to Japan was approximately 2.5 trillion yen (US $9.2 billion) in 1974. Estimated GNP of Japan was 1.36 trillion for the year 1974 and the cost of corrosion was about 1–2% of Japan's GNP. This study included only direct costs. The total costs would be much higher if indirect costs were included.

Method to Estimate Costs of Corrosion: Japan's committee estimated the cost of corrosion by:

1. Corrosion protection products and services
2. Corrosion cost by industry sector
3. Questionnaires and interviews, which were used to gather data from industrial personnel, who are well-informed experts.

TABLE 2.4   Costs to Prevent Corrosion by Protection Method

Protection Method	Cost (Yen × Billion)	Total Corrosion Costs (%)
Paints and protective coatings	1595	63
Surface treatment	648	24
Corrosion-resistant materials	239	9
Rust prevention oils	16	1
Inhibitors	16	1
Cathodic protection	16	1
Research	22	1
Total	2551	100

TABLE 2.5   Costs to Prevent Corrosion by Industry Sector

Industry Sector	Corrosion Cost (Yen × Billion)	Total Corrosion Cost (%)
Energy	60	6
Transportation	195	19
Building	175	17
Chemical industry	154	15
Metal production	27	3
Machinery and manufacturing	433	42
Total	1044	100

The Uhlig methodology was used to determine the corrosion costs on the basis of the cost of corrosion protection products and services such as coatings, inhibitors, corrosion-resistant materials, and cathode protection. The total cost amounted to 2.5 trillion yen (US $9.2 billion). Paint and protective coatings accounted for nearly US $6.1 billion. Surface treatments and corrosion-resistant materials accounted for nearly two-thirds of the corrosion costs. Surface treatments and corrosion-resistant materials amounted to nearly one quarter and one-tenth of the costs, respectively. The remaining 5% of the cost was assigned to other corrosion control methods (Table 2.4).

The Hoar methodology was applied to determine the cost of corrosion by specific industry sector. The results are summarized in Table 2.5.

The total costs by this method were nearly 1 trillion yen. Machinery and manufacturing had the highest cost of corrosion amounting to 40% of the total costs. The figures in the table show the corrosion costs to be substantial for all the sectors.

The difference between the two cost estimates determined by the methods of Uhlig and Hoar is quite large. The estimate by Uhlig's method is 1.5 trillion yen higher than the value estimated by the Hoar method. This large difference in estimated costs may be because of some omissions of some costs in the Hoar method of the estimation of costs. In general, the estimated costs by "industry sector analysis" (Hoar method) provides a higher cost than the "materials and services" (Uhlig method).

For example, the cost to prevent corrosion in the food industry was not calculated. The Uhlig method estimated the cost of surface treatment of tin-coated steel used in the production of cans at 79 billion yen. In addition, the cost to prevent corrosion by using tin-free steel (used for soft drink cans) was not included in the Hoar method of calculation. Thus more than 100 billion yen were omitted in the food industry alone in the Hoar method of calculation.

Another significant difference between the two estimation methods involved the treatment of painting costs in the transportation industry (such as ship, railroad, and motor vehicles). The cost of painting to prevent corrosion was estimated at more than 800 billion yen by the Uhlig method. According to the Hoar method, it was estimated at 200 billion yen. Thus the difference between the two methods of costing is nearly 600 billion yen with respect to the transportation industry.

Further to this there was another significant difference of 150 billion yen involving the building industry. The estimates made by the Hoar method were lower.

Even after accounting for the differences between Uhlig and Hoar approaches, the difference of about 400 billion yen remains to be accounted, which may be ascribed to the difficulties and uncertainties in the estimation of the costs of corrosion.

2.2.4 The Battelle-NBS Report (United States, 1978)

In response to a congressional directive, the National Bureau of Standards (NBS, now the National Institute of Standards and Technology (NIST)) studied the cost of corrosion in the United States. The analysis required in the study was contracted to Battelle Columbus Laboratories (BCL). The results of this work were published as NBS reports (4–6).

The Battelle-NBS study consists of a combination of the expertise of corrosion and economics experts to determine the economic impact of corrosion on the U.S. economy. A version of the Battelle National Input/Output Model was used to estimate the total corrosion cost. This model quantitatively identified corrosion-related changes in resources such as materials, labor and energy, changes in capital equipment and facilities, and changes in the replacement lives of capital equipment for all the sectors of the economy. The input/output model accounts for both the direct effects of corrosion on individual sectors and the interactions among various sectors.

The final results of the NBS-Battelle studies, after adjustments by NBS to the Battelle report for the base year of 1975, were: the total U.S. cost of metallic corrosion was estimated to be $70 billion, which amounts to 4.2% of GNP in 1975, of which $10 billion (15%) was estimated to be avoidable using the available corrosion prevention technology.

An uncertainty of ±30% for the total corrosion costs and still greater uncertainty in avoidable costs were estimated.

The final results were based on NBS analysis of the uncertainty in the Battelle input/output model estimates and adjustments to the Battelle results based on uncertainty analysis. For reference, Battelle estimated the total cost of corrosion to be $82 billion (4.97% of $1.677 trillion GNP in 1975). About 40% or $33 billion (2% of GNP) was estimated to be avoidable.

2.2.4.1 Method to Estimate Corrosion Costs The Battelle-NBS study (4–6) used an input/output methodology to estimate the cost of corrosion for the U.S. economy.

The U.S. economy was divided into 130 industrial sectors in the input model. For each industrial sector, experts were asked to estimate the costs of corrosion prevention such as the use of coatings, inhibitors, and the cost of repair and replacement of corroded parts.

The input–output (IO) analysis model was invented by Wassily Leontief who was awarded the Nobel Prize in 1973. The IO model is a general equilibrium model of an economy showing the extent to which each sector uses inputs from the other sectors to produce its output, and thus showing how much each sector sells to each other sector. The IO model shows the increase in economic activity in every other sector that would be required to increase the net production of a sector by, for example, $1 million. In the case of $1 million worth of paint required for corrosion prevention, the IO model would show the total activity in all sectors would amount to the $1 million worth of paint. The IO matrix was constructed by the U.S. Department of Commerce on the basis of the census of manufacturers in 1973 and represents the actual structure of the U.S. economy at that time. The IO model has been very invaluable for planning. The IO framework has also been useful in estimating the total economic activity that will result from net additional purchases from a sector and the total economic loss because of closure of an industry.

Economic IO analysis accounts for direct (within the sector) and indirect (within the rest of the economy) inputs to produce a product or service by using IO matrices of a national economy. Each sector represents a row or a column in the IO matrix. The rows and the columns are normalized to add up to one. When selecting a column (industrial sector P) the coefficients in each row would tell how much input from each sector is needed to produce $1 worth of output in industry P. For example, an IO matrix might indicate that producing one dollar worth of steel requires 15 cents worth of coal and 10 cents of iron ore. A row of matrix specifies to which sectors the steel industry sells the product. For example, steel might sell $0.13 to the automobile industry and $0.06 to the truck industry of every dollar of revenue.

Elements were identified within the various sectors that represented corrosion expenditures such as coatings for pipelines. The coefficient of coatings for the steel pipelines was modified so that, for instance, pipelines spend nothing on coatings, where the purpose of coatings is to prevent corrosion. After the coefficients in the steel pipeline column are modified, the column is normalized to add to one. This new matrix represents the world without corrosion. With the new matrix, the level of resources used to produce GNP in a world of corrosion would result in higher GNP than in a world without corrosion.

The Battelle-NBS study collected data on corrosion-related changes in the following:

1. Resources (materials, labor, energy, value added required to produce a product or service).
2. Capital equipment and services.
3. Replacement rates for the capital stock of capital items.
4. Final demand for the product.

On the basis of these data, coefficients in the IO model were adjusted. Data were gathered from interviews with experts in the field, review of literature, and use of technical judgment.

The total cost of corrosion was defined as that increment of total cost incurred because of corrosion. It was also noted as to what cost would not be incurred if corrosion did not exist. Thus it developed into three scenarios:

1. Existence of corrosion.
2. Nonexistence of corrosion.
3. Practice of economically most effective corrosion prevention method.

The IO model was constructed involving all three scenarios. The total national cost of corrosion corresponds to the difference in GNP of scenarios one and two.

The corrosion costs were divided into avoidable and unavoidable costs.

1. Avoidable corrosion costs are those because of the use of corrosion control.
2. Unavoidable costs are those that are not amenable to reduction by the presently available technology.

The following direct costs are included in the calculation of the total cost:

1. Replacement of equipment or buildings.
2. Loss of product.
3. Maintenance and repairs.
4. Excess capacity.
5. Redundant equipment.
6. Corrosion control (inhibitors, organic and metallic coatings).
7. Engineering research and development testing.
8. *Design*: material of corrosion not for structural integrity; material of corrosion for purity of product; corrosion allowance and special processing for corrosion resistance.
9. Insurance.
10. Parts and equipment inventory.

2.2.4.2 Sector Costs on the Basis of the Input–Output Model
The Battelle-NBS study used the IO model to estimate the cost of corrosion to sectors. In addition to the IO model, the report focused on four areas: federal government, personally owned automobiles, the electric power industry and loss of energy and materials.

On the basis of the experts' judgment, industry indicators (coefficients in the IO matrix) were calculated to obtain the cost of corrosion for any given sector. These indicators were based on expert judgment as to how much purchases can be avoided if there were no corrosion. The effects of corrosion reflected in: (i) changes to

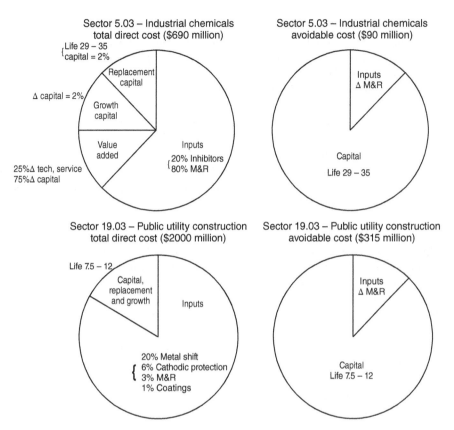

Figure 2.1 Breakdown of industry indicators into its components (2).

material inputs such as coatings, corrosion inhibitors, corrosion-resistant materials; (ii) changes in capital equipment; and (iii) changes in technical services.

The breakdown of industry indicators into its components is shown in Figure 2.1 for industrial chemicals and public utility sectors.

Inputs. These include the costs of coatings and plating for corrosion control, corrosion inhibitors, maintenance and repair, corrosion-resistant metals, and cathodic protection.

Capital Replacement. Replacement of capital equipment and associated facilities in industry is affected by corrosion.

Growth Capital. The costs of capital equipment are affected by corrosion through replacement lives of capital equipment.

Value Added. Activity of industry is affected by corrosion through inputs such as costs of research and development and technical services. When the coefficients (industry indicators) in the IO matrix are modified to reflect the absence of corrosion, the IO matrix can be used to indicate the inputs needed to produce the same bundle of goods and services that consumers purchased in the world with corrosion. The IO matrix will indicate cost-savings because of the absence of corrosion. One needs less input to produce the same output if there was no corrosion. The savings because of the difference between the world with corrosion and the world without corrosion are indicated in dollars.

The impact of corrosion is that one has to spend more money in the world with corrosion than in the world without corrosion. This difference is the cost of corrosion. In the Battelle-NBS study, the cost of corrosion was determined as a percentage of sales and on a dollar basis. The highest total costs of corrosion were attributed to mining, manufacturing, public utilities, and construction. The highest corrosion costs were in wholesale and retail trade, auto manufacture, livestock, and petroleum refining.

The source of corrosion costs varies considerably from one industry to another. Replacement of equipment accounts for corrosion costs in wholesale and retail sections in the livestock sector. In the case of industrial chemicals, the largest part of total corrosion costs is because of inhibitors/water treatment and maintenance and repair. In the case of public utility construction, the largest portion of total corrosion cost is because of inputs such as corrosion-resistant materials, cathodic protection, and coatings.

Battelle-NBS studies centered on corrosion cost analysis of four special areas of the economy, such as:

1. Federal government.
2. Personally owned automobiles.
3. Electric power industry.
4. Loss of energy and materials.

2.2.4.2.1 Federal Government The four agencies that had considerable capital equipment are: (i) Department of Defense; (ii) National Aeronautics and Space Administration (NASA); (iii) U.S. Coast Guard; (iv) U.S. Government Services Administration (GSA).

The total cost of corrosion to the U.S. federal government was estimated to be $8 billion out of which capital cost is $6 billion and maintenance cost is $2 billion. The total corrosion costs represent 2% of the total federal budget of $400 billion.

The total capital and maintenance costs of government assets are listed in Table 2.6.

2.2.4.2.2 The Automobile Sector The total corrosion cost of automobiles was determined to range from $6 billion to $14 billion. The avoidable cost was estimated between $2 billion and $8 billion.

The principal areas of automotive corrosion were associated with the degradation of iron and steel components, which amount to 80% of the weight of the automobile. The elements of automobile costs are:

TABLE 2.6   Capital and Maintenance Costs

Category	Total Capital ($× billion)	Maintenance Cost ($× billion)
Aircraft	195	0.99
Ships	56	0.40
Buildings and Real estate	144	0.655
Total	395	2.045

1. The cost of built-in corrosion protection included in the purchase price of the car.
2. The portion of maintenance and operating costs attributable to corrosion.
3. The cost of premature replacement of automobiles because of corrosion.

The built-in costs of corrosion of automobiles were identified as corrosion protection for steel body panels such as metallic zinc coatings, paint, adhesives and sealants, nonferrous metals, corrosion-resistant materials, rust-proofing heat exchanger components, mufflers, and tail pipe corrosion. The greatest impact on the cost of corrosion for automobiles was the adverse effect of corrosion on the cost of replacement of the automobile. Both the IO model and focused sector study showed that in both models the cost of replacement of the automobiles dominated the total cost and avoidable cost estimation.

2.2.4.2.3   The Electric Power Industry   The total corrosion costs in the electric power industry in the generation and distribution of power were estimated to be $4 billion and the corrosion-related expenditure was estimated at $1.1 billion. A significant portion of excess capacity of power plants was attributed to corrosion, where corrosion-related excess capacity was assumed to be approximately 10% of the total capital investment.

The two main segments of the electric power industry are: (i) generation and (ii) transmission/distribution of electricity. The five types of plants are fossil fuel, hydroelectric, nuclear, geothermal, and solar. Corrosion costs varied depending on the type of the plant. It was found that corrosion greatly increased the frequency and duration of outages, resulting in significant costs. For the transmission and distribution of electricity, atmospheric corrosion, and underground corrosion of buried structures were found to be primary contributors to corrosion costs.

The output of the Battelle-NBS analysis was used to estimate the additional energy and materials consumed because of metallic corrosion. Approximately 3.4% of the country's energy consumption ($1.4 billion) was related to corrosion. Within the energy sector, the impact of corrosion was greater on coal usage than on petroleum or natural gas usage. Nearly 0.6% of energy consumption or 0.23 billion dollars was estimated as avoidable. Nearly 17% of the nation's demand for metallic ores ($1.4 billion) resulted from corrosion and about one-eighth of that (2.1%) of metallic ore demand or $180 million was judged to be avoidable. The accomplishments of the Battelle-NBS study are the following:

1. A measure of the severity of corrosion costs.
2. An indication of where and how the impacts of corrosion are felt.
3. A useful method for the analysis of corrosion costs.
4. Bibliography and database on corrosion economics.
5. Reference point for the impact of corrosion on other factors affecting the economy.
6. Assessment of the economic effect of proposed means to reduce corrosion costs.
7. Identification of sectors where affordable and unavoidable corrosion costs are encountered.

2.2.5 The Economics of Corrosion in Australia

Since the 1950s, many countries have attempted to assess their national corrosion costs. In 1973, R.W. Revie and H.H. Uhlig estimated that the direct losses to the Australian economy caused by corrosion (7) were "in the order of A\$470 million," which was equivalent to 1.5% of the GNP for Australia in 1973. However, no element for indirect losses caused by corrosion was included. Revie and Uhlig believed that the indirect losses were high, but not easily estimated. Since indirect costs may equal or exceed the direct costs, total corrosion costs to Australia could have been estimated to be about 3% of GNP in 1973. Assessments of uncertainties or a separation of costs into avoidable and unavoidable components were not quantified.

The 1973 estimate of A\$470 million was noted to be about double the 1955 estimate of \$240 million made by Worner (8) and was further compared with an estimate of the annual cost of corrosion to Australia of \$900 million published in 1972 by E.C. Potter in 1972. Potter (9) used a published estimate of corrosion costs to Great Britain (2) equal to 3.5% of GNP and included corrections relevant to Australia to arrive at A\$900 million corrosion costs. Potter's cost estimate is higher than that of Revie and Uhlig because some elements for the indirect costs of corrosion were included in the Potter report.

2.2.5.1 The NBS/B.C.L. Economic Model in Relation to the Australian Economy
The methods of N.B.S./B.C.L. econometric model in relation to the Australian economy was considered appropriate to determine the costs of corrosion and evaluate the potential economic savings that might accrue to the Australian community with the improved application of corrosion mitigation technology. By applying the B.C.L. methodology to the Australian economy, a figure for the potential economic savings to the Australian community has been computed. As in the B.C.L. model, it has been calculated by forming a judgment of the extension of the useful life of capital assets by the application of an input/output econometric analysis.

2.2.5.2 Concepts and Definitions Relating to Corrosion as Developed by N.B.S./B.C.L. and as Applied to the Australian Economy
Corrosion was restricted to metals and was defined as the degradation of metals where the environment – aqueous or gaseous – contributed to the mode of degradation. This included

general or local metal wastage by dissolution, stress corrosion cracking, corrosion fatigue, erosion corrosion, oxidation, and sulfidation. Phenomenon such as creep, mechanical damage, and stress rupture were excluded.

Elements of the cost of corrosion were identified to provide a convenient checklist for the consistent treatment of costs in all segments of the economy.

Capital Costs

Replacement of equipment and buildings

Excess capacity

Redundant Equipment

Control Costs

Maintenance and repair

Corrosion control

Design Costs

Materials of construction

Corrosion allowance

Special processing

Associated Costs

Loss of product

Technical support

Insurance

Parts and equipment inventory

Costs incurred for coatings of buried steel pipelines are corrosion costs while painting of automobiles are ascribed partly to corrosion and partly to aesthetics. In this case, one half of the cost of painting of automobiles is ascribed to corrosion. Similar judgments were made in other areas where multipurpose operations were encountered.

Several items have been excluded in the B.C.L. model from the cost of corrosion. The loss of life and goodwill were not considered. Catastrophic and one-time costs were not included. Costs because of oil spill, advertising, and marketing costs related to corrosion resistance were not included.

Total corrosion costs were defined by B.C.L. as all costs above those incurred in the absence of corrosion. Avoidable corrosion costs were defined as those costs that are amenable to reduction by the most economically efficient use of presently available corrosion control technology.

Determination of avoidable costs was difficult as the best practice was difficult to measure in economic terms and there was limited database available for avoidable

TABLE 2.7   Best Practice Ratings

Industry	Incentive	Response
Industrial chemical	High	High
Livestock products	Moderate	Low
Wholesale and retail trade	Low	Low

costs of corrosion. Economic parameters such as return on investment, discounted cash flow, present and future worth of money had to be incorporated into the data.

To determine avoidable corrosion maintenance costs, B.C.L. determined a best practice rating for each industrial sector. Best practice ratings were determined by a qualitative comparison of industrial sectors as to their incentive or urge to use the best corrosion practice and their response or capacity to use the best corrosion practice. Each industrial sector was rated as high, medium, or low in both categories.

Qualitative ratings for incentive and responsiveness were based on: (i) relationship of profits to corrosion cost; (ii) quality of product; (iii) awareness of corrosion; (iv) regulation; (v) safety; (vi) personal responsibility; (vii) consequences of failure.

The factors considered in rating are: (i) size of the company (ii) level of technology; (iii) availability of corrosion expertise; (iv) time frame over which costs are incurred; (v) complexity of the problem; and (vi) administrative system through which practices are implemented. Priority is of the utmost importance in cases of the release of toxic gases.

Table 2.7 lists three of the best practice ratings according to B.C.L.

On the basis of these ratings, maintenance and repair costs were reduced from 5% for high ratings to 45% for low ratings. These ratings are considered by B.C.L. to be realistic.

2.2.5.2.1   Rationale of the Model to Determine the Cost of Corrosion   Three economic scenarios representing the U.S. economy were presented. The total cost of corrosion was defined as the difference between the GNP of a world of corrosion and a hypothetical world devoid of corrosion. To estimate the avoidable costs of corrosion another construct (World IV) was formulated in which best corrosion prevention practice was used by everyone. The difference in costs between World I and III represents the total national avoidable costs of corrosion.

To determine the costs of corrosion, B.C.L. identified the following four data requirements:

1. Inputs required to produce a product.
2. Capital equipment to produce a product.
3. Replacement lives of the capital equipment.
4. Final demand for the product.

The data were gathered by a comprehensive survey of U.S. industry and government, a literature survey, and interviews with industry experts. All the information gathered was reduced to coefficient and range changes so that adjustments could be made in the I/O model to account for corrosion costs on moving from economy I to World II (no corrosion) and World III, a world of best corrosion practice where all avoidable costs are eliminated.

In the B.C.L. model, World II and World III flow coefficient adjustments were stated as percentages of the corresponding World I coefficient. Capita/output coefficient adjustments were expressed as a percentage of excess capacity for an entire industry or as a percentage of a specific World I coefficient for redundant equipment adjustments. Replacement adjustments were expressed as change in years such as 10–30 years for World I and 20–30 years for World II. Final demand changes were expressed as a percentage change of World I values.

Thus, the input/output model used by N.B.S./B.C.L. captured all direct and indirect costs of corrosion through the creation of a total economic model for each chosen scenario.

Corrosion lifetimes of capital assets deserve particular attention, as replacement lives of capital items are affected by corrosion. The items are replaced at a greater rate than the rate when corrosion is absent. Corrosion costs of this type were treated by B.C.L. in the I/O model by adjusting the replacement life of the corroded capital. Capital replacement is also necessary because of wear and obsolescence.

Premature failure because of corrosion is well defined but proved to be difficult to be treated quantitatively in the B.C.L. model. The two difficulties encountered are:

1. Paucity of data on replacement lives of capital equipment.
2. Each capital-producing sector in the I/O model produces a bundle of goods and as a consequence of this, each capital sector produced diverse products with respect to the effects of corrosion.

As a starting point, in World I, a range of single values for replacement life in years was assigned by B.C.L. to each capital-producing sector on the basis of U.S. Internal Revenue Service data for depreciation rates. These replacement lives provided a useful starting point, although the replacement lives did not represent actual lives in service.

B.C.L. used the following procedures to estimate changes in replacement lives from the base range because of corrosion.

1. Products by industrial sectors that are affected by corrosion were identified.
2. The effect of corrosion on replacement life was estimated as "minor, moderate, or major."
3. The base ranges for replacement lives in World I were changed to reflect the relative impact of corrosion (World II).
4. The best corrosion practice (World III) was identified.

The procedure used by B.C.L. permitted changes in a consistent manner with the identification of relative magnitude of corrosion effects even in the absence of quantitative service life information. The judgment of the magnitude of corrosion effects was made on the basis of the general knowledge of the types of capital, their materials of construction, and the severity of corrosion in service.

The discussion in B.C.L. stated that although the ranges of replacement lives and changes from World I and World II or World III did not reflect actual service lives, they did represent the magnitude of the effects of corrosion.

Because of the lack of quantitative data for replacement lives and the effect of corrosion on replacement lives, it was recognized that estimates of costs associated with this element have a greater degree of uncertainty than that of other elements.

Another factor of importance with respect to the corrosion costs is the disruption cost of corrosion. When a structural system corrodes, the costs consequences are the sum or all of: (i) repairs; (ii) replacement; (iii) disruption to third parties dependent on that system.

Thus, if a bridge corrodes, there will be costs of repair or replacement. In addition, there may be disruption costs to the users of the bridge as a consequence of reduced opportunities of use. Similarly, if a high pressure gas pipeline fails because of corrosion, the disruption costs to a third party incurred as a result of diminished or halted gas supplies would far outweigh the immediate incurred costs of repair and replacement of the gas pipeline. It is known that disruption costs to third parties can exceed by a factor of 10 times the direct costs of repair and replacement. If the direct costs are of the order of $10 million, which may be considered as reasonable for the gas pipeline case, the total costs including disruption costs are therefore likely to be large. If the cause of the corrosion failure had been avoidable because of the available existing information, then the disruption costs incurred by the community would have been avoidable.

In such cases, the third parties in question have little or no control over the use of suitable corrosion control measures. Thus, in such cases where the savings from corrosion prevention measures accrue to different persons or bodies from those who would have to bear the costs of the corrosion prevention measures, then the government alone can tax such beneficiaries for the overall benefit of the community.

Another specialized cost of relevance occurs when design codes require excessive expenditure on measures to protect against corrosion because of an expected low level of capability for anticorrosion maintenance. Thus, for example, for a construction project, if it is incorrectly assumed that the capability for suitable anticorrosion maintenance is inadequate such that unnecessary overspecifications for corrosion protection measures are selected, then the costs of the construction project would be greater than necessary. If, under this context, adequate capabilities for anticorrosion maintenance had in fact existed, then the extra costs incurred would have been avoidable.

In Australia, unnecessary overspecification of corrosion protection measures occurs because of concern over sufficient long-term anticorrosion maintenance because of a lack of awareness of the extent of likely corrosion effects.

The costs incurred unnecessarily either by an industrial organization or by the community could be probably reduced by an improvement in the available information dissemination process for corrosion prevention measures.

Some community benefits cannot be considered purely as a function of market forces. The perception of time-value money of an individual is different from that of a community. The time-value of money is very high and the consumer will not spend money at the present time for a benefit occurring in the future unless the benefit is clearly apparent. If the values of corrosion prevention are not perceived at the time of purchase, then there may be insufficient incentive to pay for those corrosion prevention incentives that would preserve community resources in the long term.

The purchase of corrosion prevention measures in consumer goods is similar to the purchase of safety features in automobiles such as seat belts in cars. The purchaser may not pay for this feature although the long-term benefits in saving lives and money are present. In the case of automobile safety measures, the dilemma for the community is resolved by the government regulation of mandatory use of seat belts in automobiles. In the case of corrosion prevention measures, government intervention could lead to improved corrosion prevention awareness and practice.

Capital works are undertaken by local community forces, and these organizations must also take an active role in propagating the benefits of corrosion prevention.

With regard to the importance of corrosion to the community at large, a significant market imperfection is that the community's conception of corrosion is different from the effects of corrosion in reality. In general, the community at large perceives "corrosion" as unimportant.

In many cases, the consequences of corrosion are not even attributed to "corrosion." When a building collapses because of the corrosion of reinforcing steel bar in concrete, it is attributed to "faulty materials" instead of corrosion.

Corrosion is often perceived as a slow, inevitable degradation process. It is generally viewed as "rust," and it is plainly visible in rusting cars or corrugated iron sheeting. But many people are unaware of the significance of insidious corrosion effects, for example, on modern, technically advanced aircraft.

There is a widespread view that the effects of corrosion are insignificant and hence very little attention is given to the likely effects of corrosion and how these effects can be avoided. This erroneous perception of the neglect of corrosion effects may explain why it is surprising to note the total and avoidable costs of corrosion to be significant.

Another significant market imperfection is the lack of technology transfer that leads to the lack of awareness about corrosion and the available methods to combat corrosion. There are many instances where unnecessary costs are incurred because of incorrect specification of materials or where inadequate corrosion protection measures have been applied to proprietary equipment. The information that is required to avoid the corrosion problem was available and could have been used to control the corrosion problem.

Unnecessary costs are incurred for repair and replacement because of improper selection of materials such as the stainless steel components for environments where they fail to perform satisfactorily. Concrete spalling because of the corrosion of reinforcing steel may be a consequence of a lack of awareness of the reasoning used in construction codes of practice. Thus the costs incurred in the case of premature degradation of public buildings by corrosion must be met by the community and financed by third parties for the repairs.

An important aspect of corrosion is public safety, which may be at stake. In such cases, governments do intervene in market forces under certain circumstances. In the absence of corrosion mitigation technology, public expenses will be large. Here the concept of "internalized and externalized" liabilities may be important. For instance, governments regulate by prescribing maintenance schedules for aircraft and rigorously control standard safety. Unchecked, corrosion can cause an aircraft crash. Thus maintenance standards are put in place to avoid crashes.

If corrosion is the cause of an aircraft crash into the sea, the costs are effectively, partly internalized as the airlines obtain their money for anticorrosion measures from fare-paying passengers. However, if the aircraft crashes on a city suburb, third parties are involved, and the costs are externalized. Internalized costs represent an argument for industry support for corrosion information and dissemination and research facilities. Externalized costs refer to government support.

2.2.6   Kuwait (1995)

In 1992, Kuwait conducted an economic assessment of the total cost of corrosion to its economy using a modified version of the Battelle-NBS IO model. The resulting report containing the assessment can be found in "Economic Effects of Metallic Corrosion in the State of Kuwait" authored by F. Al-Kharafi, A. Al-Hashem, and F. Martrouk, Final Report no. 4761, KSIR Publications, December 1995.

The total cost of corrosion was estimated at about $1 billion 1987 dollars representing 5.2% of Kuwait's 1987 gross domestic product (GDP). Avoidable corrosion costs were estimated at $180 million or 18% of the total cost.

The method to estimate corrosion costs was patterned after Battelle-NBS IO method. Data gathered and information required for the model's adjustments came from three sources: (i) data compiled from a survey specifically designed for industries in Kuwait; (ii) judgment of experts in corrosion in Kuwait; (iii) experience of other countries and previous studies.

The questionnaire or survey was the choice for data compilation. The elements considered were: replacement of equipment and buildings, excess capacity and redundant equipment, loss of product, maintenance, and repair and corrosion control.

The costs of corrosion for two years, 1987 and 1992, were analyzed and the costs for 1992 were less than those in 1987 because of the following factors:

1. The economy in 1992 was smaller than in 1987.
2. The 1992 model assumed a more efficient economy with respect to cost of corrosion for each output in the IO model.
3. The study assumed that the economy in 1992 was operating below capacity.
4. The study assumed that only more efficient equipment was used in the production process.

On the sector level, the estimates for total cost of corrosion in the oil sectors (crude petroleum and petroleum refining) were $65 million in 1987. The avoidable cost was estimated to be $10 million in 1987. The commercial services sector, the government, and the social and household services sectors were responsible for the largest share (70%) of the total cost.

2.2.7 Costs of Corrosion in Other Countries

There is a brief account of the costs of corrosion in other countries such as West Germany (1969), Finland and Sweden (1965), and India (1961).

2.2.7.1 West Germany A study of the costs of corrosion was conducted at the end of the 1960s. The total cost of corrosion was estimated to be 19 billion Deutsch Marks (DM) (U.S. $6 billion) for the period 1968–1969. Of this cost, 4.3 billion DM (U.S. $1.5 billion) was estimated to be avoidable. Thus, the total cost of corrosion equivalent to 3% of the West German GNP for 1969 and avoidable costs were roughly 25% of the total corrosion costs (10).

2.2.7.2 Finland and Sweden Finland conducted a study of the cost of corrosion in 1965, and the cost of corrosion was estimated to be between 150 and 200 million markka (U.S. $47 million and U.S. $62 million) for the year 1965. Linderborg referred to these losses in the paper describing factors that must be taken into account in assessing corrosion cost (11) to Finland.

Linderborg quotes a partial study of the corrosion costs in Sweden, in which painting expenses to combat corrosion were analyzed for the year 1964. These costs amounted to 300–400 million crowns (U.S. $58–75 million) of which 25–35% were found to be avoidable.

2.2.7.3 India (1961) India conducted a study of corrosion in 1961. The cost of corrosion was estimated (12) at 1.54 billion rupees (U.S. $320 million) for the period 1960–1961. This estimate was based on calculations of expenditures for certain measures to prevent or control corrosion, including direct material and labor expenses for protection, additional costs for increased corrosion resistance and redundancy, cost of information transfer, and funds spent on research and development.

Corrosion control costs consisted of:

25% for paints, varnishes, and lacquers
20% for metallic coatings and electroplating
55% for corrosion-resistant metals

2.2.7.4 China (1986) Preliminary cost of corrosion was conducted in 1980 although no nationwide corrosion losses have been reported (13). A total of 148 enterprises in the chemical sector were surveyed. The comprehensive results of 10 enterprises showed an average corrosion of 4% of the annual income. The results of another survey of an iron and steel complex indicated corrosion costs of 1.6% of their annual income.

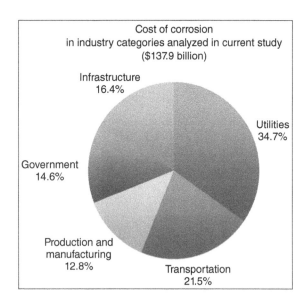

Figure 2.2 Percentage contribution to the total cost of corrosion for the five sector categories (15).

2.2.7.5 Corrosion Costs by Industry Sector in the United States The U.S. economy was divided into five major sector categories for the analysis in corrosion cost studies, and these were further divided into 26 sectors. The categories were infrastructure, utilities, transportation, production and manufacturing, and government. When added together, the total direct costs for the categories were $137.9 billion (Fig. 2.2). This figure was then extrapolated to the total U.S. economy ($8.79 trillion) for an annual cost of corrosion of $276 billion.

2.2.7.5.1 Infrastructure Aging infrastructure is one of the most serious problems. In the past, attention was focused primarily on new construction involving specification of materials and designing corrosion prevention and control systems for buildings, bridges, roads, plants, pipelines, tanks, and other elements of infrastructure. At present, aging infrastructure is nearing the end of its design lifetime, which requires attention on maintaining and extending the life of these valuable assets. The annual direct cost of the infrastructure was estimated to be $22.6 billion (Fig. 2.3).

2.2.7.5.2 Highway Bridges There are nearly 583,000 bridges in the United States out of which 200,000 are constructed of steel, 235,000 are conventional reinforced concrete, 108,000 are constructed using prestressed concrete, and the remainder is made of other construction materials. Nearly 15% of these bridges are structurally deficient because of corroded steel and steel reinforcement. Annual direct cost estimates total $8.3 billion including $43.8 billion to replace deficient bridges over 10 years, $2 billion for maintenance and capital costs for concrete bridge decks,

Infrastructure, $22.6 billion

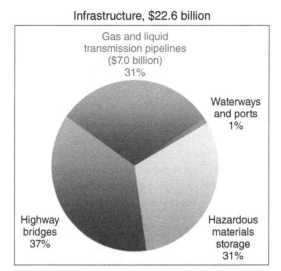

Figure 2.3   Annual cost of corrosion of gas and liquid transmission pipelines (15).

$2 billion for their concrete substructures, and $0.5 billion for painting steel bridges. Indirect costs to the user, such as traffic delays and lost productivity were estimated to be as high as ten times that of direct corrosion costs.

2.2.7.5.3   Gas and Liquid Transmission Pipelines   Corrosion is the primary factor affecting the longevity and reliability of pipelines that transport crucial energy sources throughout the country. There are more than 528,000 of natural gas transmission and gathering pipelines, 119,000 km (74,000 miles) of crude transmission and gathering pipelines, and 13,200 km (82,000 miles) of hazardous liquid transmission pipelines. The average annual corrosion-related cost estimated at $7 billion to monitor, replace, and maintain the assets and the corrosion-related cost of operation and maintenance amounted to $5 billion.

2.2.7.5.4   Waterways and Ports   Waterways and ports play a vital role in moving people and commerce in the United States. There is a total of 40,000 km (25,000 miles) of commercial navigable waterways that serve 41 states with hundreds of locks that permit travel. Corrosion is found to occur on piers and docks, bulkheads and retaining walls, mooring structures and navigational aids. On the basis of information from the U.S. Army Corps of Engineers and the U.S. Coast Guard, the expenditure is estimated at $0.3 billion. This estimate is considered to be low since corrosion costs for harbor and other marine structures were not included.

2.2.7.5.5   Hazardous Materials Storage   There are nearly 8.5 million aboveground and underground storage tanks (USTS) that contain hazardous material. Government has strict regulations on leaks of hazardous material from storage tanks into the environment. Tank owners must comply with the requirements of the U.S. Environmental Protection Agency for corrosion control and overfill and spill protection or face substantial costs toward cleanup and penalties. The total annual direct cost of corrosion for hazardous materials storage is $7 billion and $4.5 billion for aboveground storage (AST) and $2.5 billion for USTS.

2.2.7.5.6 Airports The world's most extensive airport system operates with 5324 public-use and 13,774 private-use airports. The airport infrastructure components consist of natural gas and jet fuel storage and distribution systems, natural gas feeders, vehicle fueling systems, dry fire lines, parking garages, and runway lighting. Since each of the systems is generally owned or operated by different organizations, the impact of corrosion on airports could not be quantified.

2.2.7.5.7 Railroads The U.S. railroad operates 274,399 km (170,508 miles) of railways all over the country including freight in regional and local railroads. The materials subject to corrosion are rail and steel spikes. An accurate cost of corrosion could not be determined with respect to railroad components.

2.2.7.5.8 Utilities Utilities consist of gas, water, electricity, and telecommunications services and account for the largest portion of industrial corrosion costs. The total direct corrosion costs amount to $47.9 billion. These costs are partitioned into sectors of gas distribution, drinking water and sewer systems, electrical utilities, and telecommunications (Fig. 2.4).

2.2.7.5.9 Gas Distribution The nation's natural gas distribution has 2,785,000 km (1,730,000 miles) of relatively small-diameter, low-pressure piping that includes 1,739,000 km (1,080,000) miles of distribution mains and 1,046,000 km (650,000 miles) of services. Many mains (57%) and service pipelines (46%) are made of steel, cast iron, or copper that are prone to corrosion. The total annual direct cost of corrosion was estimated to be $5 million.

2.2.7.5.10 Drinking Water and Sewer Systems According to the American Water Works Association (AWWA) industry data base, there are approximately 1,483,000 km (876,000 miles) of municipal water piping in the United States. The sewer system consists of 16,400 publicly owned treatment facilities that release about 155 million m^3 (41 billion gallons) of waste per day. The total annual direct cost of corrosion for drinking water and sewer system is $36 billion, which includes the costs of replacing aging infrastructure, lost water from unaccounted leaks, corrosion inhibitors, internal mortar linings, external coatings, and cathodic protection.

2.2.7.5.11 Electrical Utilities There are seven generic types of electricity-generating plants such as fossil fuel, nuclear hydroelectric, cogeneration, geothermal, solar, and wind. The major sources in use in the United States are fossil fuel and nuclear supply systems. The direct cost attributed to corrosion was $6.9 billion with the largest amount for nuclear power ($4.2 billion), fossil fuel (1.9 billion), hydraulic power ($0.15 billion), transmission, and distribution (0.6 billion).

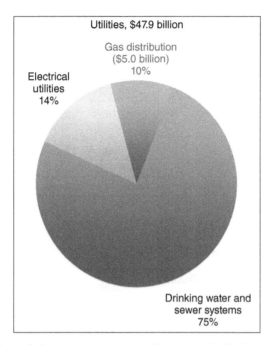

Figure 2.4 Annual cost of corrosion of gas distribution (15).

2.2.7.5.12 Telecommunications The telecommunications infrastructure includes hardware such as electronics, computers, and data transmitters, as well as equipment shelters and towers used to mount antennas, transmitters, receivers, and television and telephone systems. Towers and shelters are commonly painted or galvanized for corrosion protection. Costs are also associated with corrosion of buried copper grounding beds and galvanic corrosion of grounded steel structures. No corrosion cost was determined for the telecommunications sector because of the lack of information in this rapidly changing industry. In addition, many components are being replaced before failure because their technology quickly becomes obsolete.

2.2.7.5.13 Transportation The transportation category includes vehicles and equipment, such as motor vehicles, aircraft, rail cars, and hazardous material transport (HAZNAT). The annual corrosion cost in this category is 29.7 billion (Fig. 2.5).

2.2.7.5.14 Motor Vehicles U.S. consumers, business, and government organizations own more than 200 million registered motor vehicles. Car manufacturers have dramatically increased the corrosion resistance over the past two decades by using corrosion-resistant materials, better manufacturing processes, and more effective engineering and design. It is also thought that individual components need further improvement.

   The total annual direct cost of corrosion was estimated at $23.4 billion with $14.46 billion attributed to corrosion-related depreciation of vehicles. An additional $6.45

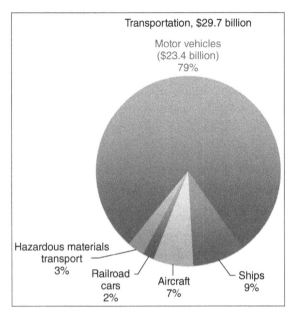

Figure 2.5   Annual cost of corrosion in the motor vehicles industry (15).

billion is spent on repairs and maintenance made necessary by corrosion and $2.56 billion represents increased manufacturing costs from corrosion engineering and the use of expensive corrosion-resistant materials.

2.2.7.5.15   Ships   The number of ships in the United States consists of: (i) 737 vessels on the Great Lakes; (ii) 33,668 inland and 7014 ocean vessels; (iii) 12.3 million recreational boats and 122 cruise ships serving North American ports. The total cost of corrosion in shipping industry is $2.7 billion out of which $1.1 billion is for new ship construction, $0.8 billion for maintenance and repairs, and $0.8 billion for corrosion-related downtime.

2.2.7.5.16   Aircraft   In 1998, the combined commercial aircraft fleet operated by U.S. Airlines numbered more than 7000 airplanes. Airplanes aging beyond their 20-year design life are of greatest concern because only recent designs have incorporated improvements in corrosion prevention during engineering and manufacturing. Total direct annual corrosion costs are estimated to be $2.2 billion including cost of design and manufacturing ($0.2 billion), corrosion maintenance ($1.7 billion), and downtime ($0.3 billion).

2.2.7.5.17   Railroad Cars   There are approximately 1.3 million freight cars and 1962 passenger cars in operation in the United States covered hoppers (28%) and tanker cars (18%) make up the largest segment of the freight car fleet. The transported goods range from coal, chemicals, ores, and minerals to motor vehicles and farm and food products. Railroad cars suffer from both internal and external corrosion with a total estimated corrosion cost of 0.5 billion. This estimated cost is divided equally between the use of external coatings and internal coatings and linings.

2.2.7.5.18 Hazardous Materials Transport According to the U.S. Department of Transportation, there are approximately 300 million hazardous material shipments of more than 3.1 billion metric tons annually in the U.S. bulk transport over land including shipping by tanker truck and rail car and by special containers on vehicles. The total annual direct cost of corrosion for this sector is more than $0.9 billion that includes cost of transporting vehicles ($0.4 billion) and of specialized packaging ($0.5 billion) as well as costs associated with accidental releases and corrosion-related transportation incidents.

2.2.7.5.19 Production and Manufacturing This category includes industries that produce and manufacture products of crucial importance to the U.S. economy and its residents' standard of living. These consist of oil production, mining, petroleum refining, chemical and pharmaceutical production, and agricultural and food production. The total annual direct cost of corrosion for production and manufacturing is estimated to be 17.6 billion (Fig. 2.6).

2.2.7.5.20 Oil and Gas Exploration and Production Crude oil and natural gas production has seen a dramatic rise in the United States. Direct corrosion costs associated

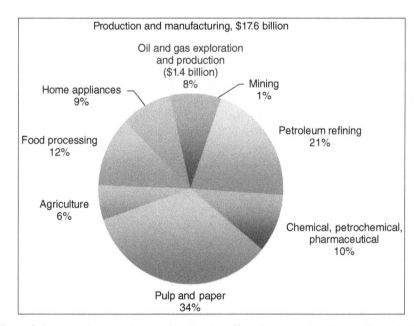

Figure 2.6 Annual cost of corrosion in the oil and gas exploration and production industry (15).

with this activity were determined to be about $1.4 billion with 0.6 billion attributed to surface piping and facility costs, $0.5 billion to downhole tubing and 0.3 billion to capital expenditures related to corrosion.

2.2.7.5.21 Mining Corrosion in this sector is not as significant a problem as in other sectors. The primary life-limiting factors for mining equipment are wear and mechanical damage. Maintenance painting, however, is heavily relied on to prevent corrosion, with an estimated annual expenditure of $0.1 billion.

2.2.7.5.22 Petroleum Refining Approximately 23% of the world's petroleum refineries are in the United States. The nation's 163 refineries supplied more than 18 million barrels of refined petroleum products per day in 1996 with a total corrosion-related direct cost of $3.7 billion. Maintenance expenses amount to $1.8 billion of this total; vessel expenses are $1.4 billion; and fouling costs are approximately $0.5 billion annually.

2.2.7.5.23 Chemical, Petrochemical, and Pharmaceutical Production The chemical industry includes manufacturing facilities that produce bulk or specialty compounds from chemical reactions between organic and/or inorganic materials. The petrochemical industry includes facilities that manufacture substances from raw hydrocarbon materials such as crude oil and natural gas. The pharmaceutical industry formulates, fabricates, and processes medicinal products from raw materials. Annual direct costs total $1.7 billion for this sector, which amounts to 8% of capital expenditures. This does not include corrosion costs related to operation and maintenance. Acquiring detailed data from individual companies and processing it can help assess the corrosion costs of operations and maintenance.

2.2.7.5.24 Pulp and Paper This industry costs $165 billion and provides 300 kg of paper per person each year. There are more than 300 pulp mills and 550 paper mills that support the production. The harsh processing environments of this industry make corrosion control costly as well as challenging. The direct annual cost for this sector is $6 billion, calculated as a fraction of the overall maintenance costs.

2.2.7.5.25 Agricultural Production According to the National Agricultural Statistics Service, about 1.9 million farms produce livestock and crops in the United States. The primary reasons for replacing machinery or equipment include upgrading and damage because of wear and corrosion. The estimated corrosion cost in this sector is $1.1 billion on the basis of the assumption that corrosion costs represent 5–10% of the value of all new equipment.

2.2.7.5.26 Food Processing This is one of the largest manufacturing industries in the country, accounting for approximately 14% of the total manufacturing output. Stainless steel is extensively used in this industry because of food quality requirements. The total estimated corrosion cost is $2.1 billion including stainless steel used for beverage production, food machinery, cutlery and utensils, commercial and restaurant equipment, appliances, aluminum cans, and the use of corrosion inhibitors.

2.2.7.5.27 Electronic Computers, integrated circuits, and microchips are exposed to a variety of environmental conditions, and corrosion manifests itself in several ways. It is also insidious and cannot be readily detected. Therefore, when corrosion failure occurs, it is often dismissed as a product failure and the component replaced. The cost of corrosion in this industry could not be estimated. It is believed that a significant part of all electric component failures is caused by corrosion.

2.2.7.5.28 Home Appliances This is one of the largest consumer products industries. The cost of corrosion in home appliances includes the cost of purchasing replacement appliances because of premature failures caused by corrosion. For water heaters alone, the replacement cost was estimated at $460 million per year, with at least 5% being corrosion related. The cost of internal corrosion protection for all appliances includes the use of sacrificial anodes ($780 million per year), corrosion-resistant materials, and internal coatings. The annual cost of external corrosion protection with coatings was estimated at $260 million. The total direct corrosion costs are at least $1.5 billion.

2.2.7.5.29 Government Although federal, state, and local governments own and operate significant assets under various departments, the U.S. Department of Defense was selected for analysis in the study because it strongly affects the U.S. economy. The other sector of importance, the nuclear waste sector, was also analyzed.

2.2.7.5.30 Defense Corrosion of military equipment and facilities has been an ongoing problem that is becoming more prominent as the acquisition of new equipment slows down. Corrosion is potentially the foremost cost driver in the lifecycle costs in this sector amounting to approximately $20 billion per year.

2.2.7.5.31 Nuclear Waste Storage Nuclear wastes are generated from spent nuclear fuel, dismantled nuclear weapons, and products such as radio pharmaceuticals. The most important design consideration for safe storage of nuclear waste is effective shielding of radiation. A 1998 total lifecycle analysis by the U.S. Department of Energy for the permanent disposal of nuclear waste in Yucca Mountain, Nevada, estimated the repository cost by the construction phase in 2002 to be $4.9 billion with an average annual cost of $205 million through 2116. Of this cost, $42.2 million is corrosion related.

2.3 TRIBOLOGY

2.3.1 Economies of Wear and Corrosion in the Canadian Industry

The estimated losses because of friction and wear in various Canadian economic sectors are given in Table 2.8.

The total estimated loss in Canada because of friction and wear amounts to more than $5 billion annually. The costs given in the table refer to 1986. The costs because of friction and wear would be much higher in 2013.

124   CORROSION COSTS

TABLE 2.8 Estimated Losses Because of Friction and Wear in Canadian Economic Sectors

Category	Friction Losses ($million/year)	Wear Losses ($million/year)	Total Losses ($million/year)
Agriculture	321	940	1261
Electric utilities	54	189	243
Forestry	111	158	269
Mining	212	728	940
Pulp and paper	105	382	487
Rail transport	283	467	750
Trucks and buses	126	860	986
Wood industries	14	189	203
Total	1226	3913	5139

Tribology combines the many branches and specialties in science and technology concerned with the understanding of design and combatting friction and wear. Specifically, tribology is concerned with the design and manufacture of sliding and rolling bearings, piston rings, gears, cutting tools, and other devices in which surfaces interact, in sliding or other relative motion. Tribology is concerned with the correct use of materials for interacting solid surfaces, with the interaction of surfaces with lubricants with chemistry and physics. It is concerned with the reduction of friction and wear and the avoidance of associated failures in machinery. Similar tribological components such as bearings are used in different technologies, and similar tribological problems such as abrasive wear occur in different industries. Thus, tribology is in itself a generic science and technology concerned with the generation and dissemination of knowledge on friction, wear and lubrication, and the application of the knowledge to a wide range of industrial products and equipment, from ploughs to space equipment, from prosthesis to engines, from washing machines to railways, from computers to power stations, from paper mills to mines.

Avoidable losses because of friction and wear occur because many who are involved in the design, specification, operation, and maintenance of industrial equipment are unaware of current technology in tribology. There is also an inadequate level of research and development in tribology, through which the reliability and economy of industrial processes and products can be improved.

The estimates of losses and potential savings are in general agreement with the estimates of this type produced in the United Kingdom, the Federal Republic of Germany, and the United States, increasing the confidence that can be placed in them.

2.3.2 Strategies Against Wear and Friction

Significant tribology programs were instituted in the United Kingdom, Germany, and the United States and structured around considerable financial contributions from the respective governments, while involving industry at the same time. "Tribology Centres" were established in the United Kingdom with the financial support of the

TABLE 2.9   Corrosion Costs in the Mining Sector

Country	Year	Corrosion Cost ($millions)	Avoidable Cost
United States	1978	8600	572
Canada	1991	1950	97

TABLE 2.10   Corrosion Costs in Various Countries

Country	Year	GNP (billions)	Corrosion Costs (billions)
United Kingdom	1970	91	32
Australia	1982	–	–
China	1986	–	–
United States	1986	3404	160
U.S.S.R.	1969	335	6.7
West Germany	1969	200	6.0
Canada	1994	724	3.5

TABLE 2.11   Corrosion Costs in Farming in the United Kingdom

	Investment (£ million)	Corrosion Losses (£ million)
Total	12,025	188

government, which decreased as the income of research and development increased. Most of the tribology projects in West Germany are cost shared between government and industry. Government involvement in the United States concerns with new major technologies such as advanced engines or military applications and a strong emphasis on promoting information systems in tribology.

Corrosion costs in the mining sector of the United States and Canada (14) are given in Table 2.9.

The data show extensive corrosion costs, and by taking necessary steps, corrosion losses can be reduced.

Corrosion costs in various countries are tabulated in Table 2.10.

The corrosion costs given so far are only direct costs of corrosion. The indirect costs because of corrosion-associated plant shutdown, lowered efficiency of equipment, and overdesign are additional to the costs cited. Loss of production during plant shutdown for repairs can cost millions of dollars per day. Leaks in pipelines and tanks result in loss of costly product. The leaks can contaminate the groundwater and cause an environmental problem. The costs in fixing this problem can be enormous.

Accumulation of undesirable corrosion products on heat exchanger tubing and pipelines decrease the heat transfer efficiency and reduce the pumping capacity. Soluble corrosion products can contaminate a system, and decontamination of the system results in additional cost. An example of this is the expensive shutdowns of nuclear reactors during the decontamination process.

Corrosion costs in the well-known old sector, namely, farming in the United Kingdom amount to 1.6% of the investment as evidenced by the data given in Table 2.11.

Although farming is a capital intensive activity and one of the largest and essential sectors of industry, very little attention has been paid toward corrosion and its control.

REFERENCES

1. HH Uhlig, *Chemical Engineering News* **27**:2764 (1949); *Corrosion* **6**:29 (1950).

2. Report of the committee on corrosion and protection: a survey of corrosion protection in the United Kingdom, Chairman TP Hoar, HMSO, London, 1971.

3. Report of the committee on corrosion and protection: a survey of the cost of corrosion to Japan, Japan Society of Corrosion Engineering and Japan Association of Corrosion Control, Chairman, G Okamoto, 1977.

4. Economic effects of metallic corrosion in the United States, NBS Special Publication 511-1, SD stock no. SN-003-003-01926-7, 1978.

5. Economic effects of metallic corrosion in the United States, Appendix B, NBS Special Publication 511-2, SD stock no. SN-003-003-01927-5, 1978.

6. JH Payer, WK Boyd, DG Lippold, W Fisher, NBS-Battelle cost of corrosion study ($70 billion), Part 1-7, Materials Performance, May–November 1980.

7. RW Revie, HH Uhlig, *Journal of the Institute of Engineers Australia* **46** (1):3–5 (1974).

8. RK Worner, Committee report of the symposium on corrosion, University of Melbourne, 1955, pp. 1–14.

9. RC Potter, *Australian Corrosion Engineering* **162**:21–29 (1972).

10. D Behrens, *British Corrosion Journal* **10**, no 3:122 (1967).

11. S Linderborg, *Kemiam Teollisuus (Finland)* **24**, issue 3:234 (1967).

12. KS Rajagopalan, Report on metallic corrosion in India, CSIR, 1962.

13. R Zhu, Approaches to reducing corrosion costs, NACE, 1986.

14. VS Sastri, M Elboujdaini, *CIM Bulletin* **90**:63 (1997).

15. Corrosion cost and preventive strategies in the United States, FHWA-RD-01-156, US Dept. of Transportation, Federal Highway Administration, McLean, VA, Mar. 2002.

3

CORROSION CAUSES

3.1 INTRODUCTION

The general causes of corrosion of material objects vary with the nature of the atmosphere to which the metallic object or component is exposed. The corrosives can be mineral acids, such as hydrochloric, sulfuric or carbonic acid, or gases such as carbon dioxide, sulfur dioxide, or hydrogen sulfide. In certain cases, it is possible that one might encounter hydrogen fluoride. Anions such as chloride and sulfite can be quite deleterious.

3.2 CORROSION IN CONVENTIONAL CONCRETE BRIDGES

The main cause of reinforced concrete bridge deterioration is chloride-induced corrosion of the black steel reinforcement, resulting in expansion forces in the concrete that produce cracking and spalling of the concrete. The source of chloride can be from either marine exposure or the deicing salts used in snow and ice removal. The use of deicing salts is likely to continue, if not increase. Very little can be done to prevent bridge structures from getting exposed to corrosive chloride salts.

3.3 CORROSION OF PRESTRESSED CONCRETE BRIDGES

Most of these bridges are relatively new, and the overall economic impact is not as significant as for conventional reinforced concrete bridges. It is estimated that the

Challenges in Corrosion: Costs, Causes, Consequences, and Control, First Edition. V. S. Sastri.
© 2015 John Wiley & Sons, Inc. Published 2015 by John Wiley & Sons, Inc.

damage because of deicing salts alone is between \$325 and \$1000 million per year to reinforced concrete bridges and car parks in the United States of America. The Department of Transport in the United Kingdom estimates total repair cost of ~\$1 billion following corrosion damage to motorway bridges. These bridges represent ~10% of the total number in the United Kingdom and hence the total cost may amount to \$10 billion (1).

3.4 REINFORCEMENT CORROSION IN CONCRETE

As concrete is porous and both moisture and oxygen move through the pores and microcracks in concrete, the basic conditions for the corrosion of mild or high-strength ferritic reinforcing steels are present. The reason that corrosion does not occur readily is because of the fact that the pores contain high levels of calcium, sodium, and potassium hydroxide, which maintain a pH of between 12.5 and 13.5. This high level of alkalinity passivates the steel, forming a dense gamma ferric oxide that is self-maintaining and prevents rapid corrosion.

In many cases, any attack on reinforced concrete will be on the concrete itself. However, there are two species, namely, chloride and carbon dioxide that penetrate the concrete and attack the reinforcing steel without breaking the concrete. The chloride and carbon dioxide penetrate concrete without causing significant damage and then promote corrosion of the steel by attacking and removing the protective passive oxide layer on the steel created and sustained by the alkalinity of the concrete pore water.

3.5 MECHANISM OF CORROSION AND ASSESSMENT TECHNIQUES IN CONCRETE

There are many studies covering the mechanism of corrosion in concrete and assessment techniques (1–6). Specifications and recommended practices on how to select and apply repair methods (NACE SP1290 and 0390, BSEN 12696, BSEN 1504 and ACI 222R-01) are given in the literature.

The separation of anodes and cathodes is an important part of the understanding, measurement, and control of corrosion of steel in concrete. Corrosion of steel in concrete is basically an aqueous corrosion mechanism where there is very poor transport of corrosion product away from the anodic site. This usually leads to the formation of voluminous corrosion product and cracking and spalling of concrete, with delamination forming along the plane of the reinforcing steel. In the absence of oxygen at the anodic site, the ferrous ion will stay in solution or diffuse away and deposit elsewhere in pores and microcracks in the concrete leading to severe section loss without the advanced warning given by concrete cracking and spalling.

3.5.1 Chloride Ingress and the Corrosion Threshold

Chloride can be present in concrete for various reasons such as:

1. Contamination because of (i) deliberate addition of calcium chloride set accelerators; (ii) deliberate use of seawater in the mix; (iii) accidental use of inadequately washed marine source aggregates.
2. Ingress consisting of (i) deicing salts; (ii) sea salt; (iii) chlorides from chemical processing.

Until the later 1970s, it was widely thought that the chlorides cast into concrete existed as chloroaluminates and would not cause corrosion. It was later found that large numbers of structures were corroded because of the presence of chloride in the mix, and the binding was not as effective as initially believed.

ACI Report 222 R-01 reviews the national standards and laboratory data. The consensus is that 0.4 wt% of chloride in cement is a necessary but not sufficient condition for corrosion and in variable chloride and aggressive conditions, corrosion can occur at lower chloride levels, down to ~0.2% of chloride by weight of cement.

According to the literature review (5), it is thought that whether chlorides are bound or not, the chloroaluminates decompose releasing chloride ions, which cause breakdown of passivity. It is also suggested that the amount of chloride bound in the cement paste is not very important. The amount of calcium hydroxide available to maintain the pH has a profound effect on the initiation of corrosion.

Chloride ingress into concrete follows Fick's second law of diffusion, forming a chloride profile with depth into the concrete:

$$\frac{d[Cl^-]}{dt} = D_t \cdot \frac{Cl^2[Cl^-]}{dx^2}$$

where $[Cl^-]$ is the chloride concentration at depth x and time, t. D_t is the diffusion coefficient (about 10^{-8} cm^2/s).

The solution to the above differential equation is:

$$\frac{(C_{max} - C_{x,t})}{(C_{max} - C_{min})} = \mathrm{erf} \frac{x}{(4D_c t)^{1/2}}$$

where C_{max} is surface or near surface concentration; $C_{x,t}$ is the chloride concentration at depth x; and time t, C_{min} is the background chloride concentration erf – is error function.

The parameter C_{max} must be a constant as surface measurement is used to avoid fluctuations in surface levels on wet and dry surfaces.

3.5.2 Carbonation of Concrete and Corrosion

Carbonation is a simpler process than chloride attack. Atmospheric carbon dioxide reacts with pore water to form carbonic acid. Carbonic acid reacts with calcium hydroxide and forms solid calcium carbonate. The pH drops from about 13 to $\sim$8. The steel corrodes around pH 11.0. Carbonation rates follow parabolic kinetics:

$$d = At^{0.5}$$

where d is carbonation depth, t is time, and A is constant.

Steel bridges corrode on exposure of steel to atmospheric conditions. The extent of corrosion is greatly enhanced because of marine (salt spray) exposure and other corrosive industrial environments. The only corrosion preventive method is to provide a barrier coating such as paint.

There is considerable need for additional studies on innovative construction materials such as corrosion-resistant alloy/clad rebars (both metallic and nonmetallic) and more durable concretes with inherent corrosion-resistant properties. Further research and development in rehabilitation technologies that can mitigate corrosion with minimal maintenance requirements such as sacrificial cathodic protection (CP) systems is desirable.

According to the National Bridge Inventory Database, there are nearly 600,000 bridges in the United States of America of which nearly 300,000 were built between 1950 and 1995. The materials of construction for these bridges are concrete, steel, timber, masonry, timber/steel/concrete combinations, and aluminum. Reinforced concrete and steel bridges are in the majority and built since 1950, which can deteriorate because of corrosion. According to a report by the American Society of Civil Engineers, the condition of the bridge structures was rated as "poor" and was found to be the largest contributor to the US infrastructure cost of corrosion.

Because of the specific concrete property of weak tensile strength as compared to its compressive strength, steel reinforcing is placed in the tension regions in concrete members, such as decks and pilings. The two primary forms of steel reinforcing in concrete bridges are "conventional" reinforcing bar (rebar) and prestressed tendons. The difference between conventional reinforcement and prestressed tendon reinforcement is that the prestressed tendons are loaded in tension (prestressed) either prior to placing the concrete (pretensioned) or after placing and curing of the concrete (posttensioned). It is useful to note that prestressed tendon generally has a higher tensile strength than conventional rebar steel.

The majority of concrete deterioration leading to reduced service life and/or replacement is associated with conventional reinforced steel bridge structures. This is simply because the large majority of the reinforced concrete bridges and the longer in-service times experienced by these bridges. Although conventional rebar and prestressed tendon bridge structures have specific design, construction, and corrosion-related concerns and consequences, the basic corrosion mechanism is similar, and many control methods are applicable to both.

3.5.3 Conventional Reinforced Concrete

Reinforced concrete bridges suffer from corrosion of the reinforcement and, consequently, concrete degradation because of the high tensile forces exerted by

Chloride-induced macrocell corrosion initiates on top rebar

Expansion of corrosion product produces tensile stresses in concrete

Tensile stresses in concrete lead to cracking/spalling

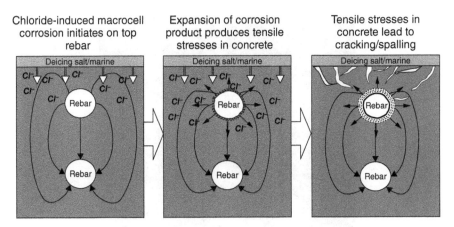

Figure 3.1   Schematic of corrosion damage to rebar (7).

the corroding steel (corrosion products have three to five times greater volume than the original steel). These high tensile forces cause cracking and spalling of the concrete at the reinforcement (Figs 3.1 and 3.2). Steel in high-pH concrete in the absence of chloride is passive and corrosion is negligible, which in theory should give reinforced concrete structures a long corrosion-free life. However, in practice, corrosion in concrete is accelerated through two mechanisms:

1. Breakdown of the passive layer on the steel by chloride ions
2. To a lesser extent by carbonation because of carbon dioxide reaction with cement phase.

In the case of highway bridges, the vast majority of the problems are caused by chloride migration into the concrete because of deicing salt application and marine exposure. On reaching the surface, the chloride ions cause the disruption of the passive film and create conditions favorable for accelerated corrosive attack on the reinforcing steel.

In addition to chloride ions, oxygen is also necessary for accelerated corrosion. Chemical, physical, and mechanical properties of concrete can have a significant effect on concrete deterioration by controlling the following:

1. The chloride and oxygen permeation in the concrete.
2. The sensitivity of passive layer to chloride attack.
3. The rate of corrosion reactions at the steel surface following corrosion initiation.
4. The rate of cracking and spalling of the concrete when exposed to the expansion forces of the corrosion products. The effect of concrete properties on the corrosion and concrete deterioration processes of bridge structures has been

Figure 3.2   Example of deteriorating bridge element (7).

documented (8, 9). It was shown that concrete mix design has a significant effect on the corrosion of rebar.

The uneven distribution of chloride ions in the concrete and at the steel surface also greatly affects the corrosion. The high level of chloride on the top layer and decreasing chloride concentration with distance results in increased corrosion rate because of macrocell corrosion.

Corrosion of steel in concrete is a very complex phenomenon. Although significant research on modeling in the corrosion processes of steel in concrete has been performed, accurate life prediction for concrete structures is difficult.

Nonmarine corrosion-related reinforced concrete bridge failures became a growing problem beginning in the 1960s in the "snow belt" regions following increased usage of deicing salts. In the worst cases, bridges began to require maintenance after a service life of 5–10 years, with the average maintenance interval being around 15 years. The quality of concrete used in bridge construction was of improved quality in the 1970s and 1980s. This together with increased thickness and the epoxy coated rebar led to increased service lives. New bridge structures built and maintained with high performance, greater cover thickness, corrosion-resistant rebar, corrosion-inhibiting admixtures, overlays, sealants, and improved CP are expected to give service lives of 75–100 years. In designing long-life bridge structures, it should be noted that changing load and capacity requirements may make the structure functionally obsolete before it becomes structurally deficient. Hence, emphasis should be placed on forecasting traffic loads and patterns and on designs to handle changes in traffic volume.

Prestressed concrete bridges are low in number (18%) in comparison to reinforced concrete bridges (40%), and the total economic impact is not as large as reinforced concrete bridges. However, on an individual basis, failure of prestressed concrete components may have a significant impact on the structural integrity of a bridge. Prestressed concrete bridges rely on the tensile strength of tendons to sustain the load, and the loss of even a few tendons may lead to catastrophic failure of a bridge component.

There were 107,700 prestressed concrete bridges in the United States of America in 1960. In 1992, the UK Ministry of Transportation imposed a ban on the commissioning of grouted, bonded posttensioned bridges. This ban is the result of the collapse of two foot bridges in 1960. The failure of posttensioned Melle Bridge across the Scheldt River in Belgium was reported in 1956. This failure was traced to the corrosion of posttensioned strands.

Reliable nondestructive methods for providing assurance to the owners that the built structures have met construction specifications are not available. The main concern is whether the ducts in the posttensioned bridge members have been completely filled with the grouts and whether there is uniform coverage over the prestressing steel.

3.6 STEEL BRIDGES

Atmospheric corrosion of exposed steel is common. Painting is a universal solution to corrosion because of exposure to environmental conditions. Paints can deteriorate following moisture uptake, ultraviolet exposure, wear or mechanical damage, and exposure to chemicals. For instance, the performance of the same coatings will vary significantly depending on exposure to industrial, urban, rural, or marine environments. Once a coating is compromised, corrosion can initiate, and often, is accelerated under a deteriorated coating more than in the absence of coating. Thus selection of proper coating for the right application is important for a long service life. Proper and timely maintenance of the structure can extend the overall life of the coating significantly.

3.7 CABLE AND SUSPENSION BRIDGES

These bridges comprise a small percentage of the total nation's bridges. There are nearly 150 cable bridges, which range between 100 and 130 years in age. Construction and corrosion control of these bridges were reviewed (10). Corrosion problems are highly dependent on specific structural configuration, design, making general

Figure 3.3   Golden Gate Suspension Bridge (7).

statements very difficult. The corrosion problems are highly dependent on specific structural configurations, maintenance, and operational practices and local environmental conditions.

Corrosion concerns on cable-supported structures have been present from the early design. For example, galvanized zinc coating of the wires was used on the Brooklyn Bridge, which was completed in 1883. At that time, it was a standard practice to coat the wire with linseed oil, circumferentially wrap the assembled cable with soft galvanized wire laid into red lead paste, and to paint the finished cable.

Some of the oldest and well-known bridges in the United States, such as the Golden Gate and Brooklyn bridges, are suspension bridges (Figs. 3.3 and 3.4). The maintenance costs of these bridges are high. These cannot be replaced or taken out of service for any length of time because of historic reasons or strategic location. The specific concern with these bridges is the condition of strands. The strands are susceptible to corrosion, stress corrosion cracking, and hydrogen embrittlement, which can result in premature failure of the strands.

A more recent design of a cable bridge is a cable-stayed bridge as shown in Figure 3.5. In these bridges, the integrity of the cables is critical to the structural integrity of the entire bridge. The inspection of the cables is very difficult and these bridges are built with due consideration of corrosion protection (W Podolny, FHWA, Office of Bridge Technology, Personal Communication, 1999).

Figure 3.4   Brooklyn Suspension Bridge (7).

Figure 3.5   Cable-Stayed bridge in Tacoma (7).

3.8   CORROSION OF UNDERGROUND PIPELINES

The gas and liquid transmission pipelines sector is part of the oil and gas industry. The size of the sector is large as shown below. This sector consists of 217,000 km (135,000 miles) of hazardous liquid transmission pipelines, 34,000 km (21,000 miles) of crude oil gathering pipelines, 483,000 km (300,000 miles) of natural gas transmission pipelines, and 45,000 km (28,000 miles) of natural gas gathering pipelines. Figure 3.6 summarizes the transmission pipeline sector. In the United States, there are nearly 60 major natural gas transmission pipeline operators and 150 major

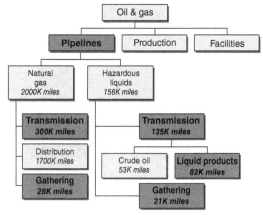

1 mi = 1.61 km

Figure 3.6   Chart describing transmission pipeline sector (7).

hazardous liquid pipeline operators (12). Figure 3.7 illustrates the different components of natural gas production, transmission, storage, and distribution system. The
components consist of production wells, gathering lines within the production fields,
processing plants, transmission pipelines, compressor stations located periodically
along the transmission pipelines, storage wells and associated gathering pipelines,
metering stations and city gate at distribution centers, distribution piping and meters
at residential or industrial centers. Hazardous liquid systems include production
wells and gathering lines for crude oil production, processing plants, transmission
pipelines, pump stations, valve and metering stations, and aboveground storage
facilities.

3.8.1   Types of Corrosion of Underground Pipelines

Types of corrosion encountered are the following:

1. General corrosion.
2. Stray current corrosion.
3. Microbiologically influenced corrosion (MIC).
4. Stress corrosion cracking (SCC).

Corrosion of the pipe wall can occur either internally or externally. Internal
corrosion occurs when corrosive fluids or condensates are transported through the
pipelines. Depending on the nature of corrosive liquid and the transport velocity,
different forms of corrosion may occur, namely, uniform corrosion, pitting/crevice
corrosion, and erosion–corrosion. Figure 3.8 shows an example of internal corrosion
that occurred in a crude oil pipeline because of high levels of salt water and carbon
dioxide (CO_2).

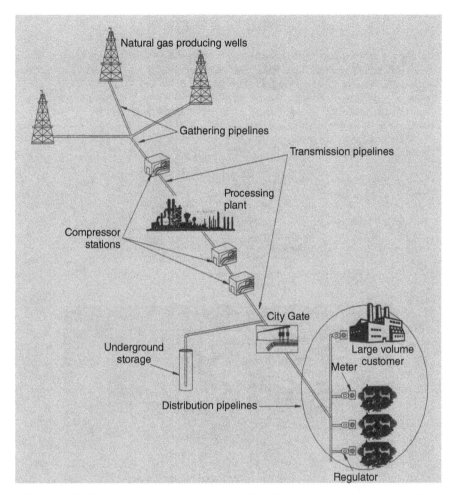

Figure 3.7   Components of natural gas production, distribution, and transmission (7).

There are several different modes of external corrosion identified on buried pipelines. The primary mode of corrosion is a macrocell form of localized corrosion following the heterogeneous nature of soils, local damage of the external coatings (holidays), and/or the disbondment of external coatings. Figure 3.9 shows typical external corrosion on a buried pipeline. The 25 mm (1-in.) grid pattern was placed on the pipe surface to permit sizing of the corrosion and nondestructive evaluation (NDE) of the wall thickness measurements.

3.8.1.1 Stray Current Corrosion   Corrosion can be accelerated through ground currents from dc sources. Electrified railroads, mining operations, and other similar industries that utilize large amounts of dc currents sometimes allow a significant portion of current to use a ground path return to their power sources. These currents

Figure 3.8   Internal corrosion of a crude oil pipeline (7).

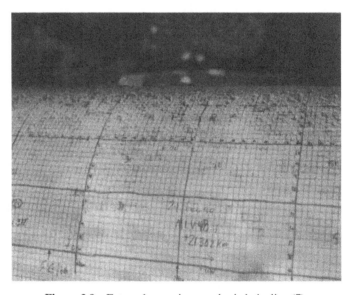

Figure 3.9   External corrosion on a buried pipeline (7).

often utilize metallic structures such as pipelines in close proximity as a part of the return path. This "stray" current can be picked up by the pipeline and discharged back into the soil at some distance down the pipeline close to the current return. Current pick-up on the pipe is the same process as CP, which tends to mitigate corrosion. The process of current discharge off the pipe and through the soil of a DC current accelerates the corrosion of the pipe wall at the discharge point. This type of corrosion is known as stray current corrosion and is illustrated in Figure 3.10.

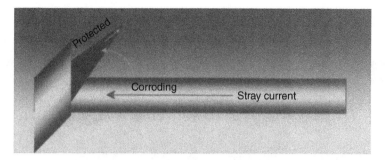

Figure 3.10 Example of stray current corrosion (7).

3.8.1.2 Microbiologically Influenced Corrosion MIC may be defined as corrosion that is influenced by the presence and activities of microorganisms, including bacteria and fungi. It is estimated that 20–30% of all corrosion on pipelines is MIC-related. MIC can affect either the external or the internal surface of the pipeline. Microorganisms located at the metal surface do not directly attack the metal or cause a unique form of corrosion. The by-products from the organisms promote several forms of corrosion, namely, pitting, crevice corrosion, and underdeposit corrosion. Typically, the products of a growing microbiological colony accelerate the corrosion process by either: (i) interacting with the corrosion products to prevent natural film-forming characteristics of the corrosion products that would inhibit further corrosion, or (ii) providing an additional reduction reaction that accelerates the corrosion process.

A variety of bacteria has been implicated in exacerbating the corrosion of underground pipelines, and these fall into the broad classification of aerobic and anaerobic bacteria. Obligate aerobic can survive only in the presence of oxygen, while obligate anaerobic bacteria can survive only in the absence of oxygen. A third type of aerobic bacteria that prefer aerobic conditions can live under anaerobic conditions. Common obligate anaerobic bacteria implicated in corrosion are sulfate-reducing bacteria (SRB) and metal-reducing bacteria. Common obligate aerobic bacteria are metal-oxidizing bacteria, while acid-producing bacteria are facultative aerobes. The most aggressive attacks generally take place in the presence of microbial communities that contain a variety of bacteria. In these communities, the bacteria act cooperatively to produce conditions favorable to the growth of each species. For instance, obligate anaerobic bacteria can thrive in aerobic environments when they are present beneath biofilms/deposits in which aerobic bacteria consume the oxygen.

In the case of underground pipelines, the most aggressive attack has been associated with acid-producing bacteria in such bacterial communities (Fig. 3.11).

3.8.1.3 Stress Corrosion Cracking One of the serious forms of pipeline corrosion is SCC. This form of corrosion consists of brittle fracture of a normally ductile metal by the conjoint action of a specific corrosive environment and a tensile stress. In the case of underground pipelines, SCC affects the external surface of the pipe, which is exposed to soil/ground water at locations where the coating is disbonded.

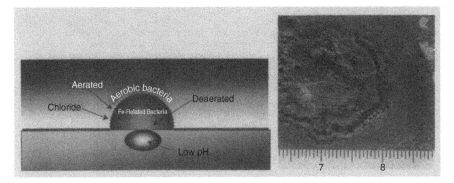

Figure 3.11   Iron-related bacteria reacting with chloride-producing acidic environment (7).

The primary component of the tensile stress on an underground pipeline is in the hoop direction and results from the operating pressure. Residual stresses from fabrication, installation, and damage in service contribute to the total stress. Individual cracks initiate in the longitudinal direction on the outside surface of the pipe. The cracks typically occur in colonies that might contain hundreds or thousands of individual cracks. Over time, the cracks in the colonies interlink and may cause leaks or ruptures once a critical size flaw is achieved.

The two types of SCC on underground pipelines that have been identified are classical or "high-pH" cracking (pH 9–10), which propagates intergranularly, and "near-neutral pH" cracking, which propagates transgranularly. Each form of SCC initiates and propagates under unique environmental conditions. Near-neutral pH SCC (<pH 8) is most commonly found on pipelines with polyethylene tape coatings that shield the CP current. The environment that develops beneath the tape coating and causes this form of cracking is dilute carbonic acid. Carbon dioxide from the decay of organic material in the soil dissolves in the electrolyte beneath the disbonded coating to form the carbonic acid solution. High-pH SCC is most commonly found on pipelines with asphalt or coal tar coatings. The high-H environment is a concentrated carbonate bicarbonate solution that develops as a result of the presence of carbon dioxide in the ground water and the CP system.

3.8.2   Replacement/Rehabilitation

Decisions for pipeline replacement versus pipeline rehabilitation are often difficult, with several important considerations. Rehabilitation includes repairing existing flaws in the pipeline and recoating the pipeline. To make the most effective decisions on replacement versus rehabilitation of a pipeline or a segment of a pipe, it is necessary to understand the extent of corrosion existing on the line and the coating condition of the pipeline. For example, excessive cutouts and replacements rapidly increase the cost of coating rehabilitation. In addition, hidden costs must be taken into account, including such items as shorter coating lives of in situ coatings. The following three conditions make replacement or rehabilitation necessary:

1. Severe corrosion damage of a pipeline not properly cathodically protected
2. Severe coating deterioration leading to increased CP requirements.
3. SCC along a large area of the pipeline.

3.8.3 Pipeline Integrity Management Programs

Pipeline integrity management programs are used by pipeline operators to determine the locations in which corrosion defects present a threat to safe operation. Repairs at these locations can vary from the installation of a reinforcing sleeve to the implementation of a large-scale pipe rehabilitation or replacement program. In the case of localized corrosion flaws, the repairs might involve composite sleeves, full-encirclement steel sleeves, or replacement of a pipe segment. For local flaws, decisions regarding the repair process can be usually handled by company procedures and criteria. In the case of large-scale corrosion and/or coating deterioration problems, the replacement/rehabilitation decision must take into account both operational and economic factors.

3.8.4 In-line Inspections

In-line inspections (ILIs) are widely used to generate a profile of defects present in a pipeline. The high-resolution UT and MFL ILI tools may be used to determine the geometry and the orientation of corrosion defects. These routine inspections may be used to determine the number and the locations of near-critical flaws that should be immediately examined such as a dig program to verify the flaw followed by repair. By using appropriate corrosion growth models, predictions can be made on future dig/repair/and/or reinspection requirements for the ILI-inspected line. If the density of corrosion defects is high or the potential exists for continued increase in dig/repair frequency, the affected pipe section may be repaired or replaced.

3.8.5 Aging Coating

Another concern related to corrosion assessment is the cost of maintaining the required level of CP. The effectiveness of the CP system can be verified using corrosion surveys. An increased number of coating defects requires an increased amount of CP current. This is accomplished by increasing current output of the impressed current rectifiers, installing impressed current rectifiers at more locations along the pipeline, or installing additional sacrificial anodes. Coating defects can be identified by conventional potential surveys or by specific coating defect surveys and verified by direct visual inspection (dig program). Under certain circumstances, coatings fail in a manner that makes assessment of the corrosion condition of pipe through conventional surveying methods difficult. Aging coating and the associated defects can make the continuous need for CP upgrading uneconomical.

3.8.6 Stress Corrosion Cracking

The presence of extensive SCC may qualify a pipeline for replacement or rehabilitation. As SCC depends on unique environmental conditions, a large-scale recoating program may protect against these environmental conditions and permit continued operation of the pipeline. On the basis of severity and density of the stress corrosion cracks, however, pipe replacement may be the most economical option.

Replacement/rehabilitation decisions involve several considerations. These considerations are terrain conditions, expected or required life, excess capacity and throughput requirements, internal versus external corrosion, and so on. A comprehensive list of considerations for pipeline rehabilitation is given below (13).

1. The location of the pipeline is important for repair considerations. A pipe in swampy clay would exclude recoating versus a repair option. On the contrary, prairies are conducive to recoating, with firm footing for the equipment and good accessibility.

If the expected life of a section of pipeline is relatively short, the pipeline operator must decide whether recoating and repair would extend the life of the pipe section to match the rest of the pipeline. If not, replacing the pipe section may be the best solution.

3.8.7   Corrosion-Related Failures

If corrosion is allowed to continue unabated, the integrity of the pipeline will eventually be compromised. In other words, the pipeline will fail. Depending on the flaw size, pipeline material properties, and the pipeline pressure, the failure refers to either a leak or a rupture. Typically, rupture of a high-pressure natural gas pipeline results in the release of so much stored energy (compressed gas) that the pipeline is blown out of the ground. A leak results when the flaw penetrates the pipeline but is not of sufficient size to cause a rupture. The leaks in natural gas pipelines are detected by either periodic inspections or third-party reporting and are repaired without any significant event. However, leaks can result in substantial problems if not detected promptly. Natural gas leaks can fill enclosed or confined spaces, and if an ignition source is present, explosions and/or fires can result, causing significant property damage and injuries or deaths. For natural gas leaks or ruptures, the immediate environmental impact is minimal.

A liquid pipeline has less stored energy than a gas pipeline, and a rupture does not cause an explosion. However, an explosion can occur on ignition of an explosive product. In the case of a hazardous liquid product pipeline, the environmental impact can be as serious as an explosion. The risk of an oil leak from the Trans-Alaska Pipeline System has continued to be the primary driver for the aggressive corrosion prevention and inspection program maintained by the operator. Of major concern is the risk of oil leakage into water streams and thereby contaminating water supplies.

Corrosion-induced pipeline failures can result in the following:

1. Loss of product
2. Property damage
3. Personal injury or death

4. Cleanup of product in the case of hazardous liquid pipelines
5. Pipeline repair and back-to-service program
6. Legal issues
7. Loss in throughput.

To prevent failures, an aggressive maintenance and integrity program is necessary.

3.9   WATERWAYS AND PORTS

The United States has more than 7750 commercial water terminals, 192 commercially active lock sites with 238 chambers, and 40,000 km (25,000 miles) of inland, intercoastal, and coastal waterways and canals (14). About 41 states, 16 state capitals, and all states east of Mississippi River are served by commercially navigable waterways (15). Both public and private works associated with waterways and ports have corrosion-related problems in both freshwater and seawater environments.

Public Works Waterway Structures operated and maintained by the US Army Corps of Engineers include locks, dams, navigational aids, levies, and decks. A typical example of a steel-reinforced concrete dam is shown in Figure 3.12. These structures are on primarily freshwater lakes and rivers. Many freshwater public works related to irrigation and flood control are owned, operated, and maintained by state and local agencies such as the Tennessee Valley Authority (TVA) or the California Aqueduct.

Public docks, piers, and bulkheads are mostly owned and maintained by port authorities. These public agencies have structures in both freshwater and seawater; however, most of the larger ports are in marine locations.

There are also a significant number of private terminals for loading grain and coal owned by shipping companies and railroads. These private terminals are located in both freshwater and marine environments. The large size of most structures at port facilities requires that they be built with steel-reinforced concrete, steel, or a combination of both. The seawater environment is significantly more severe than river or lake waters because of the high chloride content in the seawater.

3.9.1   Areas of Major Corrosion Impact

Reinforced concrete structures exposed to marine environments suffer premature corrosion-induced deterioration by chloride in seawater. Corrosion is typically found in piers and docks, bulkheads and retaining walls, mooring structures and navigational aids.

The marine environment can have varying effects on different materials depending on the specific zones of exposure. Atmosphere, splash, time, immersion, and subsoil have very different characteristics and, therefore, have different influences on

Figure 3.12   Example of a steel-reinforced concrete dam (7).

corrosion (16). Figure 3.13 shows the relative metal loss for steel piling after 5 years of exposure to seawater at Kure Beach, North Carolina.

Atmospherically exposed submerged zones and splash zones typically experience the most corrosion. The zones are found on piers and docks (ladders, railings, cranes and steel support piles, bulkheads and retaining walls sheet steel piping, steel-reinforced concrete elements, backside, and anchors on structures retaining dredged fill), and mooring structures and dams (steel gates, hinges, intake/discharge culverts, gates, and debris booms). Stationary navigational aids suffer from corrosion of support piles and steel-reinforced concrete pile caps. Floating steel buoys are also prone to corrosion.

3.9.2   Fresh Water

Airborne or splash zone attack is not normally a problem at freshwater facilities; how-ever, air pollution can cause potential problems. Under certain flow conditions, such as turbulent flow or cavitation, fresh water can cause severe corrosion to submerged metallic elements. Ice damage also can limit the effectiveness of coatings on bulkhead walls and support piling.

Piers and docks, bulkheads and retaining walls, locks, dams, and navigational aids exposed to freshwater environment experience corrosion-related problems. The most common areas of attack are submerged and splash zones on support piles (piers, docks, and navigational aids) and steel sheet piling (bulkheads and retaining walls). The zones are also found on locks (steel gates, hinges, intake/discharge culverts, valves, and sheet pile walls), dams (steel gates, hinges, intake/discharge culverts, grates, and debris booms), and navigational aids (anchorages).

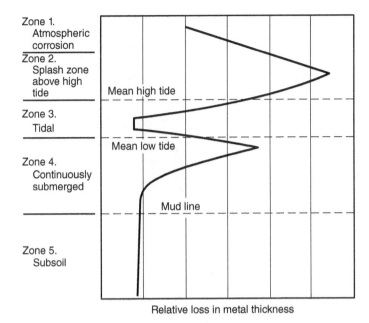

Relative loss in metal thickness

Figure 3.13   Corrosion profile of steel piling after 5-year exposure to seawater (7).

3.10 HAZARDOUS MATERIALS STORAGE

Bulk storage of hazardous liquid and gaseous materials is usually done in large steel tanks. The largest aboveground tanks are used at refineries and manufacturing plants. These range from 15 m (50 ft) to more than 61 m (200 ft) in diameter with a capacity of more than 3785 m^3 (1 million gallons). Transportation and distribution terminals of storage facilities for these materials can have a mix of aboveground and underground tanks. Liquid petroleum products at the point of sale and at the point of use are stored in direct buried underground tanks ranging from 1.9 to 114 m^3 (500–30,000 gallons) in capacity. Gases are typically stored in similarly sized aboveground tanks at the point of use. Hazardous chemicals are usually stored in vaulted underground tanks or aboveground facilities. Storage tanks for pressurized materials can be spherical in shape while storage tanks for unpressurized materials can be constructed from welded plate (Fig. 3.14a, b).

3.10.1 Aboveground Storage Tanks

Large steel aboveground storage tanks (ASTs) are located on large tank farms of oil producers (Fig. 3.15). Maintenance teams take care of external painting and both internal and external corrosion inspections. Corrosion inspection and protection of ASTs is important for the preservation of large capital investments, the reduction of maintenance and inspection costs, and the assurance of system integrity for release

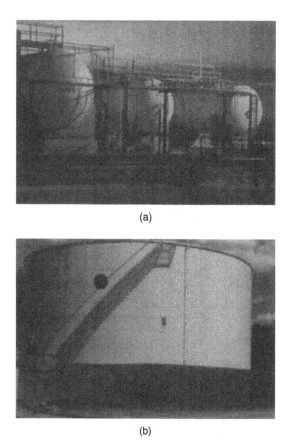

(a)

(b)

Figure 3.14   (a) Pressurized storage tanks; (b) Unpressurized storage tank (7).

prevention. ASTs are prone to a variety of modes of corrosion mechanisms. Internal and external modes of corrosion that may occur on an AST are shown in Fig. 3.16.

3.10.1.1 Internal Corrosion   There are several different corrosion conditions in the interior parts of an aboveground tank. Vapor-phase corrosion can occur in areas exposed to vapor above the stored products and includes general, crevice, and pitting corrosion, depending on the temperature and the characteristics of the material. Product side corrosion can occur on the internal wall plate when corrosive materials are stored. This type of corrosion includes general and pitting corrosion. At the interface of liquid and gas in a tank, the corrosion rate is often accelerated because the oxygen or moisture concentration gradient at the interface varies with depth in the liquid. Aqueous phase corrosion can occur when water contamination and settling in petroleum products result in a layer of water on the bottom of the tank. Although the product may be noncorrosive, the presence of contaminants such as sludges and deposits may result in internal bottom and wall general corrosion, crevice corrosion,

Figure 3.15 Oil storage tank farm (7).

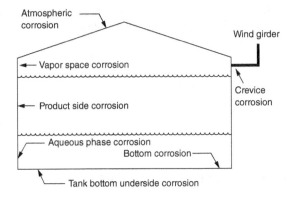

Figure 3.16 Internal and external corrosion modes of oil tanks (7).

and pitting corrosion damage. In addition, MIC can be a problem under anaerobic conditions. Internal corrosion problems are exacerbated by the stresses and flexing that the metal undergoes during fluctuations in product levels.

3.10.1.2 External Corrosion Atmospheric corrosion of the external wall and the roof is a result of general corrosion and crevice corrosion damage. Aboveground tanks suffer from external corrosion as a result of the tank bottom sitting on a grade with a variety of corrosive padding materials or on a back-filled concrete ring wall. Both types of tank bottom supports can cause external pitting of the bottom steel plate. Small aboveground tanks suffer from external atmospheric corrosion, but to a lesser degree because they can be supported off the ground, and the rounded surface minimizes crevice corrosion problems.

3.10.2 Underground Fuel Storage Tanks

These are large tanks and form a major portion of the hazardous materials storage sector. Corrosion is thought to be responsible for nearly 65% of tank failures while 35% is attributed to other causes such as third-party damage. It is generally agreed that the vast majority of underground storage tanks (USTs) and piping failures are because of external corrosion while a small percentage may be attributed to internal corrosion.

One of the primary causes of external corrosion is exposure to corrosive soils. The electrical and chemical characteristics of soil and water are closely related to corrosivity. Variations in soil characteristics because of soil type, fill compaction, amount of moisture, bacteria, chloride concentration help establish corrosion cells. Over a period of time, if untreated, the corrosion process can result in wall thickness reduction and can lead to leaks. The 6 o'clock position of the USTs is one of the most critical locations because that is the rest point where the tank bottom touches the bottom of the hole dug for the tank. At such a location, the layer of backfill is relatively thin; therefore, the soil characteristics can be different from the adjacent soil, setting up conditions for macrocell corrosion.

Analogous to the aboveground tank phenomena, internal corrosion can occur because of contaminants that settle on the tank bottom, under the stored product. Vapor-phase corrosion is usually limited because of the relatively constant temperature. A particular tank failure type, which is sometimes reported for gasoline service stations, is localized internal corrosion at the location where the internal lining is damaged. The inspector's level stick may cause mechanical damages to the lining, resulting in corrosion. Generally, a wooden pole is used to check the gas level in the UST. Lining damage occurs at the location where the pole hits the bottom of the UST.

3.11 CORROSION PROBLEMS IN AIRPORTS

The United States has the world's most extensive airport system, which essentially consists of national transportation, commerce, and defense. A typical airport infrastructure is relatively complex, and components that might be subject to corrosion are natural gas distribution systems, jet fuel storage and distribution systems, vehicle fueling systems, natural gas feeders, dry fire lines, parking garages, and runway lighting. Generally, each of these facilities is owned or operated by different organizations and companies, and the impact of corrosion on an airport as a whole is not known or documented; however, the airports do not have any specific corrosion-related problems that have not been described in other sectors such as corrosion in water and gas distribution lines, corrosion of concrete structures, and corrosion in aboveground and underground storage tanks.

Because of the diversity of airport facilities and different accountabilities, the corrosion costs cannot be assessed in a simple manner. to reduce and control corrosion costs, it is recommended that the airports establish databases that will allow engineers to track corrosion and corrosion costs and increase awareness.

A typical airport infrastructure is relatively complex, and the components that might be subject to corrosion are the following:

1. Natural gas distribution systems
2. Jet fuel storage and distribution systems
3. Deicing storage and distribution systems
4. Water distribution systems
5. Vehicle fueling systems
6. Natural gas feeders
7. Dry fire lines
8. Parking garages
9. Runways and runway lighting

Each of these infrastructure components is owned and operated by different organizations and companies. Airports do not have any specific corrosion-related problems that cannot be found in other sectors of the national economy. For example, corrosion of heat, ventilation, and air conditioning systems; corrosion of reinforced concrete floor in a parking garage; corrosion of buried metallic structures. Corrosion of buried metallic structures consists of USTs or buried fuel lines transporting fuel from the tank farms. Larger airports generate considerable volumes of wastewater during the deicing season and may have wastewater treatment facilities, which do not belong to the airports.

The problem with USTs is of concern as the Environmental Protection Agency (EPA) regulation mandates installation of corrosion protection on existing regulated USTs.

3.12 RAILROADS

America's first common-carrier-railroad, the Baltimore and Ohio (B&O), was chartered in Maryland on February 28, 1827. Some data on the basic facts of North American railroads (1999) are shown in Table 3.1.

Published information on corrosion-related issues in this industry is scarce. The elements of construction subject to corrosion are metallic objects such as rail, steel spikes for wooden ties. An area where corrosion has been reported is electrified rail

TABLE 3.1 North American Railroads (1999)

Number of railroads	561
Kilometers operated	274,399
Miles operated	170,508
Number of employees	200,906
Freight revenue	$35,295 \times 10^6$

systems. Barlo and coworkers (15) conducted a study on the corrosion of electrified trains that covered a number of transit systems. It was estimated that the damage to the rail system is primarily caused by a stray current that occurs on the electrified rail systems.

Transit systems in Chicago, Jersey City, New York City, Washington, DC, San Francisco, and Los Angeles were subjected to inspection, and it was found that corrosion-related problems exist, as manifested by the accelerated corrosion of the insulators of the rail fasteners in Jersey City and New York City, or in the wooden tie spikes in Chicago. For instance, wooden tie spikes need to be replaced after 6 months instead of the anticipated 25 years. In many instances, there was no formal tracking of corrosion-related costs.

While ostensibly there is corrosion damage to other related railroad-owned property, such as bridges and rail yard structures, from exposure to the elements, the railroad systems apparently do not consider it to be a major expense, and therefore do not track the data. No estimate of the cost of corrosion to railroads was possible from available data.

3.13 GAS DISTRIBUTION

The gas distribution pipeline sector is a part of the oil and gas industry. Figure 3.17 illustrates the different components of a natural gas production, transmission, storage, and distribution system. The components consist of production walls, gathering lines within the production fields, processing plants, transmission pipelines, compressor stations located periodically along the transmission pipelines, storage wells, and associated gathering pipelines, metering stations and city gate at distribution centers, distribution piping, and meters at both residential or industrial sites.

In 1998, the distribution pipeline included 2,785,000 km (1,730,000 miles) of relatively small-diameter, low-pressure natural gas distribution piping, which is divided into 1,739,000 km (1,080,000 miles) of distribution main and 1,046,000 km (650,000 miles) of services. There are nearly 55 million services in the distribution system. Figure 3.18 shows the distribution pipeline sector with regard to the oil and gas industry.

Several different materials have been used for main and service distribution piping. In earlier times, distribution mains were primarily carbon steel pipe. Since the 1970s, gas distribution plastic mains have replaced carbon steel. Steel mains are installed in small sections in certain "downtown" locations where the use of plastic is restricted in some large-diameter applications.

Table 3.2 summarizes the miles of gas distribution main and the number of services by material.

A large percentage of mains (57%) and services (46%) are metallic systems (steel/cast iron/copper), and corrosion is a major problem. For distribution pipe, external corrosion is the primary problem and internal corrosion has also been observed in some instances. The methods of corrosion monitoring on cathodically protected piping are similar to the methods used in the case of transmission pipelines

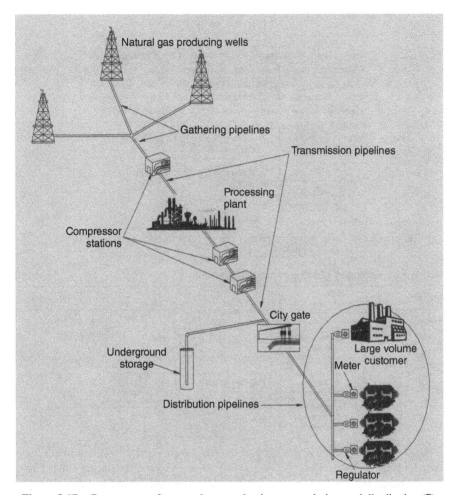

Figure 3.17   Components of a natural gas production, transmission, and distribution (7).

including pipe-to-soil potential and coating surveys. One difference is that in distribution systems, leak detection is considered to be an acceptable method of corrosion monitoring for the pipelines without CP.

For gas distribution piping, corrosion mitigation is primarily sacrificial CP. ILI is not feasible for the relatively complex network of distribution mains and services. This makes integrity assessment of the piping somewhat difficult.

Replacement costs of the infrastructure were calculated to obtain funding for corrosion control in maintaining the existing metallic piping.

3.13.1   Pipe Failures

Low-pressure gas distribution pipeline failures result in leaks that undetected might ignite and cause an explosion. The number of leaks because of corrosion was: mains

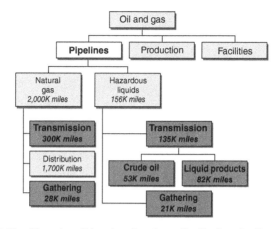

Figure 3.18   Chart describing the oil and gas distribution pipeline sector (7).

TABLE 3.2   Summary of Miles of Gas Distribution

Material	Miles of Mains	Number of Services
Steel	569,908	23,814,222
Plastic	461,433	28,506,127
Cast iron	46,023	51,090
Copper	52	1,497,638
Other	7,983	1,099,929
Total	1,085,399	54,969,006

83,864; services 99,024. The majority of the leaks were repaired without major incidents. Only 26 major incidents caused by corrosion were reported during 1984–1999, which resulted in $4,923,000 in property damage, 4 fatalities, and 16 injuries.

3.14   DRINKING WATER AND SEWER SYSTEMS

The forms of internal corrosion in water systems are uniform corrosion, galvanic corrosion, localized corrosion, concentration cell corrosion, MIC, and erosion– corrosion.

Major internal corrosion can occur in pipes made of cast iron, ductile iron, steel, galvanized steel, and cement-based materials. Table 3.3 summarizes the types of corrosion occurring in different piping materials and the possible tap water quality problems caused by them, as described by the AWWA Research Foundation in 1996 in a reference book on internal corrosion of water distribution systems (17).

Negative health effects can result from corrosion of lead, corrosion of copper alloys and solder in water supply systems, and corrosion of copper plumbing in potable water systems.

TABLE 3.3   Corrosion and Water Quality Problems Caused by Materials in Contact with Drinking Water (17)

Material	Corrosion Type	Tap Water Quality Deterioration
Cast iron	Uniform corrosion	Rust tubercles (blockage of pipe)
Ductile iron	Graphitization and pitting under unprotective scale	Iron and suspended particles release
Steel	Pitting	Rust tubercles (blockage of pipe). Iron and suspended particles release.
Galvanized steel	General corrosion	Excessive zinc, lead, cadmium, and iron release causing blockage of pipe.
Asbestos cement	Uniform corrosion	Calcium dissolution, possible asbestos fibers, and increased pH
Concrete	Uniform corrosion	Calcium dissolution and increased pH
Cement	Uniform corrosion	Calcium dissolution; increased pH
Mortar	Uniform corrosion localized attack	Copper release
		Perforation of pipe and leakage
Copper	Microbiologically induced corrosion (MIC)	Leakage from pipes
Copper	Corrosion fatigue	Rupture of pipe and leakage
	Erosion–corrosion	Leakage from pipe
Lead pipe	Uniform corrosion	Lead release
Lead-tin solder	Uniform corrosion	Lead and tin release
Brass	Erosion and impingement	Penetration failures
	Dezincification	Blockage of pipe
	Stress corrosion cracking (SCC)	Lead and zinc release
Plastic	Degradation by sunlight and microorganisms	Taste and odor

3.14.1   External Corrosion in Water Systems

External corrosion of water systems may be caused by general corrosion, stray current corrosion MIC, and/or galvanic corrosion. Corrosion mitigation techniques include the application of protective coatings, wrapping pipe in a plastic cover, and the application of CP. The areas of major external corrosion impact are generally those where localized attack may take place, such as in the proximity of other systems like galvanic corrosion or in areas where stray currents may occur.

Both DC and AC stray currents on a water line can cause corrosion. Stray current studies (18) show that the corrosion rate because of the dc current is generally greater than the corrosion rate because of ac current. General external corrosion can be a problem in corrosive soil, in particular, when the soil is of low resistivity, high moisture content, and has corrosive chemical species. When piping is electrically continuous (welded steel piping), CP can be applied; however, this is not generally applicable in the case of discontinuous pipe made of ductile iron or cast iron.

Plastic piping [polyvinyl chloride (PVC)] does not show corrosion as in the case of metal piping, but the properties of plastic piping deteriorate over time. In severely corrosive soils PVC piping may be selected rather than a metallic piping because it is inert to the chemical conditions. PVC has a lower density than steel and iron and hence it is relatively easy to handle in the field. However, PVC has lower strength and traditional welding is not possible. PVC has been used for a relatively short time, compared with steel and iron water lines. Thus, there is limited data on the expected service life of PVC pipelines, and calculations of comparative total life-cycle costs are not possible.

Cement-based piping deteriorates by corrosion of the reinforcement steel, which is accelerated by chloride from salt-treated icy roads during winter. Corrosion occurs when the passive surface film that naturally forms on the steel in high-pH concrete/cement breaks down in the presence of chloride. The corrosion product has greater volume than the original steel, creating internal stresses that cause cracking and spalling of the concrete/cement pipes.

3.15 ELECTRICAL UTILITIES

Electricity generation plants can be divided into seven generic types: fossil fuel, nuclear, hydroelectric, cogeneration, geothermal, solar, and wind. The majority of electric power in the United States is generated by fossil and nuclear steam supply systems.

Two types of light water reactors, namely, the boiling water reactor (BWR) and the pressurized water reactor (PWR) are in use in the United States of America. The fuel for these reactors consists of long bundles of 2–4 wt% of enriched uranium dioxide fuel pellets stacked in zirconium–alloy cladding tubes.

The BWR design (Fig. 3.19) consists of a single loop in which the entering water is turned directly into steam for the production of electricity. As operating temperatures must remain below the critical temperature for water, steam separators and dryers are used with a "wet-steam" turbine.

The PWR design (Fig. 3.20) is a two-loop system that uses high pressure to maintain an all-liquid-water primary loop. Energy is transferred to the secondary steam loop through two to four steam generators. The PWR design also uses a wet-steam turbine.

3.15.1 Fossil Fuel Steam Supply Systems

The electric power industry uses three types of fossil fuel power plants: coal-fired steam, gas turbine, and combined cycle power plants. The most common and widely used is the pulverized coal-fired steam power plant. Fuel oil can be used in place of

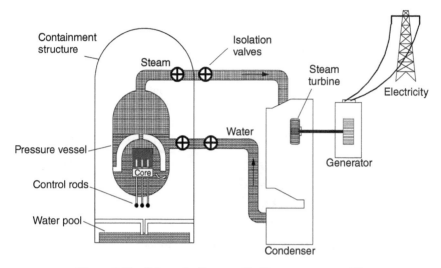

Figure 3.19   Schematic diagram of boiling water reactor (7).

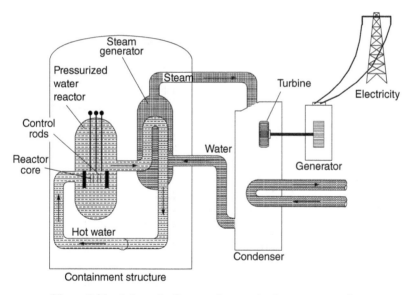

Figure 3.20   Schematic diagram of pressurized water reactor (7).

coal. Figure 3.21 shows the basic operation of a steam-generating plant. Gas turbines are usually smaller units that are used for peak loads and operate for only a few hours per day. Combined cycle plants using both steam and gas turbines are generally used for base-load service but must be capable of handling peak loads (Fig. 3.22).

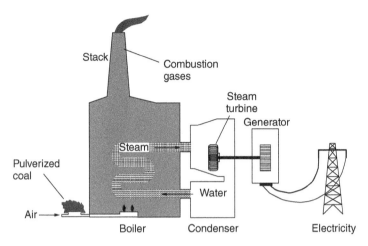

Figure 3.21   Schematic diagram of fossil fuel plant (7).

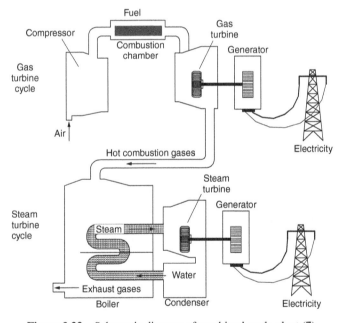

Figure 3.22   Schematic diagram of combined-cycle plant (7).

3.15.2   Hydraulic Plants

Hydraulic power systems include both hydroelectric and pumped storage hydroelectric plants. In both cases, water is directed from a dam through a series of tapering pipes to rotate turbines and create electricity. In principle, the potential energy held in the dam converts into kinetic energy when it flows through the pipes (Fig. 3.23).

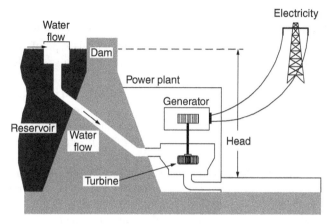

Figure 3.23   Schematic drawing of hydroelectric plant (7).

TABLE 3.4   O&M Costs for 1996

Category	Percent	1998 O&M Costs ($× billion)
Fossil fuel	64.4	117.6
Nuclear	21.1	38.6
Hydraulic	1.3	2.4
Other power generation	3.3	6.0
Transmission	3.0	5.4
Distribution	6.9	12.6
Total	100	$182.6

3.15.3   Areas of Major Corrosion Impact on Electric Utility Systems

The impact of corrosion on electric utility systems can be divided into the fraction of utility costs for depreciation, operation, and maintenance that are attributable to corrosion. The estimated costs discussed for this sector are based on detailed analysis of facilities and work activities, using input from Duke Power, an energy company serving more than two million people in North and South Carolina and the Electric Power Research Institute (EPRI) reports, technical literature, and other utilities.

The operation and maintenance (O&M) costs of various categories of energy producing sources are given in Table 3.4.The data indicate that the highest percentage O&M cost is for the fossil fuel category and the smallest percentage for hydraulic utilities.

3.16   TELECOMMUNICATIONS

This sector describes the impact of corrosion on the telecommunications infrastructure. Telecommunications hardware consists of switch boards, electronics, computers, data transmitters, and receivers. Delicate electronic components must be protected from human and weather factors in order for their operation to be executed in a reliable way over a long period of time. According to an expert in electronics, manufacturing and corrosion failures of this type of equipment are caused by environmental factors such as moisture. Corrosion of the delicate small parts can cause malfunction of the equipment.

Most of the telecommunication hardware is placed and used in buildings and thus not exposed to corrosive atmosphere. In addition, electronic hardware has a limited lifetime and becomes obsolete in a few years. The actual service life of consumer is often limited by rapid technological changes rather than material degradation issues.

Telecommunication equipment with a longer design life consists of cables, connectors, and antennas for the transmission and reception of electronic signals. These components may be placed and buried so that they become exposed to environments such as soils and water or they may be exposed to air and moisture conditions. There is no report on the percentage of failures because of corrosion for each category.

A specific corrosion problem is a possibility at telephone facilities that maintain backup power systems in case of power outages. These facilities may have diesel fuel generators supplied by USTs. A leaking UST system can lead to contamination of ground water supplies and can cause fires, explosions, and vapor hazards. Under the Resource Conservation and Recovery Act (RCRA) for USTs, the owners and operators of USTs must have upgraded, replaced, or closed existing substandard UST systems by December 22, 1998. Upgrading may involve adding spill, overfill, and corrosion protection to the UST system.

3.16.1 Shelters

Telecommunications equipment is usually housed in a shelter to protect it from wind and foul weather. Shelters are structures without windows that can be climate controlled and contain a large amount of electronics, computers, and other equipment such as transformers.

Shelters are generally located in the immediate vicinity of power stations and communication towers. Many antennas and towers are placed at high locations; therefore, shelters are located on rooftops. Shelters are also placed at locations on the ground. The mobile telephone system requires many support antennas spread throughout the landscape. Along a major interstate highway, it is possible for one to count nearly one communication tower per one mile of highway. Each tower is protected with a shelter and a fence around it, both of which protect the infrastructure.

Four construction materials are used for communication shelters: steel, aluminum, fiberglass, and concrete. Wooden blocks or concrete block are used for the foundations of shelters.

Carbon steel shelters need to be painted to protect them from corrosion. Stainless steel shelters do not require to be painted, but are more expensive than carbon steel shelters. Aluminum has favorable weight-to-strength ratio. On roof tops and on other mounted structures, the dead weight of the shelter can be important for structural purposes. Aluminum is generally corrosion-resistant in nonmarine environments. Aluminum is costlier than carbon steel and requires no painting.

The fiberglass shelters consist of a foam core with two skins of fiber-reinforced plastic. The fiberglass exterior is corrosion-resistant and requires low maintenance. The price of fiberglass shelters is higher than the corresponding aluminum and carbon steel shelters.

The largest and strongest shelters are those constructed using concrete. These are usually secured shelters with steel doors. All concrete shelters have temperature and humidity controlled environments; therefore, corrosion is not an issue for the equipment placed in concrete shelters.

Degradation of equipment because of heat and humidity is possible for cellular telephone equipment. A refrigerator-sized cabinet may be placed near the antenna. The cabinets are made of steel or aluminum. Corrosion protection is applied for cosmetic purposes as this type of technology generally becomes obsolete and is replaced before corrosion becomes a structural issue. In some cases, a double system of galvanizing and painting is applied for corrosion protection of steel cabinets. Surface preparation through grinding and application of zinc chromate primer is essential for galvanizing. The outdoor cabinets are built more robustly, that is, of thicker gauge material than the indoor cabinets and can cost twice as much. Cabinets priced in the range $1000–10,000 have an estimated corrosion cost of $200–2000. The estimated total corrosion costs of the cabinets amount to $4 million ($4000 \times 5000 \times 0.2$).

A large majority of telecommunication towers are of the self-supporting type. These have been constructed since the early 1960s and are made from hot-dipped galvanized steel. For about 40 years, there is hardly any corrosion.

Guided (wire) towers belong to the second largest group. These are made from carbon steel and were sandblasted and repainted regularly. The continued operation of these aging guided towers is a major corrosion concern because corrosion may affect the structural integrity of the towers.

The single largest corrosion problem in the telecommunications industry is the degradation of buried grounding beds and grounding rings around towers and shelters. The copper grounding systems are consumed over time by corrosive soil. Problems occur when the electrical connection between the grounding bed and the structure is interrupted or when the corrosion advances to the extent that the electrical resistance of the bed becomes too great. To prevent electrical disconnection between the grounding and the structure, the traditional mechanical connections must be replaced with CADWELD connections (American Welding Society (AWS) designation: Termit welding (TW) process). Galvanic corrosion because of connections between dissimilar metals is another factor related to copper ground beds.

The copper cables used in the telecommunications industry's electrical supply are encapsulated in plastic to prevent electrical shorts. The plastic cover also provides corrosion protection to the wires. The following anticorrosion protection measures are recommended:

1. Check galvanization condition.
2. Check paint condition.
3. Check oxidization of the structure, bolting parts, and accessories.
4. Check for masts with guide wires and oxidation of wires.

In the case of towers in salty environments:

1. Check the condition of the tower structure.
2. Wash regularly the tower structure, in the absence of rains sufficient to wash the tower free of salt.

3.17 MOTOR VEHICLES

American consumers, businesses, and government organizations own more than 200 million registered vehicles. Assuming a value of $5000 for each vehicle allows an estimate that Americans have more than $1 trillion invested in their motor vehicles, making automobiles one of the largest investments collectively among Americans.

3.17.1 Corrosion Causes

Several factors lead to various types of corrosion in motor vehicles, specifically the design process, the manufacturing process, and operating conditions.

3.17.1.1 The Design Process Motor vehicle designers make a multitude of choices that influence how susceptible a vehicle may be to corrosion. During the design of the vehicle, engineers should strive to reduce dissimilar metal contacts, crevices, stresses, poor drainage, and locations where salt and dirt can build up. The choice of materials in the design will also dramatically affect the corrosion performance of vehicles. The use of corrosion-resistant metals, coated steels, and polymers, as well as avoidance of dissimilar metal contacts, will allow vehicles to operate for many years without significant corrosion problems. One of the most critical considerations in the design process that affects corrosion performance involves the choice of primers, paints, and sealers. The use of corrosion-resistant primers over the entire body and special chip-resistant coatings for the wheel and the lower surface of the car has become standard in the industry to reduce the initiation of corrosion.

Several elements of corrosion protection added in the design phase can be rendered useless if the quality of manufacturing is below standard or low. A few elements of the manufacturing process are of specific importance. First, the quality of welding will affect the presence of crevices where corrosion can occur. Secondly, the surface pretreatment must be done properly to ensure good adhesion of the primer and the final coating. Finally, several of the special coatings, such as chip-resistant coatings and the body sealants, are applied manually by hand, and the quality of this work is very much dependent on the skill and the attention paid by the applicator.

The corrosivity of the local environment will strongly affect the corrosion performance of the vehicle (19, 20). In locations where corrosive environments are possible because of acid rain, deicing salts, or marine environments, personal driving habits and diligent maintenance of the vehicle, such as regular washing and replacement of fluids, can have a significant effect on the reduction of corrosion.

3.18 SHIPS

The size of the shipping industry can be measured by the number of miles that ships sail and the tons of cargo they haul. Corrosion of ships involves several different types of corrosion. The most common form of corrosion is general corrosion or wall thinning of the hull because of seawater attack. Studies have shown that this form of corrosion is approximately 0.1 mm (4 miles) per year (21). At this corrosion rate, it would take approximately 62 years to have a reduction of 6.4 mm (0.25 in.). Because of this slow rate, general corrosion is normally not a consideration in a ship's design life.

Galvanic corrosion occurs between two metals with dissimilar electrochemical potentials. In this form of corrosion, one of the metals is more electrochemically active and corrodes, while the second metal is protected by the corroding metal. The metals can even be of the same material if the electrochemical potential of one of the materials has been charged because of stresses or differential aeration. Previous studies have indicated that most hull corrosion is galvanic in nature (22).

Salt spray and atmospheric corrosion can severely attack external ship components. Coatings provide the primary corrosion control, and maintenance of these coatings is required at regular intervals.

Direct chemical corrosion attack occurs when certain chemicals are present in the internal holds and tanks of transport ships. Elements such as chlorine and sulfur can readily attack the steel and cause accelerated corrosion and pitting.

Corrosion in ships can also be caused by MIC. In this type of corrosion, microbial organisms present in the environment can accelerate corrosion. For example, SRB, which are present in stagnant water of many harbors, can build up on the hulls of ships. Other corrosion-causing bacteria, such as acid-producing and anaerobic bacteria, are also present in ballast tanks as well as in the liquid products that some tankers carry. The microbes cause a localized change in the environment, which can promote aggressive pitting and other types of corrosion.

Table 3.5 shows the percentage of the world's fleet by class of ship on the basis of number.

TABLE 3.5 Percentage of the World's Fleet by Class of Ship

Class	Percent
Refrigerated cargo	2
Chemical tankers	3
Bulk dry	7
Passenger/ferry	7
Oil tankers	8
Supply/tugs	15
Cargo/roll on/roll off	22
Fishing	27
Others/unknown	9

TABLE 3.6   Percentage of World's Fleet by Class of Ship on the Basis of Gross Tonnage

Class	Percent
Refrigerated cargo	1
Chemical tankers	6
Bulk dry	30
Passenger/ferry	4
Oil tankers	28
Supply/tugs	2
Cargo/roll on/roll off	15
Fishing	2
Others/unknown	12

Table 3.6 shows the percentage of the world's fleet by class of ship on the basis of gross tonnage.

3.19   AIRCRAFT

According to the data presented in the 1999 edition of the *Aviation and Aerospace Almanac*, the combined aircraft fleet operated by US airlines in 1998 totaled 7478 (23).

These include fixed-wing turboprop and turbojet and rotary wing aircraft. About 3973 are turbojet aircraft, which are divided into 29 different types for domestic and international service. Table 3.7 shows a 1997 listing of the major US carrier fleets with the number and average age of the fleets' airplanes.

3.19.1   Corrosion Modes

Corrosion in an aircraft manifests itself in several different forms. Pitting and crevice corrosion are the most common forms of corrosion in the 2000 and 7000 series aluminum alloys, which are the principal materials of construction. Pitting corrosion produces deterioration of the airframe structure in localized areas and can have high penetration rates. Pits often create stress concentrations, which may reduce the fatigue life of a component. Crevice corrosion, by itself, is more destructive than pitting corrosion. Crevice corrosion occurs when a corrosive fluid enters and is trapped between two surfaces, such as a joint, a delaminated bond line, or under a coating. Both pitting and crevice corrosion, when unchecked, can readily develop into exfoliation corrosion or intergranular SCC. Exfoliation corrosion is a form of intergranular corrosion where corrosion attack occurs along the grain boundaries of elongated grains, causing a leaf-like separation of the metal grain structure (Fig. 3.24). This form of corrosion often initiates at unprotected end grains, such as at fastener holes and plate edges. Figure 3.24 represents exfoliation corrosion around fastener holes in aluminum alloy 7075-T6 fuselage section.

TABLE 3.7 Major Carrier Jet Fleets in 1997

Carrier	Number	Age (year)
Alaska	75	7.4
America West	100	10.4
American	663	9.4
Continental	350	11.9
Delta	538	11.7
Federal Express	264	17.7
Northwest	412	18.9
TWA	200	19.9
United	592	11.5
United Parcel Service	144	16.5
US Air	408	12.3
Southwest	241	

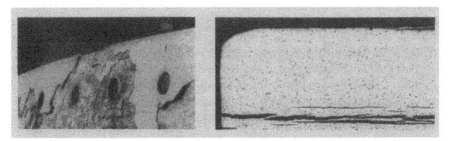

Figure 3.24 Exfoliation corrosion around fastener holes in Al alloy 7075-T6 (7).

Intergranular SCC occurs when stresses are applied perpendicular to the susceptible grain boundaries. More so than pitting and crevice corrosion, susceptibility to exfoliation corrosion, and intergranular SCC depends on alloy type, heat treatment, and grain orientation. Other common forms of corrosion are fretting corrosion, which occurs when two surfaces rub at high frequency and low amplitude in the presence of a corrosive environment. Galvanic corrosion occurs when dissimilar metals such as aluminum and steel are in direct contact. Isolation of different metals by proper design and assembly can prevent both forms of corrosion.

There are many contributing causes of corrosion in commercial aircraft. The first potential source of corrosion is in the basic design process. Materials selection, finishes, and structural configuration can have a significant impact on the corrosion performance of an airplane. During the design phase, attention must be paid to the basic principles of corrosion-conscious design, such as the selection of corrosion-resistant materials, the avoidance of dissimilar metal-contact, crevices, stresses, and poor drainage. In addition, the selection of sealants and finish systems is an important part of corrosion-conscious design. For example, the use of corrosion-inhibiting primers and sealants on fasteners and faying surfaces has become common practice for new airplanes, and the elimination of crevices is now required by "faying surface sealing" of all joints that are prone to corrosion. Corrosion-inhibiting compounds are routinely applied in the final assembly of many aircraft components, such as the inside fuselage crown and lower lobe, pressure bulkheads, pressure deck, under lavatories and galleys, wheel-wells, wing-empennage cove areas, dry bays, empennage torque box interiors, and under fairings.

Another potential source of corrosion problems is in the manufacturing process used to produce the aircraft. Specifically, the assembly and finishing processes can determine whether a specific component will be subject to premature corrosion. Of particular importance is the proper surface pretreatment and application of protective coatings and sealants, which must offer long-term durability to provide adequate corrosion protection.

Once the airplanes are in the hands of operators, many factors, including operating conditions and maintenance practices, determine the corrosion performance of the airplanes. Operational environments such as marine, tropical, high humidity, and industrial can be extremely corrosive to the external structure of an aircraft. Furthermore, during operation, the protective surface finishes can deteriorate by chipping, scratching, breaking around fasteners, abrasion, and aging. Environmental conditions inside an airplane can be even more damaging. For example, lavatory spillage, galley spillage, chemical spillage, animal waste, microbial growth, fire residue, and corrosive cargo such as fish (salt water) can create extremely corrosive conditions inside an airplane. Condensation that forms on the inside of the fuselage is also a potential source of internal corrosion. Boeing (Coating Industry Expert, Private Communication, June 2000) has conducted an inspection of airplanes with the most severe moisture problems and found that as a result of moisture uptake in insulation blankets in B-737-300 airplanes, the weight had increased by an average of 36 kg (79 lb).

3.20 RAILROAD CARS

Figure 3.25 shows a modern-day locomotive, a corroded locomotive, and reconditioned railroad cars. Commodities transported by railroad are given in Table 3.8.

The largest costs to the industry are because of corrosion of the exterior and interior of the railroad cars.

External corrosion of the cars is primarily because of atmospheric exposure. While corrosion damage is still a concern, car appearance takes precedence; therefore, the car manufacturers/lessees often choose to apply an exterior paint system to address the aesthetic issues. The paint systems are typically "direct-to-metal" (DTM) epoxy or epoxy with a urethane coat. This epoxy substrate adds protection against ultraviolet radiation.

The most common method of internal corrosion prevention is the use of coating systems and rubber linings for internal surfaces. The use of linings and coatings is aimed not only at prolonging the service life of the car fleet, but also at precluding the contamination of the transported commodity by corroding metal substrate. Considering that certain types of commodities such as chemicals are corrosive, corrosion prevention measures are an absolute necessity.

Modern-day locomotive

Corroded locomotive

Reconditioned railroad cars

Figure 3.25 Modern locomotive, corroded locomotive, and reconditioned one (7).

TABLE 3.8   Commodities Transported by Railroad Car
Loads

Commodity Group	Carloads $\times 10^3$
Coal	7027
Chemicals and allied products	1680
Motor vehicles and equipment	1546
Farm products	1404
Food and kindred products	1282
Nonmetallic minerals	1256
Metals and products	671
Lumber and wood products	645
Waste and scrap materials	581
Pulp, paper, and allied products	547
Petroleum and coke	483
Stone, clay, and glass products	475
Forwarder and shipper-associated traffic	376
Metallic ores	311
Other car loads	7421
Total car loads originated	25,705

While the largest segment of the freight has historically been coal, chemicals and allied products amount to the second largest group of transported goods, while food and kindred products make up 5% of transported goods. The latter two groups of commodities are either corrosive or sensitive to contamination. Nearly 130,000 of the covered hopper cars are used for transporting plastic pellets, which require liners to preserve product purity. The liner life is 8–10 years (Coating Industry Expert, Private Communication, June 2000).

Transportation of coal presents a problem because, when mixed with moisture, it becomes highly acidic and corrosive to the carbon steel. There are indications that a large number of cars can be significantly affected by this problem (Coating Industry Expert, Private Communication, June 2000).

Corrosion is likely to advance further by the use of thawing sheds during the winter months in cold climates, in which the cars are heated to thaw the coal. According to some estimates, there are about 100,000 cars used for coal services; therefore, the problem may be quite extensive (Coating Industry Expert, Private Communication, June 2000).

Another source of aggressive species is sodium chloride. The cars used to transport rock salt suffer from advanced corrosion attack and last for approximately only 3 years (Coating Industry Expert, Private Communication, June 2000). The high cost of rehabilitation of salt carrying cars created a trend toward using unlined, covered hopper cars previously utilized to transport grain for rock salt service. When corrosion becomes considerable, the cars are scrapped. As such a process cannot continue indefinitely, more and more rock salt is expected to be hauled by trucks and barges, as the revenue seems to be insufficient to justify the corrosion-related replacement/rehabilitation costs.

To accommodate the properties of the cargo, in addition to the use of coatings and linings, certain components of the cars, such as valves, undergo an upgrade from the lower-resistant carbon steel to the higher resistant steel grades, such as stainless steel.

Rubber linings are used for strong acids (concentrated hydrochloric and phosphoric acid). For extremely aggressive concentrated nitric acid, the entire tank car body is made from stainless steel (316L).

3.21 HAZARDOUS MATERIALS TRANSPORT

This sector includes the transportation of hazardous materials (HAZMAT) other than that of the transportation of hazardous gases and liquids by buried pipelines.

Bulk transportation of HAZMAT involves overland shipping by tanker truck and rail tank car and by specialized containers that are loaded onto vehicles. Over water, ships loaded with specialized containers, tanks, and drums are used. In small quantities, HAZMAT requires specially designed packaging for truck and air shipments.

Trucks are used for the transportation of hazardous materials. Stainless steel tanks are used in highway trucks. Storage drums and a corroded storage drum are shown in Figure 3.26.

Hazard classifications assigned for distinct HAZMAT are listed in Table 3.9.

Class 5 and 8 materials require shipping and storage containers that are resistant to corrosion to prevent internal damage. Most of the materials cited in class 1–9 can become corrosive to mild steel containers when contaminated with moisture. Depending on the environment, materials from all categories must be shipped and stored in containers that are protected from external corrosion damage.

The daily and annual number of HAZMAT shipments is listed in Table 3.10.

The Research and Special Programs Administration (RSPA) of the US Department of Transportation (DOT) published a list of the top 50 hazardous materials in a 1998–1999 summary of HAZMAT transportation incidents (24). The corrosive materials that were most involved in HAZMAT incidents in 1998 were sodium hydroxide solutions, basic inorganic liquids, hydrochloric acid, solutions, sulfuric acid, cleaning liquids, hypochlorite solutions, basic organic liquids, liquid amines, and ammonia solutions.

Table 3.11 shows hazardous materials carried by trucks as reported in VIUS.

Shippers are required to report HAZMAT incidents to the US DOT whenever there is an unintentional release of a HAZMAT. The information from all submitted forms is collected in the Hazardous Materials Information System (HMIS) incident data base, which is maintained by RSPA and includes data reported by carriers for the past 30 years. The following data shows the number of, as well as the consequences resulting from, serious incidents. In 1998, there were roughly 15,000 reported HAZMAT incidents related to HAZMAT shipments resulting in 13 deaths and 198 injuries. On DOT Form 5800.1 shippers are required to give a description of the packaging failure for each incident. Corrosion is also a contributing factor for the packaging failure.

Trucks for transport of hazardous materials

Stainless steel tanks for highway trucks

Corroded storage drum

Storage drums

Figure 3.26   Corroded storage drums (7).

TABLE 3.9 Hazard Classifications

Classification	Materials
Class 1	Explosives
Class 2	Flammable and compressed gases
Class 3	Flammable liquids
Class 4	Flammable solids
Class 5	Oxidizers
Class 6	Poisonous materials
Class 7	Radioactive materials
Class 8	Corrosive materials
Class 9	Miscellaneous hazardous materials

TABLE 3.10 Daily and Annual Number of Hazmat Shipments

Product	Annual shipped (metric tons) billion	Annual moved (metric tons) billion
Chemicals and allied	0.53	0.85
Petroleum products	2.60	3.03
Other	0.01	0.02
Total	>3.1	>3.9

TABLE 3.11 Hazardous Materials Carried by Trucks as Reported in VIUS

Classification	Hazardous Material	1997 Trucks (× 1000)
Class 1	Explosives	100.8
Class 2	Flammable, nonflammable, poisonous gas	450.4
Class 3	Combustible	127.5
Class 4	Flammable solid, spontaneously combustible, dangerous when wet	154.2
Class 5	Oxidizer, oxygen, organic peroxide	176.3
Class 6	Poison (formerly poisons A and B, solids and liquids) keep away from food	120.5
Class 7	Radioactive	19.2
Class 8	Miscellaneous hazardous materials	53.5
Class 9	Miscellaneous hazardous materials	53.5
	Hazardous materials not specified	40.6

It is essential to identify the cause of packaging failure such as "vehicle collision," "improper loading," or corrosion. It is important to realize the difference between the contents of a package involved in an incident and the root cause of an incident. Corrosive materials were the contents of 35.7% of 1998 incidents and the root cause of an incident. Corrosion was implicated as a contributing factor in 1.35% of the total incidents in 1998 (24). A total of 79 (38.3%) of the 206 corrosion-related incidents had a reported damage cost of $0, while 45 incidents had a cost between $0 and $100, and 81 incidents cost more than $100.

Internal corrosion of tankers usually requires only mitigation when an oxidizing agent or a corrosive agent is transported. Internal corrosion from settled contaminants is limited because of high throughput and movement of the product during transportation. Internal corrosion of tankers can be a problem during long periods of storage, in particular, when the tankers are not properly cleaned before storing the product in the tanker.

Steel pails and drums are used as containers for transportation. Two possible reasons for the replacement of pails and drums are damage by improper use and internal corrosion. Assuming that 50% of the damage is because of internal corrosion, the replacement costs of $145 million for pails and $342 million for drums have been estimated.

Shipping containers, such as drums and pails, are subject to internal corrosion damage and failure when corrosive materials are shipped. Internal corrosion is not a problem when materials are shipped from the manufacturer because the proper container material is used and the containers are transported in a short time. However, contaminated or corrosive materials can cause failures when stored beyond the material's shelf life. The corrosion failure of drums containing hazardous waste tends to be more of a problem. Typically, the problem occurs when the wastes are mixed or when the waste is contaminated and stored in containers made of noncompatible materials.

In the transportation industries, external corrosion of tanker trucks and railcar-mounted tanks is a common problem. Both general and pitting corrosion from the atmosphere and splash water from the roadway or rail bed can affect the tank's structured integrity and tightness. This problem is particularly severe in areas of the country with chloride sources such as road salt or airborne marine atmosphere and severe airborne industrial pollution.

3.22   OIL AND GAS EXPLORATION AND PRODUCTION

Domestic oil and gas production can be considered a "dinosaur industry" in the United States because most of the significant oil and gas reserves have been exploited. The significant recoverable reserves left to be discovered and produced in the United States are probably limited to less convenient locations, such as deep water offshore, remote arctic locations, and difficult-to-manage reservoirs with unconsolidated sands.

Oilfield production environments can range from practically zero corrosion to severely high rates of corrosion. Crude oil at normal production temperatures (less than $120\,°C$) without dissolved gases is not, by itself, corrosive. The economics of controlling corrosion in many oilfields is dependent on efficient separation of crude oil from other species. While the rates may vary, the species causing the most problems are nearly universal. Carbon dioxide and hydrogen sulfide gases, in combination with water, define most of the corrosion problems in oil and gas production. Other problems are microbiological activity and the solids accumulation.

$$H_2S \rightarrow H^+ + SH^-$$
$$Fe + SH^- \rightarrow FeS$$
$$CO_2 + H_2O \rightarrow H_2CO_3$$
$$H_2CO_3 \rightarrow H^+ + HCO_3^-$$

The mechanisms of CO_2 corrosion are generally well defined; however, the reality inside a pipeline becomes complicated when CO_2 acts in combination with H_2S, deposited solids, and other environments. H_2S is highly corrosive, but can, in some cases, form a protective sulfide scale that prevents corrosion.

Microorganisms can attach to pipe walls and cause corrosion damage. Solids such as formation sand can both erode the pipeline internally and cause problems with underdeposit corrosion, if stagnant.

Oxygen is not found in oil reservoirs and much is done to ensure that no oxygen enters the production environment; however, in many cases, a few parts per million (ppm) of oxygen enters the production pipeline, greatly exacerbating corrosion problems.

External corrosion problems in oil and gas production normally are similar to those found in the pipeline industry, but as the lines are shorter and smaller in diameter, their economic impact on the total cost of production is limited. Atmospheric corrosion of structures and vessels is a problem for offshore fields and those operating near marine environments.

Improvements in the quality of protective coatings for offshore environments have dramatically reduced the frequency of repainting platforms and tanks.

Corrosion in oil and gas production varies from location to location. Corrosion can be classified into one of three general categories of internal corrosion caused by the product fluids and gases, external corrosion caused by exposure to groundwater or seawater, and atmospheric corrosion caused by salt spray and weathering offshore. Of these, internal corrosion is the most costly as internal mitigation methods cannot be easily maintained and inspected.

A typical oil and gas production flow diagram is shown in Figure 3.27. Oil, water, and gas are produced in every oil field. Water is injected downhole to maintain reservoir pressure and stability, and often water fording from seawater or freshwater sources is used to drive oil out of the formation. As a field ages, the water cut or the ratio of water to oil in the fluids produced, increases to levels of 95% or higher depending on the economics of production. As the oil industry matures and the number of old oil fields relative to new fields increases, the amount of water produced increases and the internal corrosion increases.

Water injection from seawater or fresh water sources contributes to the "souring" of oil fields with H_2S usually resulting in an increase in the corrosion rate, which sometimes requires a complete change in corrosion strategy. These water sources may necessitate biocide injection and will require deaeration to avoid introducing a new corrosion mechanism into the existing system. Tertiary recovery techniques are often based on miscible and immiscible gas floods. These gas floods invariably contain a

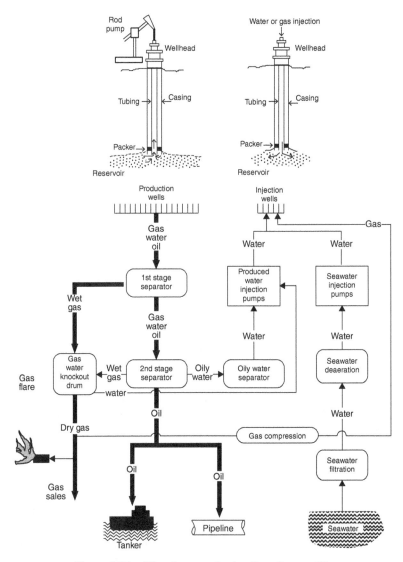

Figure 3.27   Oil and gas production flow diagram (7).

high percentage (~100%) of CO_2, which dramatically increases the corrosivity of the produced fluids.

3.23   CORROSION IN THE MINING INDUSTRY

Corrosion in the mining industry may be characterized as corrosion enhanced by abrasion. It is also difficult for corrosion engineers to plan for corrosion because mine atmospheres and waters are unique and vary widely from one mine to another; therefore, each mine experiences relatively different corrosion-related problems.

Aerobic and anaerobic microorganisms present in mine water also contribute to the extremely corrosive environments (25). Aerobic species produce sulfuric acid, making the environment very acidic. Aerobic microorganisms reduce sulfate and sulfides by using available hydrogen and producing hydrogen sulfide.

Although corrosion exists in the mining industry, the industry does not consider it to be a serious issue. Engineers from several mining companies who were interviewed could not provide any information on instances where corrosion problems were a critical issue. Past experiences and equipment suppliers provide the process engineers with enough information to keep the mining industry processing its metallic minerals, industrial minerals, and coal.

While mining engineers are not involved in serious corrosion issues, literature on the subject lists several areas of major concern because of personnel safety and continuation production. The areas of major concern are the following:

1. Wire rope
2. Roof bolts
3. Pump and piping systems
4. Mining electronics
5. Acid mine drainage.

3.23.1 Wire Rope

Wire ropes are used extensively in the mining industry to help hoist equipment. Mine workers also depend on this rope for their safety. Wire rope undergoes both corrosion and abrasion, which will degrade the mechanical properties of the wire and thus reduce its load-bearing capability and cause it to fail.

A statistical study of mine-hoist wire ropes showed that 66% of the ropes exhibited greatest loss in strength in the portion of the rope in contact with the shaft environment during its service life (26). The Mine Safety and Health Administration (MSHA) of the US Department of Labor requires that wire ropes in service be visually examined for structural damage, corrosion, and improper lubrication or dressing (27). MSHA also requires performance of careful nondestructive testing (NDT) every 6 months and cites one instance where a contractor reported that four ropes were in acceptable condition for use in an elevator shaft. However, less than 6 weeks later, one of the four 12.7 mm (0.5 in.) diameter ropes broke and another was severely corroded with several broken wires (28).

Wire ropes within the mining industry are routinely replaced every 18–36 months depending on environmental conditions and use over time. The wire ropes are made of carbon steel; however, because of their susceptibility to corrosion and wear, stainless steel and synthetic fiber ropes are becoming more widely used instead of carbon steel ropes (25).

In the mining industry, roof bolts provide support in underground mines by tying the lower layer to a stronger layer located above the main roof. In the United States, more than 120 million low-carbon steel roof bolts are used per year and are a major area of concern with regard to corrosion because a failure of the roof bolt is hazardous and could result in the loss of lives (25). In sulfide mines, roof bolts have been reported to fail within 1 year because of sulfide SCC.

Corrosion within pump and piping systems is another critical issue in the mining and mineral processing industries. The most common form of corrosion is uniform corrosion; however, pitting, crevice corrosion, intergranular corrosion dealloying, galvanic corrosion, and cavitation are each possible depending on the processing environment. Erosion–corrosion in ore grinding mills is another critical issue. Particulates are often carried in a corrosive medium through pipes, tanks, and pumps. The presence of these particulates erodes and removes the protective film of the metal and exposes the reactive alloy to high-flow velocity, thus accelerating the corrosion.

According to the Connoisseur Corporation Pty. Ltd. (29), the effects of corrosion in electrical and electronic systems in modern mines are often overlooked; however, the harsh environment of the mining industry often causes failure of electrical equipment after a short period of time.

Pyrite and other sulfide minerals on exposure to oxygen and water are oxidized to produce ferrous ions and sulfuric acid. The ferrous ions further react to form hydrated iron oxide and hydrogen ions. The hydrogen ions lower the pH of water, making it deleterious to aquatic life and corrosive to surrounding structures. The acid mine drainage can cause corrosion problems in structures such as pipes, well screens, dams, bridges, water intakes, and pumps. A survey made in 1993 by the US Forest Service estimated that 8050–16,100 km (5000–10,000 miles) of domestic streams and rivers are impacted by acid mine drainage (30).

In comparison, in 1995, the Pennsylvania Department of Environmental Protection reported that 3902 km (2425 miles) of stream in Pennsylvania did not meet EPA-mandated in-stream water quality standards because of mineral extraction (31). This significant amount affected streams in Pennsylvania, compared with the nationally estimated amount, can be explained by the fact that this state has a relatively large portion of the US coal industry.

3.24 PETROLEUM REFINING

The petroleum refining industry is undergoing intense scrutiny in the United States of America from regulatory agencies and environmental groups. The total cost of corrosion control in refineries is estimated at $3.692 billion. The costs associated with corrosion control in refineries include both processing and water handling.

Corrosion-related issues regarding processing are handling organic acids [also referred to as naphthenic acid corrosion (NAC)] and sulfur species at high temperatures as well as water containing corrosives such as H_2S, CO_2, chlorides, and high levels of dissolved solids.

3.24.1 Areas of Major Corrosion Impact

A refinery operation may have more than 3000 processing vessels of various size, shape, form, and function. A typical refinery has about 3200 km (2000 miles) of pipeline, much of which is inaccessible. Some of these pipelines are horizontal, some are vertical; some are 200 ft high and some are buried under cement, soil, mud, and water. The diameters range from 10 to 76 cm (4–30 in.).

3.24.2 Water-Related Corrosion

Crude oil desalting and distillation generates a considerable amount of wastewater. Typical wastewater flow from a desalter is approximately 8 l (2.1 gal) of water per barrel of oil processed. This water contains accelerative corrosive agents such as H_2S, CO_2, chlorides, and large amounts of dissolved solids. The wastewater also contains some crude oil that is recovered during water treatment process.

In addition to generated wastewater, cooling water (either freshwater or salt water) is used extensively in refining operations. The corrosivity of the cooling water varies considerably depending on the process. Thus, it is difficult to describe cooling water problems; however, corrosivity is highly dependent on the level and type of dissolved solids and gases in the cooling water, including chlorides, oxygen, dissolved gases, and microbes. The temperature of cooling water can also affect the corrosivity.

3.24.3 Processing-Related Corrosion

The top section of a crude unit can be subjected to a variety of corrosive agents. Hydrochloric acid formed from the hydrolysis of calcium and magnesium salts is the principal strong acid responsible for corrosion in the crude unit top section. Carbon dioxide is released from crudes typically produced in CO_2-flooded fields and crudes that contain large amounts of naphthenic acid.

Acids such as formic, acetic, propionic, and butanoic are released from crudes with a high amount of naphthenic acid. Hydrogen sulfide, released from sour crudes, significantly increases the corrosion of the crude unit top section. Both sulfuric and sulfurous acids formed by either oxidation of hydrogen sulfide or direct condensation of SO_2 and SO_3 increase corrosion.

3.24.4 Naphthenic Acid Corrosion

The high-temperature crude corrosivity of distillation units is a major problem in the refining industry. The presence of naphthenic acid and sulfur compounds increases corrosion in the high-temperature parts of the distillation units and hence equipment failure became a critical safety and reliability issue. Most of the acids have the formula $R(CH_2)_nCOOH$, where R stands for the cyclopentane ring and n is greater than 12. In addition to $R(CH_2)_nCOOH$, a host of other acidic organic compounds are also present.

Isolated deep pits in partially passivated areas and/or impingement attack in essentially passivation-free areas are typical of NAC. The damage is in the form of unexpected high corrosion rates on alloys that would normally be expected to resist sulfidic corrosion. In many cases, even very highly alloyed steels (i.e., 12 Cr, AISI types 316 and 317) have been found to exhibit sensitivity to corrosion under these conditions. NAC differs from sulfidic corrosion by the nature of pitting and impingement and the severe attack at high velocities in crude distillation units. Crude feedstock heaters, furnaces, transfer lines, feed and reflux sections of columns, atmospheric and vacuum column heat exchangers, and condensers are the types of equipment subject to pitting and impingement attack.

Sulfidic attack occurs because of the presence of various forms such as elemental sulfur, hydrogen sulfide, mercaptans, sulfides, and polysulfides. Sulfur is the second most abundant element in petroleum. Sulfur at a level of 0.2% and greater is known to be corrosive to carbon and low-alloy steels at temperatures from 230 to 455 °C (446–851 °F).

At high temperatures, especially in furnaces and transfer lines, the presence of naphthenic acids may increase the severity of sulfidic corrosion. The presence of the organic acids may disrupt the sulfide film and thereby promote sulfidic corrosion of the alloys that are normally expected to resist this form of attack. The alloys in question are 12 Cr and higher. In some cases, such as in side-cut piping, the sulfide film formed because of hydrogen sulfide is thought to offer some degree of protection from NAC.

In general, the corrosion rates of all alloys in the distillation units increase with an increase in temperature. NAC occurs primarily in high-velocity areas of crude distillation units in the 220–400 °C (430–750 °F) temperature range. No observable corrosion damage is usually found at temperatures greater than 400 °C (750 °F) probably because of the decomposition of naphthenic acids or protection from the coke formed at the metal surface.

Velocity, and more importantly, wall shear stress are the important parameters affecting NAC.

Fluid flow velocity lacks predictive capabilities. Data related to fluid flow parameters, such as wall shear stress, and the Reynolds Number are more accurate because the density and viscosity of liquid and vapor in the pipe, the degree of vaporization in the pipe, and the pipe diameter are also taken into account. Corrosion rates are directly proportional to sheer stress. Typically, the higher the acid content, the greater the sensitivity to velocity. When combined with high temperature and high velocity, even very low levels of naphthenic acid may result in very high corrosion rates.

3.24.5   Corrosion-Related Failure in Refineries

Corrosion-related failure (32) that occurred in refineries of the Union Oil Company of California resulted in a disastrous explosion and fire. An amine absorber pressure vessel ruptured and released large amounts of flammable gases and vapors. The accident resulted in 17 deaths; 17 individuals were hospitalized; and more than $100 million in damages resulted.

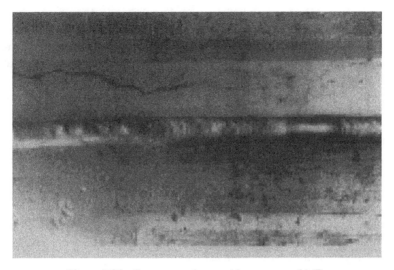

Figure 3.28   Stress corrosion cracking near a weld (7).

The National Bureau of Standards (NBS) performed a detailed investigation consisting of chemical analyses, fracture mechanics analyses, SCC susceptibility tests, and hydrogen cracking susceptibility tests. Preliminary NBS test results indicated that the subject plate material (ASTM A516, Grade 70 carbon steel) of the amine absorber was susceptible to hydrogen-induced cracking. The repair welds that were done in the field and that had not been stress relieved were especially sensitive to amine-induced corrosion and cracking. Figure 3.28 is an example of SCC both parallel and perpendicular to the weld, but not in the weld. The propagation of the crack clearly distinguishes SCC and reflects the different stresses along the weld area.

3.25   CHEMICAL, PETROCHEMICAL, AND PHARMACEUTICAL INDUSTRIES

The corrosion problems encountered are similar in the three industries (33). The most common types of corrosion encountered are: (i) caustic and chloride cracking; (ii) oxidation; (iii) sulfidation; (iv) corrosion under thermal insulation; (v) ammonia cracking, and (vi) hydrogen-induced cracking. Corrosion in the chemical industries is a well-known problem when chemicals such as hydrofluoric acid and hot sodium hydroxide solutions are dealt with on an industrial scale. There are many examples of how even a slight change in operating conditions, the presence of an unexpected impurity, or change in the concentration of process chemicals can result in a dramatic increase in the corrosion rate. The main causes of contamination in pharmaceutical manufacturing are corrosion of embedded iron particles in vessel walls, failure of glass linings, and corrosion under insulation.

According to J&H Marsh and McLennan, Inc., in the hydrocarbon processing industry (HPI) market data 2000 report (34), approximately 45% of large losses are because of mechanical failure of equipment. Equipment such as piping, tanks, reactors, process drums and towers, pumps and compressors, heat exchangers, heaters, and boilers fail regularly.

The most important raw material in the HPI is crude oil that often contains some water. Corrosive species present in oil–water mixture consist of sulfur compounds, chlorides, ammonia, sulfuric acid, hydrochloric acid, polythionic acid, carbonic acid, phosphoric acid, naphthenic acid, and cyanides. In spite of the treatment for the removal of these corrosive species, trace amounts are left behind in the water. These trace amounts of corrosive species are responsible for corrosion. Water, used in large quantities by hydrocarbon processing plants for cooling, heating, and purifying process streams, is another source of corrosives and foulants. Chlorides, sulfates, magnesium, and calcium salts dissolved in water-cooling systems can cause scale, sludge, and corrosion. Water in the process stream can accelerate corrosion as it acts as an electrolyte and dissolves certain materials.

Three types of corrosion have been identified. The first type of corrosion can be identified by visual examination. The second type of corrosion may require supplementary means of examination. The third group requires studies involving optical or electron microscopy, which may sometimes be amenable to study by the naked eye.

The first group consists of general, localized, and galvanic corrosion. Localized corrosion includes both pitting and crevice corrosion. The second group consists of velocity effects such as erosion, corrosion, and cavitation; intergranular attack (IGA) where grain boundaries are preferentially attacked; and dealloying corrosion. The third group consists of cracking phenomena, and high-temperature corrosion cracking phenomena include SCC, hydrogen-assisted cracking (HAC), liquid metal cracking (LMC), and corrosion fatigue.

The Materials Technology Institute (MTI) of the Chemical Process Industries, Inc. published a compilation of experiences of corrosion failure mechanisms in process industries (35).

Cracking was found to be the most frequent failure mode. Cracking ranged from 27% to 36% of the corrosion failure mode. General corrosion was the next most frequent (17–26%) mode followed by 12–20% of localized attack. In the case of localized attack mechanisms, pitting was the most frequent failure mode followed by intergranular corrosion. The study found that steel and stainless steel were involved in the majority (48–61%) of the SCC failures reported.

The MTI report (35) listed the corrosion failure modes along with the frequency of the occurrence as follows in Table 3.12.

The reported data on failures was collected from five companies and a total of more than 1272 failures. The failure mode of cracking includes SCC, fatigue cracking and caustic cracking.

The distribution of SCC of different materials of construction is as follows in Table 3.13.

TABLE 3.12   Corrosion Failure Modes Along with the Frequency of Occurrence

Failure Mode	Average Frequency (%)
Cracking	36
General corrosion	26
Local attack	20
Temperature effects	7
Velocity effects	5
Galvanic, stray current, and "macro cell"	3
Hydrogen effects	2
Biological	0
Total	99

TABLE 3.13   Distribution of Stress Corrosion Cracking of Different Construction Materials

Material	Average Frequency (%)
Stainless steel	61.4
Steel	30.4
Copper alloys	4.3
Nickel alloys	2.8
Titanium	0.7
Tantalum	0.3

Chlorides were involved mainly in the cracking of stainless steels, and caustics and nitrates were responsible for the cracking of steel. The studies reported that a 50–50 distribution between mechanical and corrosion failures was found to hold good.

3.26   PULP AND PAPER INDUSTRY

The paper-making process consists of various steps from pulp production, pulp processing and chemical recovery, pulp bleaching, and stock preparation to paper manufacturing.

Pulp production is done by: (i) mechanical pulping; (ii) semichemical pulping, and (iii) chemical pulping.

1. Mechanical pulping utilizes steam, pressure, and high temperature to tear the fibers. Newspaper and paperboards are typical products of the mechanical pulping process. Mechanical pulping is done in 300 series stainless steel containers to prevent corrosion.
2. Semichemical pulping involves the use of dilute solutions of sodium sulfite and sodium carbonate to digest the lignin in the pulp. In addition to this, mechanical refining is used to separate the fibers.

3. Chemical pulping uses various chemicals to produce long, strong, and stable fibers and to remove the lignin that bonds the fibers together. The two main types of chemical pulping are Kraft (sulfate) pulping and sulfite pulping. The corrosion rates in these processes can be significant depending on the chemicals and materials used. Traditionally, carbon steel has been used as a material of construction. Because of erosion from particles in solution and corrosion because of the increase in sulfur content, stainless steel type 304L and 316L have been used as cladding for the carbon steel digesters. It is advisable to check for intergranular SCC, which may occur in a heat-affected zone (HAZ) of the weldments (36). Specific stainless steel grades used in Kraft pulping digesters and connecting pipes are duplex stainless steels 2205 and 2304 and austenitic stainless steel 312 (36).

The acid pulping process uses sodium bisulfite or magnesium bisulfite in a pulp digester at a pH of 3. Type 316 stainless steel is normally used as a minimum alloy because the sulfur dioxide can be converted into sulfuric acid. Sulfuric acid can corrode the stainless steel depending on the pH, temperature, and pressure of the system. SCC has been observed in the HAZ of weldments in the pulp digesters in the presence of sodium hydroxide. Because of SCC susceptibility of the austenitic stainless steels, duplex stainless steel 2205 is often used in pressure vessels and tanks.

Kraft pulping chemical recovery consists of passage of black liquor along with the slurry passes through evaporators, recovery boilers, and causticizers to eventually produce white liquor. Corrosion on the fireside of the recovery boiler is accelerated by the presence of reduced sulfur species. The hydroxide mixtures present in black liquor are extremely corrosive to the recovery boilers made of type 304 stainless steel (37). Several phenomena in the recovery section cause different forms of corrosion such as: (i) corrosion under ash build-up; (ii) corrosion in the thin condensation layers; and (iii) high-temperature metal/gas interactions.

(i) Ash build up on the heat exchanger tubes can occur in recovery boilers when the incineration of the liquid waste is incomplete. The deposited ash decreases the efficiency of the heat exchanging process. Underdeposit corrosion such as crevice or pitting corrosion may occur.

(ii) Condensation can occur in the ductwork between the recovery boiler and the off-gas scrubbers when the off-gases cool down to a temperature below their flash point before reaching the scrubber. Localized attack in the condensate phase can be severe (>1000 mpy) and can be accelerated by alternate condensation and revaporization. The concentration of corrosive species in the thin condensed layer is the highest just before complete revaporization.

(iii) High-temperature metal/gas interactions in the recovery boiler tubes are oxidation, carburization, and sulfidation. The rates of the processes vary with the concentration of the burned black liquor waste and the temperature of the recovery boiler. High-temperature gaseous attack does not require an aqueous or molten salt electrolyte. Continued scale growth on the metal surface leads to metal consumption and decreased wall thickness of boiler tubes and boiler walls.

Recausticizing is the process used to transform the inorganics recovered from the recovery boiler into white liquor so that the chemicals may be recycled.

Sulfite pulping chemical recovery consists of greater amounts of sulfite, hydrogen sulfide, and hydrochloric acid than those used in the Kraft process. Because of these corrosive species, the internal portions of the recovery boilers and the evaporators are generally constructed of reinforced plastics, type 316L stainless steel, type 317L stainless steel, or nickel-based alloys. To prevent pitting and crevice corrosion, scale build-up should be prevented, wet-dry zones should be avoided, and chloride concentration should be kept to a minimum.

The bleaching is done in three to five stages in which the pH of the pulp is alternated between acid and alkaline conditions. During the acid cycle, the chemical reactions between the bleach and lignin bonds turn the pulp lighter in color. During the alkaline cycle, the reaction products from the acid stage are removed (38).

Chlorine dioxide is similar to chlorine with respect to its corrosivity (36). Suitable materials for bleach washers using chlorine and chlorine dioxide are super austenitic 6–7% molybdenum stainless steels such as 25–45 Mo or 25–6 Mo.

In recent years, less chlorine and more sodium hydroxide have been used for bleaching. At present, chemical pulps and the deinking of secondary fibers are the primary users of chlorine bleaching techniques. The European paper and pulp industry uses ozone, oxygen, and peroxide in place of chlorine.

Many pulp mills have a paper mill adjacent to them, thus making a transfer of the product easy. The equipment used to store and transport the pulp can undergo crevice and pitting corrosion.

The suction rolls used to remove water from the paper during the drying process experience general corrosion, fatigue failures, pitting, and MIC because of exposure to stock and white water, deposits of paper fiber present in crevices and bacterial growth. The inside diameter of the drilled holes within the suction rolls experience fatigue failure because of the presence of high stress concentrations. The sodium thiosulfate present in stock and white water systems can result in severe pitting (39).

The drilled holes in the suction roles, along with crevices and deposit sites, experience the growth of microbes as a result of high temperature (40–50 °C), dissolved organic materials, and dissolved inorganic salts. Microbiological attack occurs beneath deposits on the microbiological slime and increases with the degree of closure of the paper mill. Corrosion problems within the paper machines are chloride pitting and crevice corrosion, thiosulfate pitting, and microbiological attack.

3.27   AGRICULTURAL PRODUCTION

The major input items and farm production expenditures in 1997 are shown in Table 3.14

Corrosion problems occur in the plumbing systems of agricultural sprayers. Urea ammonium nitrate (UAN) is used as a fertilizer. Anhydrous ammonia, used in farm fertilizers, is very corrosive. Nitrate formed on oxidation of ammonia can be corrosive.

TABLE 3.14   Major Input Items and Farm Production Expenditures on 1997

Item	Percent of Total Expenditure
Chemicals, fertilizer, and seeds	15
Feed	14
Fuel	3
Farm services	13
Interest, taxes	10
Livestock	8
Machinery, vehicles	7
Labor	10
Rent	10
Supplies, repair and construction	10

Fertilizer tanks are susceptible to pitting and erosion–corrosion in the acetic acid solutions used in fertilizers.

Corrosion problems occur in milking process systems because of the moisture, sanitizing chemicals, and animal respiration. Corrosion problems occur because of hot water cycles and wash water used to clear milking equipment.

Agricultural fumigants, such as methyl bromide and phosphine, are used to disinfect food products. Phosphine in combination with carbon dioxide and humidity is corrosive to copper and electronic and electrical equipment in food processing area.

Agricultural buildings that house livestock require special care in selecting wiring materials, wiring methods because of the corrosive dust, gases, and moisture. Corrosion of metallic conduit, boxes, and fixtures lead to electrical system failure. Accelerated corrosion because of condensation occurs on electrical panels that are not properly designed.

3.28   THE FOOD PROCESSING SECTOR

The corrosion environment in the food and beverage industry involves moderately to highly concentrated chlorides, often mixed with significant concentrations of organic acids. The water side of the processing equipment can range from steam heating to brine cooling. Purity and sanitation standards require excellent corrosion resistance to pitting and crevice corrosion. Sulfating agents producing sulfur dioxide when used to treat foods are sodium sulfite, sodium bisulfite, potassium bisulfite, sodium metabisulfite, and potassium metabisulfite. All of these additives are corrosive to food processing equipment.

Underdeposit corrosion is likely to occur in cooling systems where scales or foulants exist. The presence of general fouling and scales can cause the formation of a differential cell, which begins the process of corrosion because of the difference in oxygen concentration at the metal surface beneath the deposit and the oxygen concentration in water, a differential cell forms, resulting in the corrosion reaction.

The food processing industry uses water for washing, transporting, blanching, cooking, cooling, and cleaning. In particular, heating and cooling processes require large amounts of water. Underdeposit corrosion is caused by using water in boilers, rotary cookers, and hydrostatic sterilizers.

Aluminum alloys are susceptible to underdeposit corrosion. Stainless steels are also susceptible to underdeposit corrosion as well as deep pitting. Anodic, cathodic, and filming inhibitors are used to mitigate corrosion (40).

Both aluminum alloys and stainless steels are susceptible to underdeposit corrosion and stainless steels are prone to deep pitting. Both anodic and cathodic as well as filming inhibitors are used to mitigate corrosion.

Biocides such as chlorine dioxide and bromine compounds (oxidizers) are used for sterilization. The biocides might interfere with the performance of inhibitors and hence the concentration of biocides must be controlled.

Galvanic corrosion is an accelerated attack between two dissimilar metals that are in electrical contact and exposed to an electrolyte. In hydrostatic sterilizers, the flight bars are made of aluminum or stainless steel, and the transport chain is made of carbon steel, and both are exposed to hot water and steam (40). Aluminum, the less noble metal, will corrode.

SCC in AISI 304 and AISI 316 steel piping and tanks is a problem in water lines in brewery applications (41). A common form of SCC occurs at temperatures higher than ambient in the presence of chlorides. Cracking may occur from the process or from the outside, for example, under insulation.

3.29 ELECTRONICS

The common electronic components consist of integrated circuits (ICs), printed circuits (PCs), and connectors and contacts. IC conductors are made from aluminum alloys, often alloyed with silicon and copper. PC boards such as conductors and connectors are made of copper with solders consisting of lead–tin alloys. Contacts are made from copper covered with electroplated nickel or gold for improved corrosion resistance.

Electronic devices are common and are exposed to much harder conditions than air-conditioned rooms. The small dimensions of a microchip and the silicon-based IC elements spaced less than 0.2 μm show that the tolerance for corrosion loss is as small as a picogram (10^{-12} g). Submicron dimensions of electronic circuits, high-voltage gradients, and a fairly high sensitivity to corrosion lead to a unique corrosion-related issues.

Environmental contaminants can cause failure in electronics. The contaminants are fine and coarse particles of salts such as chlorides, sulfates of sodium, ammonium, potassium, magnesium, and calcium. The environmental condition of importance is relative humidity, which in combination with sulfur dioxide or nitrogen oxide will cause a great deal of corrosion damage.

According to the Instruments System Automation Society standard there are four classes as given in Table 3.15.

TABLE 3.15   Instruments System Automation Society Classes

Class	Description	Expected Time-to-Failure
G1 (mild)	Corrosion not a factor	No corrosion-related failure
G2 (modern)	Corrosion <1000 Å per month	Failure in 3–4 years
G3 (harsh)	High corrosion <2000 Å per month	Failure in 1–2 years
G4 (severe)	Considerable corrosion <3000 Å per month	Failure in 1 year

The various forms of corrosion that may be encountered in the field of electronics are as described below.

1. *Anodic Corrosion.* The spacing between components of the ICs is small, and when a voltage is applied to a device, voltaic gradients of the order of 10^5–10^6 V/cm can exist across surfaces, accelerating electrochemical corrosion reactions and ionic migration. In ICs, positively biased aluminum metallizations are susceptible to corrosion. A combination of electric fields, atmospheric moisture, and halide contamination leads to corrosion of aluminum. Gold and copper metallization are also subject to corrosion under these conditions.

2. Negatively biased aluminum metallizations can corrode in the presence of moisture because of the high pH (basic) produced by the cathodic reaction of water reduction. The high pH can dissolve the passive oxide on aluminum along with the corresponding increase in conductor resistance possibly up to open-circuit value.

3. *Electrolytic Metal Migration.* Detected early on in electromechanical switches, this problem occurs in silver-containing compounds. In the presence of moisture and an electric field, silver ions migrate to the negatively charged cathodic surface and plate out in the form of dendrites. The dendrites grow and eventually bridge the gap between the contacts causing an electric short and an arc. Large dendrites may be formed even from small volumes of the metal. Under certain humidity and voltage gradient conditions, a 30-day exposure becomes equivalent to 4 years of service in a typical office environment (42). Other metals susceptible to metal migration are gold, tin, palladium, and copper.

4. *Pore-Creep in Electrical Contacts and Metallic Joints.* To prevent tarnishing of connectors and contacts, a noble metal such as gold is plated on the contact surface. The coverage of noble metal such as gold may not be perfect, and corrosion can occur at the site of imperfections. If the substrate is copper or silver and exposed to either chloride or sulfur-bearing environment, corrosion products can creep out from the pores and cover the gold plating with a layer of high contact resistance.

5. *Fretting Corrosion of Separable Connectors with Tin Finishes.* Fretting corrosion in electronic components occurs as the continuous formation and flaking of tin oxide from a mated surface on tin-containing contacts. As more tin is oxidized and utilized, it leaves no choice but to replace the component with a new part.

6. Galvanic corrosion can occur when two dissimilar metals, such as aluminum and gold, are coupled, as is commonly done for packaged (plastic-wrapped) ICs. The polymers used for packaging are porous and the gaskets around hermetic covers such as ceramic or metal sometimes leak; therefore, in humid environments, moisture can permeate to the IC bond pad, leading to conducive conditions for galvanic corrosion. Electronic devices dissipate a considerable amount of heat during operation, which leads to reduced relative humidity. During power-down or storage periods, the relative humidity rises leading to the likelihood of corrosion (43).

7. *Processing-Related Corrosion of ICs.* ICs are exposed to aggressive corrosive media used in reactive ion etching (RIE) or wet etching for patterning of aluminum lines, which can lead to corrosive residue. RIE of aluminum metallizations utilizes a combination of aggressive chlorine-containing gases. If removed untreated from the etcher, patterned structures are covered with aluminum chloride residue, which hydrolyzes to give hydrochloric acid in the presence of moisture (43).

8. Micropitting on Aluminum on ICs during Processing. Aluminum metallizations, alloyed with copper, can form intermetallic compounds such as Al_2Cu along the grain boundaries, which act as cathodic sites relative to aluminum adjacent to grain boundaries. This leads to dissolution of aluminum matrix in the form of micropitting during the rinsing step after chemical etching.

9. *Corrosion of Aluminum by Halogenated Solvents.* Both liquid and vapor-phase halogenated solvents used for the production of ICs and PCs corrode aluminum-containing components. Water contamination of the solvent increases the time to corrosion on the one hand and increases the corrosion rate on the other hand. Dilution of the stabilized solvents with alcohol results in the breakdown of halogenated solvents and the decomposition product, chloride ion corrodes aluminum, and aluminum–copper alloys.

10. *Solder Corrosion.* The corrosion resistance of lead–tin solder in aqueous environments is a function of the alloy composition. The corrosion resistance of lead-tin increases when the tin content of the solder increases above 2 wt%. Lead forms unstable oxides, which react with chlorides, borates, and sulfates (43).

11. *Corrosion of Magnetic and Magneto-Optic Devices.* Corrosion-related failures can occur in advanced magnetic and magneto-optic storage devices, where thin film metal discs, thin film inductive heads, and magneto-optic layers are affected. Corrosion occurs in sites where the deposited carbon overcoat is lacking because of intentional roughening of the disc and where the magnetic cobalt-based layer and nickel–phosphorus substrate become

exposed. The potential differences between the noble (positive) carbon and the metal substrate, a galvanic couple may develop, leading to rapid galvanic-induced dissolution of the magnetic material (43).

Magneto-optic devices utilize extremely reactive alloys for the recording media (because of a high terbium content). Exposure of magneto-optic films to aqueous solutions or high-humidity conditions results in a localized pitting attack even during storage in ambient office conditions (43).

While attempts have been made to mitigate corrosion of electronics by encapsulating the components in plastics, it is useful to note that polymers are permeable to moisture. Hermetically sealed ceramic packaging is more successful. However, care must be taken to prevent moisture and other contaminants from being sealed in. A useful common approach for mitigating corrosion of circuits housed inside a relatively large-size chassis includes the use of volatile corrosion inhibitors.

3.30 CORROSION PROBLEMS IN HOME APPLIANCES

There are some common areas of significant corrosion impact for major home appliances and comfort conditioning appliances. The corrosion types are internal corrosion from process water and external corrosion from wet conditions.

The most important reason for corrosion in appliances is the water that is being handled by the equipment. This type of corrosion affects the internal components of appliances and limits life expectancy. In the case of home appliances, the appliances most susceptible to internal corrosion are refrigerators, water heaters, washers, dishwashers, and water softeners. The air conditioners, humidifiers, high-efficiency furnaces, dehumidifiers, and boilers are susceptible to internal corrosion.

Internal corrosion in appliances is a problem because it limits useful life. This is a direct cost of corrosion. A unitary air conditioner has an average life of 13 years. One of the reasons for this limited life expectancy is that condensate in the air conditioner corrodes the internal metal components.

External corrosion of appliances results in deterioration in the appearance of the surface of the appliance without affecting its capability in performance. However, the value of the corroded appliance will decrease because of the corroded appearance. In addition to corrosion of noncoated surfaces, corrosion can occur when painted or coated surfaces become chipped or nicked. Some of the wet environments around appliances are a furnace or a boiler in the humidity of a wet basement, an air conditioner in the yard or window being exposed to rain and moisture, and kitchen equipment and whose exteriors are often cleaned with water or wet towels.

External corrosion of appliances results in the deterioration of their appearance and therefore affects their resale value. External corrosion of appliances generally does not limit the capability of an appliance to function properly. The reduction in the resale cost of a corroded appliance is the direct cost of corrosion.

Corrosion problems vary with the type of appliance. Corrosion can build up and destroy parts of the entire appliance. The appliances that are subject to corrosion are: (i) water heaters, (ii) boilers, (iii) high-efficiency furnaces, and air conditioners.

The heating coils in water heaters are exposed to the water in the heater tank. Common water contaminants such as chlorides, fluorides, and sulfites can cause corrosion of the heating coils, the water heater connections, the tank wall, and the tank frame. The elevated temperature of the water in the heater is further likely to increase the internal corrosion rate.

Boilers are heat exchangers constructed of carbon steel to produce hot water or steam by being heated with an oil or gas burner. The hot water or steam is transferred to radiators to provide heat. After releasing heat, the cool water or steam condensate is returned to the boiler for reheating.

A common problem in boilers is the occurrence of calcium oxide build-up on the heating elements. This is not a corrosion problem in itself, because it is caused by a chemical reaction in the water at high temperatures. However, a scale deposit present on a metal surface may cause corrosion under the deposit. This type of underdeposit corrosion can be aggravated when corrosive species such as sulfides and/or chlorides are present in the water. While scale deposits reduce the thermal conductivity of the steel, and thereby increase energy costs, corrosion of the heating element can lead to a catastrophic tubing failure, which requires costly repairs.

3.30.1   High-Efficiency Furnaces

Corrosion can occur in furnaces when condensation occurs, which can corrode the internal metal surfaces. Condensation is a problem in high-efficiency furnaces, because operating at high efficiencies means that the appliance must operate in a condensing mode. At present, these furnaces are designed with a maximum annual flue utilization efficiency (AFUE) of 90% compared to a standard minimum efficiency of 78%. To operate a high-efficiency furnace, the flue gas must be cooled to a temperature below the dew point, by which the combustion-generated moisture is condensed in the heat exchanger, and the latent heat of vaporization is recovered for utilization.

Research in the mid-1980s on the corrosion of materials used in condensing heat exchangers in furnaces indicated that the greatest probability of corrosion occurs when the appliance goes through the transition from wet to dry conditions. This is because the acidity of condensate increases as the water evaporates. The flue gas generated is a mildly acidic liquid that is corrosive to type 304 and 316 stainless steels commonly used in heat exchanger furnaces. The corrosivity of the condensate can increase because of airborne contaminants in particular chlorine-bearing compounds, present in indoor environments, and carried into the burner by the combustion process.

3.30.2   Air Conditioners

Aluminum and copper are the materials in air conditioners. Coils and cooling fins are made from aluminum, and piping is usually made of copper. Aluminum is susceptible to galvanic corrosion when in contact with copper components. Galvanic corrosion can occur when two dissimilar metals are in electrical contact in an electrolyte.

Piping or plumbing systems made of copper alloys are susceptible to erosion-corrosion in unfavorable fluid flow conditions. Erosion–corrosion can occur when erosive action of the flowing stream removes the protective copper oxide film from the metal surface, and thus exposing the bare metal surface to a corrosive environment (44).

3.31   CORROSION PROBLEMS IN THE US DEPT. OF DEFENSE

The ability of the US Department of Defense (DOD) to respond rapidly and effectively to national security and foreign policy commitments can be adversely affected by equipment-related failures. Using available resources, minimization of downtime and maximization battle readiness must be accomplished through the useful operational lifetime of the equipment. If this is done effectively, equipment can be deployed in a timely and responsive manner and maintained in the field with a minimum of downtime.

Corrosion of military equipment and facilities is a significant and ongoing problem. Large amounts of costs are incurred to protect the assets from corrosion, affecting procurement, maintenance, and operations. The effect of corrosion on various types of equipment is a problem that is becoming more prominent as the acquisition of new equipment slows down and more emphasis is placed on total care and operation of the current system. As the intention to operate aging aircraft, ships, land combat vehicles, and submarines continues into the twenty-first century, the potentially detrimental effects of corrosion on the cost of ownership, safety, and readiness must be fully appreciated. The effect of corrosion of the DOD equipment will continue to get worse unless and until new technologies can be used to reduce the cost of ownership. The total annual cost of corrosion to the DOD is approximately $20 billion for systems and infrastructure (45).

3.31.1   Weapon Systems

The available data from the services indicate that corrosion in weapons systems is the primary cost driver in life-cycle costs (46). Quantifying corrosion is difficult as neither the mechanisms nor the methodologies exist to quantify accurately. Analysis of field data reveals instances where questionable materials selection early in the acquisition process has led to enormous unanticipated increases in life-cycle costs because of corrosion (J Argento, US Army TACOM-ARDEC, Picatinny Arsenal, NJ, Personal Communication, 1999). In view of force reduction and a reduction in budgets, consideration must be given to the selection of advanced materials, processes, and designs that will require less manpower for corrosion inspection and maintenance.

The following discussion concerns specific information on corrosion and related costs incurred by the Army, the Air Force, the Navy, and the Marine Corps. It should be noted that the corrosion costs of selected components in these services do not add up to the $20 billion quoted above. It only serves to demonstrate how corrosion can significantly affect the equipment and facilities of the armed services.

3.31.2 Army

This is a major branch of the armed forces, owns and operates a range of facilities and equipment. Corrosion creates a significant burden for the Army, affecting the Army's readiness, equipment reliability, troop morale, and, in particular, the cost of maintenance of the weapons systems.

3.31.3 Vehicles

A significant portion of the corrosion cost ($2 billion) is attributed to Army ground vehicles. The major types of vehicles operated by the army are the following:

Abraham tank systems – M1 Abrams
Bradley Fighting Vehicle Systems
M2 Infantry Fighting Vehicles (IFVs)
M3 Cavalry Fighting Vehicles (CPVs)
Multiple-Launch Rocket Systems (MLRs)
Command and control vehicles (C2Vs)
Bradley Carrier Systems (BCSs)
Bradley Fire Support Vehicles
Medium Tactical vehicles
2 $\frac{1}{2}$-ton cargo trucks
5-ton cargo trucks
High-Mobility Multipurpose Wheeled Vehicles (HMMWVs)
Light armored vehicles

In general, very little attention is given to corrosion and corrosion control of army vehicles. In fact, corrosion on these vehicles is allowed to occur until it affects their load-carrying capacity. Moreover, little has been done to incorporate corrosion protection and control in the design and manufacturing of army vehicles. For example, none of the medium tactical vehicles has galvanized steel in the body. The HMMWV has several corrosion control shortcomings that result in very high corrosion maintenance costs (J Argento, US Army TACOM-ARDEC, Picatinny Arsenal, NJ, Personal Communication, 1999; 49). In designing HMMWV, several corrosion control features that are now common in commercial vehicles have not been applied. One of the most glaring faults with the HMMWV is that the frame is built of AISI 1010 steel and that no galvanizing or other prevention is applied. Another problem with the frame is that holes are drilled into the sides of the frame with no drain holes in the bottom. This allows water and dirt to enter and deposit in the frame. Other problems are the use of AISI 1010 carbon steel for fasteners, handles, and brackets. Another serious problem is the use of dissimilar metal couples such as aluminum frames bolted to steel frames. These and other omissions of corrosion control have led to costly maintenance and repairs. The following shortcomings have been identified as a result of an audit: (i) use of AISI 1010 carbon steel without galvanizing or any protective coating; (ii) the presence of many galvanic couples and the use of 2800 rivets that may act as locations for corrosion; (iii) use of painting procedures that are not state-of-the-art standards; (iv) use of paints that give very little corrosion protection such as the chemical agent-resistant coatings (CARCs) that deteriorate rapidly in the presence of a corrosive environment.

3.31.4 Case Study of HMMWV

In spite of the availability of knowledge and technology, the manufacture of the Army's HMMWV did not use available knowledge during manufacture. One of the most glaring faults was in the design and construction of the steel frame using AISI 1010 steel without galvanizing or any other form of protection. Holes were drilled in the sides of the frame with no holes at the bottom to allow drainage of water and other corrodents. This allowed water to stagnate leading to corrosion from inside out. This being the case, it is beyond comprehension as to how one can expect the HMMWV vehicle to wade through water up to 1.5 m (60 in.) deep without undergoing corrosion damage.

The main problem is the use of carbon steel 1010 for the fasteners, handles, brackets, and the frame. The corrosion data obtained on 275 vehicles are presented in Table 3.16.

Most of these problems could be eliminated or prevented by using galvanized steel or high-quality coatings. Other problems can be overcome by using polymers and other alternative materials.

Use of dissimilar metals such as aluminum and carbons steel can result in galvanic corrosion. The entire HMMWV vehicle is secured with more than 2800 rivets, and this design affords the vehicle high-strength-weight ratio: each rivet is a preferential site for corrosion.

A particular weakness of the HMMWV compared to standard commercial vehicles is the coating system used. Most of the commercial vehicles use a multistep coating process to both protect the galvanized steel and to enhance the appearance of the vehicle. Electrodeposition or E-coating is used in the coating of ordinary vehicles. Electrodeposition assures complete coverage of the vehicle. In the case of HMMWV, spraying technology was used, and the resulting coating did not turn out to be as good as the E-technology coating.

The corrosion protection of the HMMWV was to be provided by the military coating specification Mil-C-46164 and the CARC. The CARC paint system consists of a surface cleaning, epoxy primer, epoxy interior topcoat, and a polyurethane exterior topcoat. The purpose of this coating was to provide chemical resistance (penetration of the coating) and to aid in the decontamination of the vehicle in case of chemical attack. The coating was to provide corrosion protection as well as

TABLE 3.16　Number and Percentage of Corrosion-Affected Parts in HMMWV Vehicles

Vehicle Parts	Number of Vehicles Affected	Percent of Vehicles Affected
Engine Compartment		
Heads	49	18
Injectors	53	19
Engine mounts	78	28
Valve covers	87	32
Radiator assembly	131	48
Suspension and Steering		
Idler arms	48	17
Control arms	78	28
Rie rods	124	45
Axle housings	161	59
Springs	205	75
Body		
Fenders	72	26
Bumpers	105	35
Door frames	115	42
Beds	120	44
Tie downs/lift points	209	76
Underbody		
Metal brake lines	35	13
Air tanks	40	15
Drive shafts	105	38
Fuel lines	106	39
Universal joints	135	49
Other		
Welded seams	73	27
Fuel tank assemblies	135	49
Nuts, bolts, fasteners	177	64
Frame	187	68

camouflage protection as the paint was available in different colors. The protection by the CARC paint system was not complete as shown by the data given in Table 3.17.

The CARC paint system hardens after application to an extremely inelastic product. The metals to which CARC was applied were much more elastic and also expanded as well as contracted rapidly following environmental conditions. The net result is the CARC paint and is disbonded from the metal and falls off and leads to corrosion.

Another weakness of CARC paint is that it is difficult to apply, and field repair of the coating is difficult. CARC paint contains a high level of volatile organic compounds (VOCs). Strict environmental regulations allow only 0.9 L (1 quart) per day per area to be used to reduce the level of VOC emissions.

TABLE 3.17   Data on HMMVS with Deteriorated CARC Paint

Sector	Location	Number Inspected	Number of Deteriorated HMMWV	
			Coating	Percent
Army	Fort Bragg	17	4	24
Army	Fort Still	13	9	69
Army	Fort Knox	9	3	33
Army	Fort Drum	11	11	100
Marine Corps	MCLB-Atlantic	2	2	100
Marine Corps	Camp Lejeune	40	30	75
WI National Guard	Various	29	13	45

Firing Platforms. Howitzer firing platforms contribute significantly to corrosion costs in the Army. The 119 is a 105 mm Howitzer of British design, and a total of 500 were acquired. Severe corrosion was detected on the platform. An investigation by the Army indicated several deficiencies that lead to severe corrosion, including various dissimilar metal contacts resulting in galvanic corrosion in some areas of the platform where water could collect. The design of the Howitzer platform was such that it needed to be replaced at a cost of $18,000 each; the total cost to remedy the problem is estimated to be nearly $9 million.

A second Howitzer corrosion problem is experienced with M198 Howitzer of which 1800 are in service. To keep the Howitzer in readiness, an annual cost of $5300 for parts replacement is required for each M198. The total annual maintenance cost for just corrosion-related parts replacement is estimated at $10 million ($1800 \times \$5300$).

3.31.5   Helicopters

The army operates several helicopters, some of them dating back to the Vietnam era (Table 3.18).

TABLE 3.18   Helicopters and Duties

Helicopter	Duties
UH-1 Iroquois (Huey)	Personnel ferrying helicopter (900)
UH-60 Blackhawk	Personnel ferrying helicopter
CH-47 Chinook	Heavy-cargo-lifting helicopter (431)
AH-1 Cobra	Gunship (19)
AH-64 Apache	Attack helicopter (743)
OH-58 Kiowa	Reconnaissance helicopter
RAH-46 Comanche	Reconnaissance, light attack, and air combat helicopter (1213) (Deployed in 2008)
MH-16 Little bird	Light assault

In February 2000, the Army released a report indicating that 40% of the helicopter fleet is not combat ready. In addition, these problems are experienced particularly with aging equipment such as the Vietnam War–era Hueys and Cobras, which are assigned mostly to the National Guard and the Army Reserve Units. In addition, newer helicopters, such as the Apaches and Chinooks, also suffer from combat-readiness problems. Approximately 8–22% of overhaul and repair costs are because of corrosion. In fact, it was estimated that in 1998, approximately $4 billion was spent on the corrosion control of helicopters alone (J Argento, US Army TACOM-ARDEC, Picatinny Arsenal, NJ, Personal Communication, 1999).

3.31.6   Air Force

As the fleet of military aircraft and support equipment ages, the damage caused by corrosion becomes an increasing concern. The aircraft spend a longer time in depots for maintenance and repair, which leads to a decrease in readiness and an increase in cost to maintain the aircraft. Moreover, a possible loss of integrity of the structure is possible if the corrosion goes undetected and becomes severe.

Recently, a study was done for the Air Force Corrosion Program Office to determine the annual cost of direct maintenance to the Air Force (48). The Air Force study examined the cost for the fiscal year 1996 and examined costs for all Air Force systems and equipment, including all aircraft, aircraft subsystems, ground systems, vehicles, missiles, munitions, ground support equipment, and space equipment. Corrosion maintenance was defined as a comprehensive inspection for corrosion, all repair maintenance because of corrosion, washing sealant application and removal, and all coating applications and removal. Intangible or indirect costs, such as aircraft downtime, and the depreciation effects that result from corrosion maintenance, such as repeated grind-outs of skin and structure, were not addressed in the study. Other intangible or indirect costs that were not addressed include the costs of building corrosion control facilities; the cost of building and maintaining formal corrosion maintenance schools for training of maintenance technicians, and the cost to produce, distribute, and install specialized corrosion control equipment in corrosion control shops.

The total cost of direct corrosion maintenance to the US Air Force for fiscal year 1997 was estimated at approximately $800 million (48; Table 3.19).

The table clearly indicates that the major portion of the cost can be attributed to aircraft repair and paint. Significant amounts are also spent on washing and vehicle maintenance. In addition to the total cost findings, it was found that maintenance in the depot accounted for 80% of the total cost of corrosion maintenance. While the total number of aircraft in the fleet decreased by 20%, the costs decreased by only 10%, and the maintenance costs have increased.

The data presented in Table 3.20 show the effect of aging on weapon system costs. The difference in age and size has an effect on the cost of aircraft.

3.31.7   KC-135 Stratotanker

This is a strategic air refracting tanker built by the Boeing Company, which can also be used as a cargo carrier or troop transport. The first KC-135 entered the Air Force

TABLE 3.19 Elements of Total Corrosion Maintenance Cost to Air Force

	Total Maintenance Cost, Fiscal Year 1997 ($)
Repair	572,352,704
Wash	38,443,783
Paint	145,951,530
Vehicles	23,291,759
Munitions	6,247,341
Other	18,540,036
Total	794,827,153

TABLE 3.20 Corrosion Maintenance Cost for Individual Military Aircraft in 1990 and 1997

Aircraft	1990 Fleet Cost ($)	Number of Aircraft	Corrosion Cost (%)	1997 Fleet Cost ($)	Number of Aircraft	Total Corrosion Cost (%)	Change in Number of Aircraft
A-10	25,611,157	524	4.25	4,325,700	375	0.69	−149
13-1	1,267,086	76	0.21	7,326,979	95	1.17	19
B-52	95,751,947	228	15.90	39,545,321	94	6.29	−134
C-130	137,963,143	694	22.91	50,351,736	694	8.01	0
KC-135	13,554,678	644	18.86	205,561,487	602	32.72	−42
C-141	68,621,286	231	11.40	102,584,893	220	16.33	−11
C-5	17,019,858	126	2.83	104,595,003	126	16.65	0
CLS	3,286,630	180	−0.55	6,301,275	321	1.00	141
E-3	3,698,062	32	0.61	19,851,017	32	3.16	0
F-111	41,778,986	245	6.94	7,749,299	37	1.23	−208
F-15	23,325,398	749	3.87	29,194,683	737	4.65	−12
F-16	17,010,711	1,260	2.83	15,728,095	1,513	2.50	253
Helos	4,854,452	179	0.81	2,511,531	215	0.4	36
C-10	666,302	52	0.11	7,439,773	59	1.18	7
T-37	2,278,434	527	0.38	1,326,593	420	0.21	−107
T-38	13,105,291	812	2.18	23,894.508	451	3.80	−361
A-7	1,600,922	214	−0.27				−214
A-37	345,047	58	0.06				−58
F-4	26,867,597	746	4.46				−746
F-5	72,943	7	0.01				−7
OV-10	3,438,883	54	0.57				−54
Total	602,118,813	7638	100.01	628,288,893	5991	99.99	−1647

fleet in 1957 and the last one was delivered in 1965. At present, about 550 of the 732 tankers built remain in service. As a result of decreasing DOD budget, the current KC-135 fleet has been projected to remain in service until 2040. With the average KC-135 tankers being more than 40 years of age, they will be more than 80 years old in 2040 and will have been in service for more than four times their original design service life. In general the structural life of both commercial and military aircraft is based on flight hours and number of fatigue cycles. In general, the life of aircraft is fatigue-limited and corrosion is never considered to be a life-limiting factor. The minimum KC-135 structural fatigue life-limited components are the fuselage and the upper wing skin at 66,000–70,000 h, while the actual fleet hours are only 15,000. As the KC-135 utilization averaged only 300–400 flight hours per aircraft in 1 year, it appears that the fleet can easily remain in service until 2040.

However, severe corrosion has been experienced on the aluminum alloy components of the KC-135 aircraft. This corrosion is the result of low utilization, where the majority of the time is spent on the ground being exposed to the corrosive atmospheric environments. In the 1950s, the KC-135 was never designed and constructed with corrosion prevention as a primary concern. The original structural alloys were aluminum alloys 2024-T3 and 2024-T4, 7075-T6, and 7178-T6, which are all susceptible to corrosion and SCC. The original construction was without any sealant in the lap joints and fuselage skins that had spot-welded doublers attached to them. The upper wing skins, which are made of the highly corrosion-susceptible aluminum alloy 7178, were attached with high-strength steel fasteners, causing dissimilar metal corrosion in certain areas.

A particularly severe problem is the corrosion of the fuselage lap joints, where the voluminous corrosion products at the contact or faying surfaces of the lap joints cause deformation of the skin (49, 50). Because of the resulting stress fatigue and stress corrosion, cracks can nucleate near the fastener holes, jeopardizing the structural integrity of the fuselage. Other corrosion problems on the KC-135 aircraft include dissimilar metal corrosion and lap joint corrosion on the 7178 upper wing skin, lap joint corrosion on the 7075-T6 fuselage crown section, and SCC of the 7075-T6 forged frame section.

3.31.8   Navy

The Navy consists of several components, including ships, submarines, aircraft weapons, and facilities such as buildings, piers, docks, and harbor structures. An internal Navy study conducted in 1993 estimated the total cost of corrosion for all naval systems at $2 billion per year (Sedriks, Office of Naval Research, Personal Communication, July 2000).

The Navy fleet consists of various surface ship battle forces, including 11 aircraft carriers, 106 surface combatants (i.e., cruisers, destroyers and frigates), 39 amphibious warfare ships, 34 combatant logistic ships, and 31 support/mine warfare ships (total of 221 ships).

The surface ships are subject to extremely aggressive environments. An extensive corrosion control program is required to maintain the fleet during dry-dock cycles. The primary defense against corrosion is the diligent use of protective coatings. In addition to coating, CP is used for the protection of underwater hull. The cost to maintain CP systems are low compared to the cost of maintaining the various protective coating systems. Figure 3.29 shows a photograph of a destroyer, indicating the different shipboard coatings that are currently in use. The traditional coatings indicated

Figure 3.29   Destroyer with different shipboard coatings (7).

TABLE 3.21   Annual Maintenance Demand on Sailors for Coating Maintenance

Maintenance Activities	Man-Years Per Ship
Topside and freeboard (enamel, silicone alkyd)	9.0
Flight decks and topside decks	4.0
Bilges/wet space corrosion	4.5
Machinery space	2.25
Interior bulkheads and decks	3.00
Superstructure, catwalks, mixing	3.25
Total	26.00

in the figure have a design life of 10–15 years, after which the ship has to be in dry dock to completely remove the "old" coating and apply the "new" coating.

Between the major maintenance cycles, there is an annual maintenance demand for continuous corrosion maintenance. Table 3.21 shows the annual maintenance demand on sailors for coating maintenance of navy surface ships.

3.31.9   Submarines

The Navy operates 18 fleet ballistic missile submarines and 56 nuclear attack submarines. Because of the confidential nature of the submarines, no information on the corrosion aspects could be obtained.

3.31.10   Aircraft

Corrosion has a significant impact on the life-cycle costs of naval aircraft. The navy has three levels of aircraft maintenance; namely, organizational, intermediate, and depot maintenance. The organizational maintenance is performed on individual equipment and includes inspection, servicing, lubrication, adjustments and replacement of parts, assemblies, and subassemblies. The intermediate maintenance is conducted on parts after removal from equipment and includes calibration repair, or replacement of damaged or unserviceable parts and components or assemblies. The depot maintenance involves major overhaul or complete rebuild of parts, assemblies, subassemblies, and end-items, including the manufacture of parts, modifications, testing, and reclamation as required. A current estimate for corrosion maintenance cost is $200,000 per navy aircraft per year (Sedriks, Office of Naval Research, Personal Communication, July 2000). This cost is nearly twice as much as the corrosion cost for Air Force aircraft ($100,000 per year) operating under less corrosive conditions. It is needless to state that naval aircraft operate in significantly harsher conditions than Air Force aircraft.

3.32   NUCLEAR WASTE STORAGE

Nuclear wastes are generated from spent nuclear fuel from electric power plants, dismantled weapons, and products such as radio pharmaceuticals. The most important design item for the safe storage of nuclear waste is the effective shielding of radiation. To minimize the probability of nuclear exposure, special packaging is designed to meet the protection standards for temporary dry or wet storage or for permanent underground storage. The most common materials of construction consist of steel and concrete. The wall thickness of the packaging is generally thick in comparison to the contained volume.

Corrosion is a form of material degradation that results when moisture or water comes into contact with the packaging materials. A corrosion failure may not result in a large release of nuclear waste and radiation; however, a leak would be considered potentially hazardous and, therefore, would not be acceptable. Currently, nuclear waste is stored at temporary locations, including water basins in nuclear power plants and at dry locations aboveground. Deep underground storage in Yucca Mountain, Nevada, has been proposed as a permanent storage solution.

It is useful to note that when considering the total costs of nuclear material storage, it is nearly impossible to distinguish the specific corrosion cost, especially for some corrosion-related costs for nuclear waste packaging design and packaging fabrication and the costs for remediation of temporary sites that are being used for longer periods than the time periods for which they were designed.

The great majority of nuclear shipments are very small in size [less than 0.45 kg (1 lb) per shipment] and total nearly 2.8 million shipments per year (an average of 7656 shipments per day) (51). Spent fuel shipments (material only) typically weigh 0.5–1.0 metric tons for truck shipments and up to 10 metric tons for rail shipments.

TABLE 3.22   Volume of Low-Level Waste Received at US
Disposal Facilities

Year	Volume of Low-Level Waste Received (Cubic Meters × 1000)
1985	75.4
1986	51.1
1987	52.2
1988	40.4
1989	46.1
1990	32.4
1991	38.8
1992	49.4
1993	22.1
1994	24.3

In addition, the protective lead shipping casks for containment of the spent fuel weigh many more additional tons. Corrosion is not an issue in the transportation of nuclear waste because of the stringent package requirements and the short duration of the transport; however, corrosion is an important problem in the design of casks used for permanent storage. Table 3.22 shows the volume of low-level waste received at US disposal facilities in the 10-year period between 1985 and 1994 (52).

3.32.1   Transition from Interim Storage to Permanent Storage

In 2000, interim storage facilities for nuclear waste were numerous. Interim nuclear wastage storage consisted of a number of older tanks that have a radioactive leak history and have a need for remedy. Low-level radioactive waste can be liquid or solid waste in containers. Low-level waste is stored "dry" aboveground or relatively shallow underground. At present, there are a total of 250,000 cubic meters of buried low-level waste and 106,000 m^3 of stored aboveground low-level waste at US Department of Energy (DOE) facilities (53). The cost of dry storage is reported to be $1.2 million per cask.

High-level waste (HLW) from spent nuclear fuel from nuclear power plants is generally stored in water basins at the plants where it was used. At present, nearly 30,000 metric tons of spent nuclear fuel is stored at commercial reactors (53). Dry storage and wet basin storage are designed as temporary solutions. The need for a long-term storage repository is acute and is under study and development.

For example, the K West and K East basins in Hanford, Washington, are two concrete basins that were built in 1951 for the temporary storage of nuclear fuel produced at DOE's Hanford site. Although the initial plan was to terminate the storage after 20 years, the two basins continue to receive spent fuel from reactors. It has been reported that rods in open canisters have corroded in the basin, releasing radioactive isotopes into the basin water. Basin cleanup plans, waste removal, and groundwater contamination were subsequently reported. The cost of this work will be considerable.

3.32.2 Cask Design for Permanent Storage

In addition to the unavoidable material aging because of exposure to radiation from the radioactive material and hence corrosion is expected to be a concern in the long-term storage of nuclear waste. It has been further suggested that heat generation from radiation can drive the corrosion rate higher. Many cask designs have been proposed, and each with different materials of construction. The most common proposed materials are carbon steel, stainless steel, and concrete construction.

At the present time, all the nuclear waste generated is solid waste. Hence, the waste is relatively noncorrosive, which minimizes the risk for internal corrosion damage to storage and transportation banks. There is, however, a significant amount of old liquid nuclear waste in storage, which can corrode the containers internally. In addition, the presence of water in the solid waste could potentially cause corrosion problems. External corrosion is a potential problem, because the older liquid waste is stored in buried tanks, and these tanks are therefore exposed to groundwater. The consequences of leaks are numerous and hence the structural integrity of the storage containers must be assured for centuries.

The potential for corrosion of permanent storage canisters has been and continues to be under investigation. A literature review and summary of plutonium oxide and metal storage packaging failures was published (54). Metal oxidation in nonairtight packages with gas pressurization was identified as the most common mechanism of packaging failure. An example of a possible corrosion problem was further described in a study on hydrogen/oxygen recombination and generation of plutonium storage environments (55). There are also literature citations available with respect to the predicting of service life of steel in concrete used for the storage of low-level nuclear waste (56).

In a September 2000 meeting on key technical issues regarding container life, the US Nuclear Regulatory Commission (NRC) and representative of the DOE discussed the ongoing research into the effects of corrosion processes on the lifetime of the containers (57).

In the above meeting, a wide range of material issues that designers are facing were discussed. In nuclear waste containers, both corrosion from the inside and from the outside should be considered. The issues included, but were not limited to, general and localized corrosion of the waste package outer barrier; methods for corrosion rate measurements; documentation on materials such as Alloy 22 and titanium; the influence of silica deposition on the corrosion of metal surfaces; passive film stability, including that on welded and aged material; electrochemical potentials; MIC; stress distribution because of laser peening and influence on rock fall impact strength; and dead load stressing and the effects of fabrication sequence and welding.

3.32.3 Effect of Location on Corrosion of Nuclear Storage Containers

The current plans for a permanent nuclear storage repository are to build it at a relatively dry site at a depth of several hundred meters below earth's surface. It is thought that the presence of water will eventually corrode the storage containers.

Figure 3.30   Yucca Mountain site for high-level nuclear waste storage (7).

In the United States, the Yucca Mountain site (Fig. 3.30) is reported to be a good location because of the low water content in the site. The proposed design for nuclear waste disposal is for steel canisters containing the spent fuel to be stored within other steel canisters and buried horizontally in chambers 300 m below the earth's surface. The canisters were designed to last at least 1000 years, which will depend on the mountain itself to provide a natural barrier to survive the minimum 10,000 years required by the government; however, there is no guarantee that the canisters at Yucca Mountain will be free from water flow for 10,000 years.

REFERENCES

1. JP Broomfield, *Corrosion and Corrosion Protection of Steel in Concrete*, Vol. **1**, R Narayan Swamy (ed.), Sheffield Academic, Sheffield, UK, pp. 1–25, 1994.

2. JP Broomfield, *Corrosion of Steel in Concrete: Understanding, Investigation and Repair*, 2nd ed., Taylor and Francis, London, 2007.

3. K Tuutti, *Corrosion of Steel in Concrete*, Swedish Cement and Concrete Research Institute, Stockholm, 1982.

4. CL Page, KWJ Treadway, *Nature (London)* **297**:109 (1982).

5. HO Bonstedt, *Corrosion International* **20**(10):65–68 (1998).

6. JN Enevoldsen, MSc thesis, Queen's University, Kingston, ON, 1993.

7. *Corrosion Cost and Preventive Strategies in United States*, Appendices D, E, F, G, J, L, P, R, S, U, BB, CC, Report no. FHWA-RD-01-156, FHWA, 2002.

8. NG Thompson, DR Lankard, *Improved Concretes for Corrosion Resistance*, Report no. FHWA-RD-96-207, FHWA, May 1997.

9. NG Thompson, DR Lankard, *Optimization of Concretes and Repair of Materials for Corrosion Resistance*, Report no. FHWA-RD-99-096, FHWA, Sept 1999.

10. FL Stahl, CP Gagnon, *Cable Corrosion in Bridges and Other Structures*, ASCE Press, New York, 1996.

11. http://ops.dot.gov/stats, May 22, 2000.

12. OGJ Special Report *Oil and Gas Journal* (Aug 23, 1999).

13. DJ Boreman, BO Wimmer, KJ Leewis, *Pipeline and Gas Journal* (Mar 2000).

14. *An Assessment of the US Marine Transportation System, a Report to Congress*, The Marine Transportation System (MTS), www.doc.gov/mts/report, Sept 2000.

15. *Navigation Datacenter Publications and US Waterway Data*, CD, US Army Corps of Engineers, Alexandria, VA, Vol. **6**, 2000.

16. JA Beavers, GH Koch, WE Berry, *Corrosion of Metals in Marine Environments*, Metals and Ceramics Information Centre, A Department of Defense Information Analysis Center, Battelle Columbus Division, Columbus, OH, June 1986.

17. American Water Works Association Research Foundation and DVGW-Technologiezentrum Wasser, *Internal Corrosion of Water Distribution Systems*, 2nd ed., Cooperative Research Report, Denver, CO, 1986.

18. SJ Duranceau, MJ Schiff, GEC Bell, *Effects of Electrical Grounding on Pipe Integrity and Shock Hazard*, order no. 90702, AWWA Research Foundation, June 1996.

19. R Baboin, *The Automotive Environment: Automotive Corrosion by Deicing Salts*, National Association of Corrosion Engineers, 1991.

20. R Baboin, *Chemistry and Corrosivity of the Highway Environment* Paper no. 91371, National Association of Corrosion Engineers, Annual Meeting, 1991.

21. K Staumbagh, JC Krecht, SSC-348 "Corrosion Experience Requirements" Ship Structure Committee, 1991.

22. SNAME, *SNAME 0–23 Paint Panel Fundamentals of Cathodic Protection for Marine Service*, Technical & Research Report R-21, 1976.

23. GKMG Consulting Services Inc., Aviation Week Newsletter, *The Aviation & Aerospace Almanac*, 1999.

24. Office of Hazardous Materials Safety, *Summary of Incidents*, http://hazmat.dot.gov/files/summary/9899/brindex98.htm.

25. *ASM Metals Handbook, Corrosion*, Vol. **13**, 9th ed., ASM International, Metals Park, OH, pp. 1293–1298, 1987.

26. RL Jentgen, RC Rice, GL Anderson, *Canadian Institute of Mining and Metallurgy Bulletin* **77**11:50–54 (1984).

27. Mine Safety and Health Administration, 30CFR, 56.19023 MSHA www.msha.gov/regdata/msha/56.19023.htm, Oct 2000.

28. Mine Safety and Health Administration, MSHA, www.msha.gov, Oct 2000.

29. Connoisseur Corporation Pty. Ltd., Asset Protection Technology, Mining Industry, www.connaisseur.com.au/rd-ect/industry/mining.html, Sept 2000.

30. US Environmental Protection Agency, Office of Compliance, Sector Notebook Project, *Profile of the Metal Mining Industry, Standard Industrial Classification*, Washington, DC, 10 Sept 1995.

31. J Earle, TC Callaghan, *Impacts of Mine Drainage on Aquatic Life, Water Uses and Man-Made Structures*, Department of Environmental Protection, Harrisburg, PA XXX

32. *Reduced Corrosion in Amine Gas Absorption Columns*, www.hydrocarbon processing.com/archive/archive_99-10/99-1-_reduce-mogul.htm, Oct 2000.

33. V Novokshehenov, *Proceedings of the Fifth Middle East Corrosion Conference*, Manama, Bahrain, pp. 209–223, Oct 28–30, 1991.

34. 2000 Market Data Report: Special Report on Hydrocarbon Processing Industry, *Hydrocarbon Magazine*, Nov 1999.

35. RB Puyear, *Corrosion Failure Mechanisms in Process Industries: A Compilation of Experiences*, MTI Report no. R-4, MTI, St. Louis, MO, 1997.

36. PH Thorpe, Selection and Fabrication of Stainless Steels for the Pulp and Paper Industry, *Corrosion & Prevention 1998 Proceedings*, PH Thorpe and Associates, pp. 109–113, Nov 1998.

37. A Harrison, *Pulp Paper* **68**10:135–136 (1994).

38. US Environmental Agency, Office of Compliance, Sector Note Book Project, Standard Industrial classification 261–265, *Profile of Pulp and Paper Industry*, Washington DC, Sept 1995.

39. RM Uschan, LC Trick, *American Papermaker* **57** (9), (1994).

40. J Rauh, Midland Research Laboratories, Oct 2000.

41. D Larson, Anheuser-Busch, St. Louis, MO, Oct 2000.

42. RB Comizzoli, RP Frankenthal, JD Sinclair, Corrosion Engineering of Electronic and Photonic Devices, *Corrosion and Environmental Degradation*, Vol. **19**, *Materials Science and Technology Series*, Wiley-VCH, Weinheim, Germany, 2013.

43. GS Frankel, Corrosion of Microelectronic and Magnetic Storage Devices, *Corrosion Mechanisms in Theory and Practice*, P Marcus and J Oudar, (eds), Marcel Dekker, NY, 1995.

44. BJ Moniz, Copper, *Metallurgy*, American Technical Publishers, 1994.

45. *Science of Technology News*, HQ, US Army Material Command, Vol. **2**, Nov 1997.

46. H Mindlin, BF Gilp, LS Elliot, M Chamberlain, *Corrosion in DOD Systems: Data Collection and Analysis (Phase I)* MIAC Report 8, Metals Information Analysis Center, Aug 1995.

47. *Audit Report on High-Mobility Multipurpose Wheeled Vehicles*, Office of the Inspector General, US Dept of Defense, 1993.

48. G Cooke et al., *A Study to Determine the Annual Direct Cost of Corrosion Maintenance for Weapon Systems on Equipment in the United States Air Force, Final Report*, CDRI, no. A001, 1998.

49. NC Bellinger, S Krishnakumar, JP Komorowski, *Canadian Aeronautics and Space Journal* **40**(3):125 (1994).

50. GH Koch, S Styborski, CA Paul, Effect of Corrosion in Lap Joints on Fatigue Crack Nucleation, *Proceedings of Fourth Joint DOD/FAA/NASA Conference on Aging Aircraft*, Mar 2000.

51. *Hazardous Materials Shipments*, Office of Hazardous Materials Safety Research and Special Programs Administration, US Dept of Transportation (DOT), Washington, DC, Oct 1998.

52. *Radioactive Waste: Production, Storage, Disposal*, NRC Report no. NUREG/BR-0216, Office of Public Affairs, US Nuclear Regulatory Commission, Washington, DC, July 1996.

53. M Sei, G Borchert, K Chance, *An Examination of US Nuclear Waste*, Chem. 480, Apr 6, 1998, www.utm.edu/departments/artsci/chemistry/NUWaste.htm.

54. PG Eller, RW Szempruch, JW McClard, *Summary of Plutonium Oxide and Metal Storage Package Failures*, LA-UR-99-2896, Los Alamos National Laboratory, 1999.

55. JA Lloyd, PG Eller, L Hyder, *Literature Search on Hydrogen/Oxygen Recombination and Generation of Plutonium Storage Environments*, LA-UR-98-4557, Los Alamos National Laboratory, 1999.

56. C Andrade, M. Cruz Alonso, Values of Corrosion Rate of Steel in Concrete to Predict Service Life of Concrete Structures, *Application of Accelerated Corrosion Tests to Service Life Prediction of Materials, ASTM STP 1194*, G Cragnolino, N Sridhar, eds, American Society for Testing and Materials, Philadelphia, 1994.

57. Summary of Key Technical Issues (KTI) Meetings and Associated Files: Summary of the Resolution of the Key Technical Issues on Container Life and Source Term, Nuclear Regulatory Commission, Sept 12–13, 2000, www.nrc.gov/nmss/DWM/092100 agreement.htm, May 2001.

4

CORROSION CONTROL AND PREVENTION

4.1 INTRODUCTION

Before embarking on the main topic of corrosion control and prevention, it is prudent to consider the economics of corrosion control methods. In one of the methods, the total cost of corrosion was estimated by Uhlig (1), which was later adapted to estimate the cost of corrosion to the Japanese economy (2, 3).

Corrosion control methods consist of protective coatings, corrosion-resistant metals and alloys, corrosion inhibitors, polymers, anodic and cathode protection, corrosion control services, corrosion research and development, and education and training. The total annual cost of corrosion estimated with this method for the average year of 1998 was $121.41 billion or 1.381% of the $8.79 trillion gross domestic product. Table 4.1 shows the distribution of corrosion control methods and services costs.

4.2 PROTECTIVE COATINGS

Both organic and metallic coatings are used to provide protection against corrosion of metallic substrates. The metallic substrates, particularly carbon steel, will corrode in the absence of the coating, resulting in a reduction of the service life of the steel part or component.

Challenges in Corrosion: Costs, Causes, Consequences, and Control, First Edition. V. S. Sastri.
© 2015 John Wiley & Sons, Inc. Published 2015 by John Wiley & Sons, Inc.

TABLE 4.1   Costs of Corrosion Control Methods and Services

Material and Services		Range ($× billion)	Average Cost ($× billion)	Percent
Protective coatings	Organic coatings	40.2–174.2	107.2	88.3
	Metallic coatings	1.4	1.4	1.2
Metals and alloys		7.7	7.7	6.3
Corrosion inhibitors		1.1	1.1	0.9
Polymers		1.8	1.8	1.5
Anodic and cathodic protection		0.73–1.22	0.98	0.8
	Services	1.2	1.2	1.0
	Research and development	0.02	0.02	<0.1
	Education and training	0.01	0.01	<0.1
Total		$54.16–188.65	121.41	100

4.2.1   Organic Coatings

The major organic coatings are often classified by a curing mechanism, with the two basic types of cured coatings being nonconvertible and convertible (4). The nonconvertible coatings cure solely by evaporation of the solvent with no chemical change in the resin matrix. They can be redissolved in the solvent originally used to dissolve the resin. Convertible coatings, on the contrary, cure primarily by a polymerization process in which the resins undergo an irreversible chemical change.

4.2.1.1   Nonconvertible Coatings   The common types of nonconvertible coatings are the following:

Chlorinated Rubbers. Elastomers formed when natural rubber or a polyolefin is reacted with chlorine. These materials are usually modified by other resins to obtain high solid contents and to decrease brittleness.

Vinyls. These are made by dissolving polyvinyl chloride (PVC) polymers in a suitable solvent. They are generally low solid coatings applied in very thin coats. Vinyl coatings are used for their weathering ability.

Acrylics. These are made by dissolving polymers made from acrylic acid and methacrylic acid or acrylonitrile. Water-based acrylics are widely used because of their weathering properties and ease of application.

Bitumen. Generally based on residues from petroleum or coal mining processes. Bitumen coatings can also come from naturally occurring sources such as gilsonite. The presence of some aromatic hydrocarbons such as benzene in some of these coatings has limited their acceptability in recent times because of environmental and health concerns.

Flame-Spray Polymers. These are nonevaporative cure coatings. These coatings function by cooling from the molten state. The molten polymer hits the surface and cools, solidifying into a protective film. The most common flame-spray is polyethylene, which is ground into a powder and flocked through a flame that converts polyethylene into a molten state. This type of coating can be remelted or dissolved in a suitable solvent.

Coalescence Coatings. In this type of coating, tiny particles of resin are encapsulated in a soap-like material and then dispersed in water, which acts as a diluent rather than as a true solvent. This type of blend is known as an emulsion. When the water evaporates, the resin particles fuse (coalesce) to form a stable, cured coating film. These coatings, once cured, cannot be redissolved in water, although other organic solvents may dissolve them. These coatings consist of acrylic latex suspensions and epoxy emulsions.

4.2.1.2 Convertible Coatings Most convertible coatings cure by polymerization. Polymerization occurs when two or more resin molecules combine to form a single, more complex molecule. The resin molecules may be monomers or short chain polymers that react to form polymers. The four main types of polymerization used in coating technology are oxygen-induced, chemically induced, heat-induced, and hydrolysis.

Oxygen-induced Polymerized Coating: Alkyds. These are oil-based primers and topcoats. Alkyds are based on fish or vegetable oils blended with pigments and catalysts in a solvent. The film is formed when the oil reacts with oxygen assisted by the catalyst followed by the evaporation of the solvent. Most paints sold in cans are alkyds. Drying oils consist of penetrating oils and lacquers that form a thin protective film.

Chemically Induced Polymeric Coatings. Epoxies are the preferred corrosion control coating for severe environments. Epoxies are a generic class of materials based on the presence of an epoxide polymer side group. Epoxies exhibit superior adhesion and chemical resistance but are susceptible to weathering degradation (by chalking) and are often top-coated to shield them from ultraviolet light.

Polyurethanes. Polyurethanes are used for color retention and weathering. These are extensively used over steel for long-term decorative corrosion protection. Polyurethanes vary widely in chemistry and can be formulated to be very flexible elastomers, rigid foams, or dense brittle films.

Heat-Induced Polymerized Coatings. Polyesters and vinylesters are based on styrene monomers with a very reactive catalyst. These may be classified as chemically induced curing polymers; however, the actual reaction is heat-induced. The catalytic reaction generates a great deal of heat, which polymerizes the styrene monomer and the ester groups. They are used as tank linings and form the basis of many freestanding fiberglass structures.

Phenolics. These are thin films, which form by evaporation of solvent followed by baking at, or greater than, 204 °C. Phenolics form a very strong, hard chemical and temperature-resistant film used for storage of strong acids and solvents.

Silicones. Chemically, silicones vary greatly; however, the corrosion-resistant coatings based on silicones are baked to create an inorganic silicone backbone that withstands very high temperatures. In applications such as furnaces and boilers, silicone-based coatings are often the only option.

Fusion-Bonded Epoxies. These are powder-based epoxies applied to hot substrates. The powder on contact with the hot substrate melts followed by a chemical reaction. On cooling, the solid film is formed. Fusion-bonded epoxies are widely used in pipelines and concrete rebar applications.

Hydrolysis-Induced Polymerized Coatings: Inorganic Zinc. Zinc metal powder is dispersed in a zinc silicate binder, and the zinc silicate uses the moisture from the air to form a cured matrix. The zinc particles behave like individual anodes to sacrificially protect the steel from corrosion. Many steel bridges and free-standing structured steel members are coated with inorganic zinc, which has a characteristic gray-green color. For other applications, the zinc is top-coated with an epoxy and/or polyurethane to provide an excellent system for corrosion control. There are also water-based inorganic zinc coatings, which react with CO_2 to cure.

Moisture-Cured Polyurethanes. Some polyurethane coatings form their protective cured film by reaction with the moisture from the air. The properties of the resulting film are different from two-component polyurethanes, but contain a basic urethane side group that classifies them as polyurethanes.

The selection of coating for the different industrial applications is based on the intended service, application, intended service life, and cost. According to the US Department of Commerce, Census Bureau, the total amount of organic coating material sold in the United States in 1997 was 5.56 billion liters (1.47 billion gallons) at a cost of $16.56 billion.

The four types of coatings are the following:

1. Architectural coatings
2. Original equipment manufacturers' coatings
3. Special-purpose coatings
4. Miscellaneous allied paint products

Architectural coatings are applied on new and existing residential, commercial, institutional, and industrial buildings. Small percentages of these are used as primers and undercoats and may be classified as corrosion control coatings. Architectural coatings are water-based and water-thinned. Some data on architectural coatings sold in 1997 in the United States are as follows in Table 4.2.

This value of $486 million spent on corrosion-related architectural coatings amounts to nearly 8% of the total $6.2649 billion spent on all architectural coatings in 1987.

Original Equipment Manufacture (OEM) Coatings are factory-applied to manufactured goods as part of the manufacturing process. There is an element of decoration

TABLE 4.2 Value of Corrosion-Related Architectural Coatings

Sold in 1997	Cost ($1 million)
Exterior solvent-based	91
Exterior water-thinned	100
Inferior solvent-based	101
Interior water-thinned	194
Total	486

TABLE 4.3 OEM Corrosion Control Coatings

	Value ($ million)
Automotive finishes	1,128
Automotive part finishes	78
Heavy-duty truck/bus/RV finishes	369
Aircraft/railroad finishes	166
Heating/AC/appliance finishes	84
Metal building product finishes	662
Machinery and equipment finishes	241
Nonwood furniture and fixture finishes	384
Automotive powder coatings	110
General metal finishing powder coatings	311
Other OEM powder coatings	130
Product finishes for OEM equipment	134
Total	$3797

TABLE 4.4 Value of Special-Purpose Corrosion Control

Coatings Sold in 1997	Value ($× billion)
Industrial maintenance coatings, interior	139
Industrial maintenance coatings, exterior	609
Automotive refinishing	1302
Marine paints for shipping/offshore	248
Total	$2298

in OEM finishes, but for those applied to steel, their primary function is corrosion control, either for weathering resistance or flash rust protection. The market breakdown is given in Table 4.3.

Special-purpose coatings include heavy industry corrosion control coatings as well as marine and automotive refinishing. The distribution of corrosion-related special-purpose coatings is shown in Table 4.4.

The value of special-purpose corrosion control coatings represents 79% of the $2896 billion special-purpose coatings market in 1997.

Another category of total sales is miscellaneous allied paint products, which includes paint/varnish removers, thinners, pigment dispersions such as art suppliers and putties. The contribution to corrosion protection from this category consists

TABLE 4.5   Summary for Corrosion Control Coatings Sold in 1997

Corrosion Control Coatings	Cost $× million
Total architectural corrosion control coatings	486
Total OEM corrosion control coatings	3797
Total special purpose corrosion control coatings	2298
Total miscellaneous allied corrosion control points	118
Total	$6699

TABLE 4.6   Distribution of 1998 Coating Sales by End-Users

End-Use Industry	Percent Sales
Petroleum refining and chemical production	14
Bridges and highways	8
Railroads	8
Water and waste treatment	7
Offshore oil and gas production	7
Marine	7
Defense/space	7
Electric utilities/gas	5
Pulp and paper	4
Land-based oil and gas production	4
Food and beverages	3
Primary metals and mining	3
Aircraft	1
Other (not specified)	22/100%

of only thinners used in nonarchitectural solvent-based coatings. Solvent-based corrosion control coatings account for 75% of the solvent-based coating market. It is estimated that the amount of thinner used in corrosion control applications is 75% of the thinner sold at a cost of $118 million. This $118 million accounts for 7% of $1.6479 billion allied paint products market in 1997.

Extracting the corrosion coating portions from the total costs given leads to the total estimate of all corrosion markets in the paint industry (Table 4.5).

A survey by the Steel Structure Painting Council (SSPC) of industrial coatings performed in 1998 separated the coating sales by the end-use industry. Table 4.6 shows the distribution of coating sales in 1998 by the end-use industry.

The average cost of paint per gallon is estimated at $23, which is derived from a wide range of costs for high-performance coatings such as epoxies ($30–50 per gallon) and polyurethane ($80/gallon) to industrial waterborne acrylics ($12–15 per gallon) (6).

The raw material cost of any coating application, while significant, is only a portion of the cost of a coating application project. The SSPC survey (6) indicated that, for example, for a typical aboveground crude oil storage tank, the total cost of coating is distributed as shown in Figure 4.1. The figure clearly shows that surface preparation

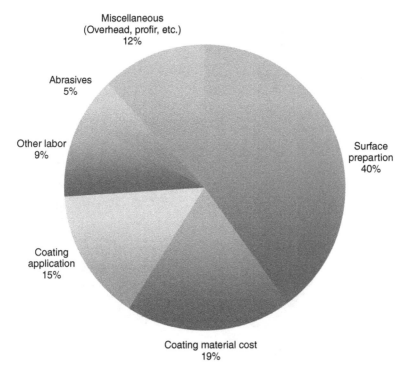

Figure 4.1 Cost distribution of coating application on aboveground storage tank (adapted from 5).

(40%) and coating application (15%) require a significant portion of the total coating budget whereas the actual coating material cost is only 9%.

According to a report (7) of the Federal Highway Administration, for a typical heavy duty maintenance job on a steel bridge structure, the cost of the coating material is only 4% of the total cost (Fig. 4.2). Large portions of the total cost are to access (20%), containment (19%), and workers' health (15%).

Using these figures, the total cost of application of the $6.699 billion in coatings is estimated to range from $35.3 to $167.5 billion for the entire coating industry in the United States. These costs do not include the costs' hard-to-define items, such as the costs of performance testing, personnel costs for time spent specifying coating products and application procedures, overhead for handling of bids and contracts and other support services that would be unnecessary if coating application is not needed. The total cost does not include downtime, lost production, or reduced capacity during maintenance painting.

Some major changes have affected the coating industry over the past 30–40 years. The first major change is minimizing the volatile organic compounds (VOC) content in the paints. High-solid coating with minimal solvents or waterborne equivalents is recommended. More than 10% of the high-performance industrial coatings in the United States are of the waterborne variety. The second major change is to

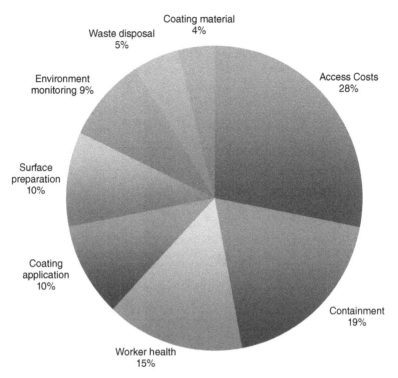

Figure 4.2   Cost distribution of coating application on steel highway bridge structure (7).

use lead-free paint. Currently, 4% of all coating sales is for the purpose of replacing lead-based coatings. A third major change is the toxicity of chromates and its presence in corrosion-inhibiting primers that are used in protecting aluminum alloys. Chromates may be replaced with molybdates or tungstates in combination with sodium nitrite.

4.2.2   Metallic Coatings

The most widely used metallic coating process for corrosion protection is galvanizing, which involves application of metallic zinc to carbon steel for corrosion protection purposes. The statistics of the Department of Commerce in 1998 state that nearly 8.6 million metric tons of hot-dip galvanized steel were produced in 1997 (8). The total market for metallizing and galvanizing in the United States, which is considered a corrosion control cost, is estimated at $1.4 billion. This figure represents the total material cost of the metal coating and the cost of processing. This does not include the cost of carbon steel that is being galvanized.

4.2.2.1   Galvanizing   Hot-dip galvanizing differs from other zinc coatings and the metallizing process in that the zinc is alloyed to the metal during galvanizing. In contrast, organic or inorganic zinc coatings including electroplated metallic coatings are

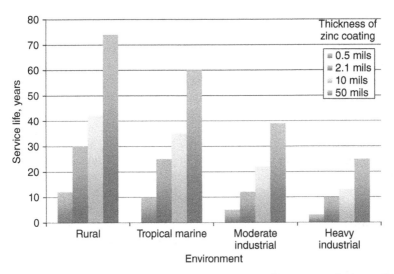

Figure 4.3 Expected service life of galvanized steel under different atmospheric conditions. (Reproduced with permission of National Association of Corrosion Engineers from *A User's Guide for Hot Dip Galvanizing*, TPC9, Houston, TX, 1983 (9.)

nonalloyed coatings, although the protection mechanism is essentially the same. Both alloyed and nonalloyed zinc coatings act as barriers to the corrosive environments and as sacrificial anodes when the barrier is breached.

The extent of protection by galvanizing depends entirely on the thickness of the galvanized layer. Galvanizing is unique in the sense that empirical data accumulated over the years provide guides for estimating the service life of galvanized coating under a wide range of exposure conditions. Figure 4.3 shows the service life predictions for several standard environments (9). Hot dip galvanizing is the most common process and consists of dipping the steel in a bath of molten zinc.

The galvanizing industry in the United States is divided into two classes; namely, fabrication and sheet galvanizing. The fabrication business deals with structural components such as piping, I-beams, poles, handrails, and other heavy duty products. The sheet business deals with galvanized sheet metal for equipment, roofing, panels, and other nonstructural steel applications.

In 1998, fabrication industry sales were nearly $750 million (American Galvanizers Association, Personal communication, Apr. 2000.). The profitability of galvanizing is directly tied to the price of zinc that remained fairly steady at $1.06–1.15/kg over the past 10 years. The improved efficiencies have offset the rising material/labor costs.

According to the American Zinc Association, sheet and strip galvanizing accounted for 540,000 metric tons of zinc (Metallized coatings supplier, Personal communication, 2000.).

The commodity price of zinc was $1076 per metric ton, with an additional average premium of $60 per metric ton paid by manufacturers.

The cost of processing sheet steel into galvanized parts is dependent on the facility and the cost ranges from $50 to $100/metric ton of zinc (R Leonard, Galvanization Information Center, Personal communication, 1990). The total sheet galvanizing sales were estimated to be $654 million.

The profitability of galvanizing is dependent on zinc commodity price and the size of the market. The two growth markets have been identified by the American Galvanizers Association. The first opportunity is the transition of utility poles from wood to galvanized steel. The elimination of wooden poles in favor of galvanized steel is estimated to add $200–300 million for the galvanizing industry.

Another growth market involves the use of galvanized metal studs for home construction. This aspect is connected with the home-building industry, which is related to the general prevailing economy.

4.2.2.2 Metallizing Metallizing may be defined as the application of a very thin metallic coating over a substrate metal for either active corrosion protection (zinc or aluminum anodes) or as a protective layer on stainless steels and other alloys. Metallizing may be done by flame spraying or electroplating. Other advanced processes, such as plasma arc spraying, can be used for exotic refractory metals for very demanding applications; however, advanced processes are not usually used for less demanding applications such as corrosion protection.

There is a market for metallizing anodes and it ranges from $5 to 10 million, which continues to grow following the recognition by government and industrial agencies that life-cycle costs are of importance (Metallized coatings supplier, Personal communication, 2000.).

4.3   METALS AND ALLOYS

Corrosion-resistant alloys are used where corrosive conditions are severe enough to prohibit the corrosion prone carbon steels and where protective coatings provide insufficient protection or are economically not good enough. The total cost for these alloys used in corrosion control applications is $8.3 billion with $7.9 billion for stainless steels, $0.28 billion for nickel-based alloys, and $0.15 billion for titanium alloys.

According to the US Census Bureau of Statistics, a total of 2.5 million metric tons of raw stainless steel was sold in the United States in 1997 (11) at an estimated production cost of $5.5 billion. About 25% of the US market is imported and the total consumption in the United States is estimated to be $7.5 billion. The amounts of stainless steel products consumed are given in Table 4.7.

The stainless steel consumed consists of sheet and strip, plate, bar, and pipe and tube. The data given in the table show that the transportation industry accounts for 23.8%, with the food equipment and construction sectors at 15.3% and 14.2%, respectively. It is interesting to note that the oil and gas sector and chemical sector only had 1.4% users.

In highly severe environments nickel-based and titanium alloys are used. Nickel-based alloys are extensively used in oil production and refinery, and the

TABLE 4.7   Stainless Steel Consumption

Industry	Total Metric Tons	Percent
Construction	343,175	14.2
Food equipment	370,119	15.3
Oil/gas – chemical	33,332	1.4
Fabricated metal products	302,607	12.5
Industrial/commercial machinery	95,297	3.9
Transportation	576,583	23.8
Electric	89,240	3.7
Furniture and fixtures	10, 327	0.4
Pulp and paper	25,851	1.1
Measure/analyze	26,642	1.1
Electrical/gas/sanitary	4243	0.2
All others/not classified	359,750	14.8
Conversion (pipe and tube)	186,078	7.7
Total	2,425,105	100.0

TABLE 4.8   Prices of Metals

Year	Price
1960	$2.20 per kg of nickel
1998	$4.40 per kg of nickel
1998	$5.00 per kg of molybdenum
1960	Cr $2.0 per kg
1998	Cr $8.0 per kg

chemical process industries where the conditions are aggressive. These alloys are extensively used in other industries where high-temperature and/or corrosive conditions exist. It is interesting to note the changes in the price of metals and alloys with time (Table 4.8).

The primary use of titanium alloys is in the aerospace and military industry where the high-strength-to-weight ratio and the resistance to high temperatures are of interest. Titanium and its alloys are corrosion-resistant to many environments such as oil production and refinery, chemical process, and pulp and paper industries. It was estimated in 1998 that 65% of mill products were used in aerospace applications and 35% were used in other applications (12). The most common form of titanium used is titanium sponge, which is produced in the United States, China, Japan, Russia, and Kazakhstan. The price of titanium increased from $2 per pound in 1960 to $4 per pound in 1980–1990. The price of titanium is very much dependent on the aerospace industry.

In 1998, 21,600 metric tons of titanium sponge was produced. The total consumption was 39,100 metric tons at a cost of $391 million. The cost of $150 million has been reported for titanium and titanium alloys for corrosion control applications.

4.4 CORROSION INHIBITORS

A corrosion inhibitor in general terms is a chemical substance that when added in a small amount to an environment effectively reduces the corrosion rate of a metal or alloy exposed to the corrosive environment. A more precise definition of an inhibitor is not possible because of the number of mechanistic and/or chemical considerations when classifying corrosion inhibitors.

In most cases, corrosion inhibition is achieved through interaction or reaction between the corrosion inhibitor and the metal surface, resulting in the formation of an inhibitive surface film. In some cases, the chemistry of the environment may be modified to make it less corrosive, such as adjusting the pH of the solution to promote passivation, scavenging dissolved oxygen or neutralizing acidic species. Anodic inhibitors such as chromates, molybdates, tungstates, phosphates, and nitrites function by interfering with the corrosion reaction occurring at the anodic site. Carbonates and arsenates inhibit the current flow by interfering with the cathodic reaction. Film forming inhibitors such as organic amines and imidazolines function as anodic or cathodic or both anodic and cathodic inhibitors.

Corrosion inhibition is used internally with carbon steel pipes and vessels as an economic corrosion control alternative to stainless steels and alloys, coatings, or nonmetallic composites. A particular advantage of corrosion inhibition is that it can be implemented or changed in situ without disruption of the process. For example, in processes that produce environments of increasing corrosivity with time, such as "souring" oil fields, corrosion can be effectively controlled with a suitable inhibitor.

The major industries that use corrosion inhibitors are: petroleum production and refining, chemical and heavy industrial manufacturing, and the product additive industry. The usage summary of corrosion inhibitors in various industries is given in Table 4.9.

The data in Table 4.9 show that the largest consumption of corrosion inhibitors is in the oil industry with the single highest amount in the petroleum refining industry.

TABLE 4.9 Consumption of Corrosion Inhibitors in the United States in 1998

End-Use Industry	Kg × Million	Lb Million	$× Million
Petroleum refining	248.1	547	246
Petroleum production and drilling	63	139	153
Petroleum storage and transport	15.4	34	31
Pulp and paper	182.8	403	198
Chemical manufacturing	272.2	600	180
Iron and steel	57.2	126	50
Miscellaneous material handling (includes food processing, utilities, and institutions)	132.9	293	88
Additives to petroleum products	54.4	120	108
Automotive and fuel additives, others	4.5	10	12
Total	1030.5	2272	1066

The use of corrosion inhibitors has increased to a great extent since the 1980s. It may be noted that the total consumption of corrosion inhibitors in the United States was nearly $1.1 billion.

Increased amounts in inhibitor usage in 1981, 1986, and 1998 for three of the largest industrial sectors, namely, the oil production, chemical, and refining industries have been noted (13). It is useful to note the inhibitor usage for petroleum production in 1986 occurred during the time that the US oil industry suffered a downfall because of the increase in oil prices to above $30 per barrel. A significant drop in prices occurred during the second half of the 1980s. With the drop in prices to values below $10 per barrel in 1998, domestic production decreased, consequently. This coincided with the drop in inhibitor usage for drilling and production.

4.4.1 Petroleum Production, Transportation, and Refining

The consumption of inhibitors in the petroleum industry is directly connected to the size of the petroleum-based production market. This in turn is tied to the price of crude oil. The overall consumption of gasoline in the United States has increased only slightly since 1980. Although there are more cars on the road, the average consumption of gasoline has decreased. Increasing quantities of crude oil on a relative basis are being imported from the Middle East.

The increased input of foreign oil results in the reduction of inhibitor used by the oil companies in the production sector but not in the refining sector. Quite often the petroleum production and refining industries run in opposite economic cycles. When the crude oil price is high, domestic production is profitable; however, the margin of product on refining is lowered. Conversely, when oil prices are low, refinery feedstock is cheap, and the production of refined products and specialty chemicals results in a higher profit as consumption of the products is only slightly tied to oil prices.

Upstream oil production uses inhibitors for oil drilling operations as well as permanent production tubulars and pipelines, where two-phase and three-phase production streams are treated with filming inhibitors. The amount used ($153 million) in production is dependent on the amount of water produced with the oil. The economics of such a system is calculated as cost per barrel of oil or water produced. This has been estimated in the lower 48 states to range from $0.02 per barrel of oil produced to $0.23 per barrel of oil produced, depending on factors such as temperature, corrosive gases present, and operating procedures.

Refinery operations utilize the highest amount of inhibitors ($246 million) because higher temperature processes are encountered in refining operations. Not only do process and boiler water streams require inhibition, but also process environments produce HCl, which must be neutralized with pH modifying inhibitors, imported oils from Middle East sources tend to be sour (high sulfur content) and are more corrosive; therefore, in terms of per barrel of oil refined, costs of inhibition for production crudes are expected to increase. An annual rate of increase in refining sector has been predicted to be approximately 2.5%.

The cost of inhibitors for petroleum storage and transportation is approximately $31 million, which is related to the price of oil as new pipelines are built when oil prices are high enough to justify the costs of construction and operation. As domestic oil exploration and production move farther offshore and to more remote areas without an existing pipeline infrastructure, the need for new pipelines will continue, provided the oil prices remain high enough to sustain the increased operating costs far offshore.

4.4.2 Pulp and Paper

The major amount of corrosion in the pulp and paper industry is in the paper-making process, which uses large amounts of process water. The amount of corrosion inhibitors consumed in the pulp and paper industry from 1996 to 1998 increased an average of 2%/year. About $198 million was spent on corrosion inhibitors in 1998.

During the past 35 years, the pulp and paper industry had moved from open-water systems to closed-loop systems, thereby increasing the severity of the environment. These "white liquors" are acidic and very corrosive, and severe crevice corrosion problems may occur whenever the process flow is halted.

The paper industry in the United States is a mature industry but is expected to maintain steady growth in the upcoming years.

The demand for paper products is closely related to the economic growth and disposal income. Most paper is consumed by packaging, printing, publishing, and business communications, which are directly tied to business expansions. Similarly, consumption of cardboard and paper shipping containers is tied to the demand for industrial production.

Competitive materials and technologies such as plastics for packaging and electronic publications/communications are encroaching on traditional markets for paper products. These are not realistically expected to replace paper but will curtail the growth in view of the expected paperless society in the future.

4.4.3 Iron and Steel

It can be stated that on average, nearly 100 million tons of raw steel are produced. The inhibitors used in 1998 amounted to $50 million, and inhibitors used in steel production are expected to increase only slightly (1–1.5%) in the future. Like most systems requiring inhibition, the water treatment piping and vessels in both cooling and boiler water systems are the most affected.

4.4.4 Additives

The applications of corrosion inhibitor additives are primarily for petroleum products such as gasoline, motor oil, and grease. Other inhibiting additives are antifreeze and coolants, brake fluids, fuel additives, and plant cleaning and metalworking fluids.

Of the $120 million market for corrosion-inhibiting additives, $90 million is spent on an additive that is used for 560 billion liters (150 billion gallons) of motor fuel (both gasoline and diesel). The consumption of gasoline is growing at a slow rate because of a steady increase in fuel-efficiency of automobiles since the early 1980s.

4.4.5   Deicers

A potentially lucrative market for corrosion inhibitors exists in deicers used on streets and bridges in northern areas. The deicers, which consist of rock salt and calcium carbonate, cause corrosion damage not only to automobiles, but also to steel and steel reinforcements in bridges, light-stanchions, and underground pipes and cables. The technical challenge is to formulate an inhibitor, which is not only nontoxic to the environment, but is also economical enough to be attractive for city budgets.

4.5   ENGINEERING COMPOSITES AND PLASTICS

In 1996, the plastics industry accounted for $274.5 billion in shipments (11). It is difficult to estimate the use of plastics for corrosion control, as in many cases, plastics and composites are used for a combination of reasons, including corrosion control, light-weight, economics, strength to weight ratio, and other unique properties.

Certain polymers are used largely, if not exclusively, for corrosion control purposes. The significant markets for corrosion control by polymers are composites (primarily glass-reinforced thermosetting resins) PVC pipe, polyethylene pipe, and fluoropolymers.

4.5.1   Composites

In the context of corrosion, composites refer to glass or other fiber or flake-reinforced thermosetting resins. Composite products use for their anticorrosion properties are fiberglass-reinforced pipe and storage tanks, fiber-reinforced plastic grating, handrails, I-beams, and other shapes equal to these that are made of steel.

The Composites Institute, a division of the Society of Plastics Industry, Inc., estimates that composite shipments in 1998 were 1.63 billion kg and have increased 53% (K. Walshon, Society of Plastics Industry, Composites Institute, Personal communication, Oct. 1999) since 1991. Table 4.10 shows the distribution of composite shipments according to industry sectors and indicates that the largest shipment percentage is to the transportation and construction sectors. Corrosion-resistant applications account for only 11.8%.

The cost of composites was estimated by one major manufacturer of fiber-reinforced plastics (FRP) for corrosion-resistant applications to be $9.70/kg ($4.41/lb). This works out to $1.864 billion spent on composites in the United States for industrial corrosion-resistant applications.

Fiberglass pipe is representative of composites used for corrosion control purposes. The fiberglass pipe market in the United States is estimated to be $350 million. About a third of this market is in oil and gas production, 25% is in gasoline transportation and storage, and 15% in the petrochemical industry. Another advantage is the light weight of fiberglass pipe aside from the major aspect of corrosion protection.

The difference in installed cost between steel and fiber reinforced plastic component must be used in estimating the corrosion costs, as the composites are replacement for steel.

TABLE 4.10   Distribution of Composite Shipments (K. Walshon, Society of Plastics Industry, Composites Institute, Personal communication, Oct. 1999.)

Industry Sectors	Percentage of Shipments
Transportation	31.6
Construction	20.8
Marine	10.1
Electrical/electronic	10.0
Appliances and business equipment	5.5
Consumer goods	6.3
Aircraft	0.6
Corrosion-resistant applications	11.8
Other	3.3
Total	100.0

For a pipe less than 20 cm in diameter, the installed cost of fiber-reinforced pipe (FRP) is 50% greater than steel pipe, while for 20–40 cm pipe, the installed costs of FRP and steel are equal. In the case of pipe larger than 40 cm in diameter the installed cost of FRP is less than that of steel. Overall, the installed cost of an FRP pipe is about 30% higher than the installed cost of steel pipe (Ameron Fiberglass, Personal communication, Aug. 1999.). The 30% extra cost is the actual cost of corrosion, and the annual contribution to the total cost of corrosion by composites is $1.864 billion × 30%, which equals $559 million.

PVC pipe first developed by German scientists during World War II has grown by leaps and bounds. A total of 6.6 billion kg of PVC resin was produced in 1998 in the United States of which 907 kg are used toward manufacture of PVC pipe. For buried pipes, 10.2 cm in diameter and larger, which includes water, sanitary, and storm sewers, 137,500 km of PVC were produced in 1997 at a total worth of $1 billion.

PVC pipes have many advantages over steel pipes such as light weight, ease of fabrication (no welding required), and ease of installation (no torch cutting required). Thus, the total amount of PVC pipe is not the direct cost of corrosion. The PVC industry is found to play a significant role in the context of corrosion resistance. The cost attributable to corrosion is approximately $500 million.

4.5.2   Polyethylene

This is by far the most used polymeric material in the United States. More than 12.2 billion kg (27 billion lbs) of polyethylene resin was produced in 1998. Polyethylene is chemically inert and hence finds application in the form of polyethylene pipe in the corrosion-resistant market. The use of polyethylene pipe by industry in the United States in 1998 as reported by the Conduit Plastic Pipe Institute is given in Table 4.11.

The commodity price of polyethylene pipe is $1.32/kg and translates into a total cost of $461.4 million of polyethylene pipe sold in the United States. This is considered as corrosion-related cost.

TABLE 4.11   The Use of Polyethylene Pipe in the United States in 1998 (14)

Application	Amount kg × million
Potable water	30.4
Irrigation/agriculture	15.4
Gas distribution	94.3
Oil/gas production	51.7
Industrial/sewers	94.8
Other	62.1
Total	348.7

TABLE 4.12   Fraction of Polymers Used for Corrosion Control in 1997

Polymers	Cost
Composites	$559 million
PVC (pipe)	$500 million
Polyethylene (pipe)	$461 million
Fluoropolymers	$560 million
Total	$2,080 million

4.5.3 Fluoropolymers

All the polymers that contain fluorine side groups in the molecular structure have excellent high-temperature stability and chemical resistance. Polytetrafluoroethylene, known as Teflon, is the best known fluoropolymer. Other fluoropolymers have been used in petrochemical and chemical properties. Cost is a prohibitive factor in their use. On a weight basis, they cost 50–65 times that of polyethylenes. The fraction of fluoropolymers used for mitigation of corrosion in 1997 was estimated at $560 million.

In summary, Table 4.12 shows the fraction of polymers used for corrosion control in 1997.

4.6   CATHODIC AND ANODIC PROTECTION

The cost of cathodic protection (CP) and anodic protection of metallic structures prone to corrosion can be divided into the cost of materials and the costs of installation and operation. Industry data have provided estimates for 1998 sales of various hardware components amounting to a total of $146 million as shown in Table 4.13.

The largest share of CP market is taken up by sacrificial anodes at $60 million of which magnesium has the greatest market share. The costs of installation of a CP system vary to a significant extent depending on the location and specific details of the construction. The range of cost for labor, materials, and the number of installations for various systems in 1998 are given in Table 4.14.

TABLE 4.13　Total Cost of Components for Cathodic and Anodic Protection

Components	Cost ($× Million)
Rectifiers	15
Impressed current cathodic protection anodes	25
Sacrificial anodes	60
Cable	6
Other accessories	40
Total cost	146

TABLE 4.14　Cost of Installation of Cathodic Protection Systems

Installation	Cost Range Per Installation	Estimated Number	Total Cost ($× billion)
Rectifier (replacement)	1.5–2.5	800	0.0012–0.002
Impressed current CP ground bed including rectifier and 10 anodes per bed	8–12	6000 ICCP ground beds	0.048–0.072
Galvanic ground bed with magnesium anodes (10 anodes per bed)	0.35–0.6	1 million anodes	0.35–0.6
CP on underground storage tank (3 USTs with 1 ICCP system)	6–10	50,000 UST CP systems	0.3–0.5
CP on aboveground storage tank (37 m diameter AST)	15–25	2,000 AST-CP	0.03–0.05
Total			$0.73–1.22
Average			$0.98 billion/year

A major market for sacrificial anodes not included in Table 4.14 is the domestic water heater market. About 120 million water heaters are in use in the United States. Assuming that 5% of all water heaters get their anodes replaced each year and the cost of a magnesium anode is $150, the annual expenditure of $780 million can be estimated. Annually, about 9.2 million water heaters are replaced. Assuming that 5% of water heater failure is because of corrosion, and an average replacement cost of $1000, an annual expenditure of $460 million can be estimated. Addition of CP cost of $0.98 billion and water heater anode cost of $1.24 billion results in a total cost of $2.22 billion/year.

4.7　SERVICES

This may be defined as companies, organizations, and individuals that are providing their services to corrosion control. When taking the National Association of Corrosion Engineers (NACE) international membership as a basis, a total number of engineers and scientists that provide corrosion control services may be extrapolated. In 1998, NACE membership was about 16,000; 25% of those are providing consulting and engineering service both externally and internally. Assume that the average revenue generated by each is $300,000. This amount includes salary, overhead, and benefits for the NACE member, as well as the cost to persons who are nonmembers involved in performing corrosion control activities. The total control services cost can be estimated as $1.20 billion. This figure is a conservative estimate as not all engineers involved in corrosion control are NACE members.

4.8   RESEARCH AND DEVELOPMENT

Over the past few decades, less funding has been made available for corrosion research and development. This is significant in light of the cost and inconvenience of dealing with leaking and exploding underground pipelines, bursting water mains, corroding storage tanks, and aging aircraft. In fact, several government and corporate research laboratories have significantly reduced their corrosion research capabilities or have even closed down. Moreover, research and development funding has been reduced by both government and private agencies.

Corrosion research may be divided into academic and corporate research. According to NACE International, there are 114 professors performing corrosion research with a total annual budget of less than $20 million.

4.9   CORROSION CONTROL OF BRIDGES

The methods utilized for corrosion control on bridges are specific to the type of bridge construction and whether its intended use is for new construction or maintenance/rehabilitation of existing structures. The present discussion is focused on the following:

1. Conventional reinforced concrete
2. Prestressed concrete
3. Steel.

In the present discussion, reinforced concrete and prestressed concrete corrosion control methods are combined. Although prestressed concrete bridges have special concerns such as anchorage in both posttensioned and pretensioned structures and ducts for post-tensioned structures, the general corrosion control methods are applicable to both prestressed and conventional reinforced bridges.

4.9.1   Reinforced Concrete Bridges

Conventional reinforced concrete bridges refer to those with superstructures made of reinforced concrete. In general, prestressed concrete and steel bridges will have conventional reinforced concrete decks or substructures. Thus, corrosion control practices for conventional reinforced concrete are applicable to components of many other bridge structures.

Corrosion protection can be incorporated into new bridge structures by proper design and construction practices, including the use of high-performance concrete such as silica fume additions, low-slump concrete, and an increase in concrete cover thickness. Each of these attempts to impede migration of chloride and oxygen or other corrosive agents through the concrete to the steel rebar surface. However, eventually, the corrosive agents penetrate through the concrete cover and cracks, necessitating other corrosion control processes. A widely used method of corrosion prevention is the use of coated carbon steel rebar and to some degree, corrosion-resistant alloy/clad rebars. The typical organic rebar coating is fusion-bonded epoxy, while the metallic rebar coating is galvanizing (very limited use in bridge structures). Rebar cladding with a corrosion-resistant alloy such as stainless steel is relatively new. Solid rebars made out of stainless steel alloys have been used to a limited extent. In addition, nonmetallic composite materials have been used. Another corrosion control practice involves the addition of corrosion-inhibiting admixture to the concrete.

4.9.1.1 Epoxy-Coated Rebars A technical note prepared by the FHWA provides a discussion of the use of epoxy-coated rebar in bridge decks. Epoxy coatings (also referred to as powders or fusion-bonded coatings) are 100% solid, dry powders. The dry epoxy powders are electrostatically sprayed over clean, preheated rebar to provide a tough, impermeable coating. The coatings achieve the toughness and adhesion to the metal sample because of a reaction caused by heat. As epoxy powders are thermosetting materials, they are not affected by temperature. The epoxy coating provides a physical barrier between aggressive chloride ions and the steel bar.

For some time, bridge deck deterioration because of corrosion of reinforcing bars has been the main problem. Prior to 1970, Portland cement itself was considered to give protection from corrosion. Later, it was found that deicing salts caused corrosion of reinforcing steel bars. Many thousands of bridge decks containing corroded black reinforcing steel showed signs of spalling in 7–10 years after construction. Corrosion of the substructure was also observed because of the leakage of deicing salt solution.

Epoxy coated rebar was introduced in the 1970s as a means of extending the life of reinforced concrete bridge components by minimizing concrete deterioration caused by corrosion of reinforcing steel. The epoxy coatings prevent the moisture and chlorides from reaching the surface of the reinforcing steel and reacting with the steel. Since the 1970s, the highway industry has widely used epoxy coatings. Just as any protective system, the coatings will degrade over time, leading to corrosion of rebar.

In the case of many substructures exposed to a severe corrosive marine environment epoxy-coated rebars did not perform as well as in bridge deck applications. This was encountered in bridges located in the Florida Keys. Significant corrosion was observed in substructure members of the bridges in 6–9 years. These members are subjected to salt spray in the splash zone which is highly corrosive. The amount and the degree of corrosion observed in the Florida Keys bridges raised questions concerning the durability of epoxy-coated rebar. The performance of epoxy coated rebar decks as evaluated by transportation agencies was good. The epoxy-coated rebar did not perform as well in cracked concrete as it did in uncracked concrete. Corrosion was observed on epoxy-coated rebars extracted from locations having heavy cracking, shallow concrete cover, high concrete permeability and high chloride concentrations. Reduced adhesion and softening of coating occurs because of prolonged exposure to a moist environment. The number of defects in the epoxy coating had a strong influence on the adhesion and performance of epoxy-coated bar. There was no significant premature concrete deterioration that can be attributed to corrosion of the epoxy-coated rebar. Thus use of good-quality concrete cover, adequate inspection finishing and curing of concrete and the use of epoxy-coated rebar resulted in effective corrosion for bridge decks since 1975.

At present, epoxy-coated rebar is the most common corrosion protection system and is used in 48 states. At present, there are nearly 20,000 bridge decks using fusion-bonded epoxy-coated rebar as the preferred protection system. This amounts to nearly 95% of new deck construction since the early 1980s.

The data from the Concrete Reinforcing Steel Institute (CRSI) shows that more than 3.6 billion kg (4 million tons) of epoxy-coated rebar (150 million m^2 reinforced concrete) were used as of 1998. A significant amount of the epoxy-coated rebar was used in bridge decks. Over time, the formulation of epoxy has been modified to achieve better performance of the epoxy coating.

4.9.1.2 *Metal-Coated/Clad Rebars and Solid Corrosion-Resistant Alloy Rebars*
The most promising corrosion-resistant rebars are galvanized (zinc-coated) rebars, stainless steel-clad rebars, and solid stainless steel rebars. Titanium has also been considered as a rebar metal, but its cost is prohibitive although it is highly corrosion-resistant.

4.9.1.2.1 Galvanized Rebars Hot-dipped galvanized coatings for reinforcing steel in concrete have been used since the 1940s. ASTM A767 "Standard Specification for Zinc-Coated (Galvanized) Steel Bars for Concrete Reinforcement" specifies the requirements for the galvanized coating. Class I coating has 1070 g of zinc per m^3, and class II has 610 g/m^3. The effectiveness of galvanized rebars in extending the life of reinforced concrete structures is questionable. In other applications, galvanized steel has been useful in extending the life of structures exposed to atmospheric conditions and low-chloride underground environments though not high-chloride environments. Although galvanized rebar may prove to be beneficial in certain chloride media it may not withstand the degree of corrosivity of deicing salts.

4.9.1.2.2 Stainless Steel Rebars Research in stainless steel rebars consists of clad stainless steel over a carbon steel substrate and solid stainless steel rebar. The primary concerns of cladding when stainless steel alloy is chosen are the following:

1. Adherence to rebar substrate.
2. Defects formed after bending.
3. Uniform cladding thickness (typical value is 0.5 mm)

4. Metallurgical changes because of the cladding process that may affect the corrosion resistance.

It must be noted that the chloride threshold for pitting in a nonhomogenous environment such as concrete can be significantly less than the value in the same aqueous environment. Thus, it is necessary that any work must involve a realistic concrete environment. For example, the use of stainless steel piping in underground pipelines has been discontinued because of pitting and subsequent perforation of the pipe in nonhomogenous unsaturated soil environments with relatively low chloride contents. Pitting in reinforced concrete bridge components is not as significant as decreasing average corrosion rate (overall metal loss).

McDonald et al. studied the performance of solid stainless steel rebars (types 304 and 316) and found that they performed well while ferritic stainless steels (types 405 and 430) developed pitting (15). Studies by McDonald et al. reported investigations on a 10-year exposure of 304 stainless steel in Michigan and Type 304 stainless steel clad rebar in a bridge deck in New Jersey and found no corrosion (15). In a study by Virmani and Clemena, the type 316 stainless steel-clad rebar extended the estimated time to the cracking of the concrete beyond 50 years, but not as much as solid types 304 and 316 stainless steels (100 years) (16).

In addition, McDonald et al. (15) reported on two highway structures constructed with stainless steel rebar. No corrosion was observed for solid 304 stainless steel rebar in a bridge deck in Michigan as well as in New Jersey. The chloride levels in both bridge decks were below or at the threshold level for corrosion initiation in black steel rebars. It is estimated that the use of solid stainless steel rebar provides an expected life of 75–100 years (15, 16). McDonald et al. estimated the costs, at three installations, of the use of solid stainless steel and found the overall cost to be 6–16% higher than black steel (17).

Stainless steel rebars have been reported to be used in several projects in the United States, including Michigan and Oregon (17). The expected life of structures using stainless steel rebars was stated to be 120 years.

Fluctuation in the cost of raw materials used in the production of stainless steel impacts on the economic viability of the use of stainless steel rebars in concrete decks. The rebar cost also depends on the grade of stainless steel used.

From the point of view of cost, it is preferable to use stainless steel-clad rebar instead of stainless steel rebar. It is estimated that the use of stainless steel-clad rebar provides an expected life of 50 years. The cost of stainless steel cladding can vary depending on the raw material market prices just like solid stainless steel, but also depends on the cladding manufacturer, cladding thickness, and the chosen grade of stainless steel. With proper quality control, stainless steel-clad rebar promises to be an effective means of control for bridge deterioration because of corrosion of reinforcing steel.

4.9.1.3 Alternative Means of Protection   In addition to the use of coated or alloy rebar, other means to mitigate corrosion of reinforcing steel in bridge structures consist of use of high-performance concrete, corrosion-inhibiting admixtures, or a combination of both. High-performance concretes were developed to impede the ingress of chloride to the rebar by reducing concrete permeability. This is obtained by using lower water-to-cement ratio concrete and adding mineral admixtures such as silica fume and fly ash (pozzolanic materials) to the concrete mix. Low water-to-cement ratios are achieved using high-range water reducers.

In addition to low chloride permeability, mineral admixtures impart other properties to the concrete depending on the admixture selected such as:

1. Higher corrosion resistance (higher chloride threshold for corrosion and low corrosion rate following initiation).
2. Greater cumulative corrosion prior to cracking.
3. Higher resistivity to minimize macrocell corrosion.

An FHWA study by Thompson and Lankard (19, 20) reviewed the effect on the corrosion of steel in the concrete of several variables, including cement type, mineral admixtures, water-to-cement ratio, and aggregate type. This study showed silica fume to be the most effective mineral admixture in the mitigation of corrosion of steel rebar. It also suggested that careful selection of the concrete mix components could extend the life of a concrete bridge member. It is estimated that use of a silica fume admixture provides an increase of expected life of 10 years beyond that provided by black steel rebar in conventional concrete.

4.9.1.4 Corrosion-Inhibiting Concrete Admixtures In the past two decades, the use of corrosion-inhibiting concrete admixtures has become a promising method for delaying the onset of corrosion of prestressing and conventional reinforcing steel (21). Inhibitors are generally used with permeability-reducing pozzolanic additives such as fly ash or silica fume. As the concrete has low permeability, and the corrosion inhibitor essentially increases the chloride concentration required for corrosion initiation. The inhibitor may also reduce the corrosion rate during the postinitiation period, leading to less corrosion-induced concrete deterioration.

Corrosion inhibitors can be either inorganic or organic compounds and reduce the corrosion rates to acceptable levels when present at low concentrations. Organic inhibitors function by the formation of a protective film, metal–inhibitor complex, on the metal surface. Inorganic inhibitors function by either the oxidation or reduction reactions at the steel surface.

There is extensive literature on the corrosion inhibition behavior of calcium nitrite as an inhibitor. Calcium nitrite can function as an oxidizing inhibitor in relatively high concentrations of chloride. The suitable organic inhibitors are amides and esters. A National Cooperative Highway Research Program (NCHRP) project reviewed the performance of corrosion inhibitors used in concrete and assessed the performance of commercial inhibitors (22).

The economics of the use of calcium nitrite in corrosion-inhibiting admixtures with and without the addition of silica fume has been documented (23). The cost of a calcium nitrite protection system was estimated to be $5.40/m^2, and the cost increase to construct the deck using calcium nitrite inhibitor is 1.1%. It is estimated that the use of inhibitors may increase the expected life of 20–25 years more while compared to the use of black steel rebar and conventional concrete.

4.9.1.5 Multiple Protection Systems Corrosion inhibitors are used in multiple corrosion protection systems in conjunction with epoxy-coated rebars and low-permeability concrete, in particular, for marine applications. Epoxy-coated seven-wire strands are not commonly used for prestressed concrete bridge members. Corrosion inhibitors are used in place of coated seven-wire strands in the prestressed highway construction industry.

4.9.1.6 Methods of Controlling Corrosion in New Bridge Construction Methods based on current practice in research, field performance, and emerging technologies may be described as follows (16): the preferred primary corrosion protection system is fusion-bonded epoxy-coated rebars that have been used in 20,000 reinforced-concrete bridge decks and nearly 100,000 total structures. Epoxy-coated rebar has performed very well in combating the corrosion-induced deterioration of concrete bridge decks. This does not perform well in severe marine applications.

With constant updates in the American Association of State Highway and Transportation Officials (AASHTO) and American Society for Testing and Materials (ASTM) specifications for epoxy-coated rebar, this corrosion protection system will continue to improve. The specifications involve all aspects of fabrication of epoxy-coated rebar. All these will result in improved performance of epoxy-coated rebar and more durable new concrete structures.

To provide longer service life to the concrete decks of the order of 75–120 years without the need to repair corrosion-induced concrete damage, a number of solid and clad corrosion-resistant 304 and 316 stainless steel rebars have been developed. Both alloys provide excellent corrosion protection but at higher cost. Type 316 stainless rebar requires more detailed studies.

Epoxy-coated rebar together with calcium nitrite may be a favorable corrosion protection system especially for marine applications such as piles. The long-term stability is still not confirmed and research is in progress to identify new inhibitors. Temperatures such as 38 °C and 75% humidity or moisture have been identified as conditions leading to high corrosion rates for steel in concrete. Use of a low water-to-cement ratio concrete, use of mineral admixtures, proper selection of cement type and aggregates contribute to the production of high-performance concrete with significant corrosion resistance (19).

For the protection of high-strength, seven-wire strands encased in ducts, mix designs for corrosion-resistant grout for filling the grouts have been developed. A new rapid method for evaluation of grouts has been developed (24). These developments led to new grout specification documented by the Post-Tensioning Institute in 2001.

With the recent collapse of two posttensioned bridges in the United Kingdom and one in Belgium the impact-echo nondestructive evaluation (NDE) technique was developed to detect voids in posttensioned ducts. In addition, a complementary magnetic-based nondestructive technique (NDT) for assessing section loss in

TABLE 4.15 Summary of Costs and Life Expectancy for New Construction Corrosion Control

Practice	Cost of Bar $1 kg	Bar Weight Per Deck kg/m²	Cost Per Deck $1 m²	Increase in Comparison to Baseline $1 m²	Percent Increase %	Estimated Service Life Year
Black steel (baseline)	0.44	26.4	11.60	NA	–	10
2-layer epoxy-coated rebar	0.66	26.4	17.40	5.80	1.2%	40
2-layer solid SS bar	3.85	26.4	101.64	90.04	18.6	75–120
2-layer SS clad bar	1.54	26.4	40.66	29.00	6.0	50
Calcium nitrite	–	–	–	5.40	1.1%	30
Silica fume	–	–	–	4.30	0.9	20

high-strength steel strands in the ducts also has been developed. A combination of impact-echo and magnetic-based technique allows the inspection of posttensioned systems reducing the likelihood of any sudden collapse of posttensioned bridges. Continued development of these techniques will lead to increased reliability, accessibility around trumpet locations, and resolution.

Table 4.15 gives the costs of new construction alternatives for bridge structures.

4.9.1.7 Rehabilitation The following is a summary of current practices in rehabilitation technologies (16).

Rehabilitation is achieved by overlays such as latex-modified concrete, low-slump concrete, high-density concrete, and polymer concrete. They are commonly used for the rehabilitation of bridge decks. This procedure extends the life of a bridge deck by about 15 years. Impressed-current CP systems on bridge decks are now a routine rehabilitation technique because of the cooperative research with industry and states in the development of durable anodes, monitoring devices, and installation techniques. Titanium mesh anode, used in conjunction with a concrete overlay to distribute protective current, serves as a durable anode for use in impressed-current CP of reinforced concrete bridge decks and widely accepted by state and other transportation agencies.

Several promising sacrificial anodes such as thermal-sprayed zinc, thermal-sprayed Al–Zn–In, zinc hydrogel, and zinc mesh pile jacket have been developed for use in CP of substructure members especially in marine environments. Industries in some states in cooperation with FHWA carried out some developments and identification of some anodes suitable for impressed current CP of inland concrete structures.

Significant advances have been made in the technology of CP of prestressed concrete components through extensive fundamental research and evaluation of CP systems that have been installed. Concerns about a loss of bond between

TABLE 4.16   Summary of Costs and Life Expectancy for Rehabilitation Methods

Type of Maintenance	Average Cost ($/m²)	Range of Cost ($/m²)	Average Life (Years)	Range of Expected Life (Years)
Impressed current (CO) (deck)	114	92–137	35	15–35
Impressed current CP (substructure)	143	76–211	20	5–35
Sacrificial anode CP (substructure)	118	108–129	15	10–20
Electrochemical removal (deck)	91	53–129	15	10–20
Electrochemical removal (substructure)	161	107–215	15	10–20

the prestressing steel and concrete and possible hydrogen embrittlement (from overprotection of the prestressing steel) have been alleviated by the establishment of criteria for the qualification of prestressed concrete bridge components for CP.

The costs of electrochemical rehabilitation alternatives for bridge structures are given in Table 4.16.

In general, electrochemical methods are in competition with rehabilitation utilizing an overlay such as low-slump, high-performance, or latex-modified concrete. The deck condition is often the controlling factor in the selection of the rehabilitation method. In some instances, a combination of these methods is selected. For example, electrochemical removal of chloride followed by an overlay or an overlay in conjunction with CP to mitigate any further corrosion.

Salt-induced reinforcing steel corrosion in concrete bridges is a serious problem and an economic burden. Although the positive effect of corrosion protection measures can be seen on individual cases, there are many bridges (thousands) without corrosion control.

It is useful to note that even the latest corrosion control methods are not likely to prevent all corrosion for the life of the bridge structure. Therefore, there is a need for repair/rehabilitation of bridge structures, and the mitigation of existing corrosion will draw the attention of bridge engineers for years to come.

There are several methods for the rehabilitation of concrete structures that have deteriorated because of chloride-induced corrosion of the reinforcing steel. The problems in concrete structure are generally found after significant deterioration has resulted in the cracking and spalling of the concrete, and the majority of the remedial measures are applied after removal and patching of the damaged concrete. The available methods are based on one of the following principles (16).

1. Provision of a barrier on the surface of the concrete to prevent future ingress of chloride (overlays, membranes, etc.).

2. Control of the electrochemical reactions at the steel surface to mitigate the corrosion reactions by imposing proper voltage field on the rebar CP.

3. Altering the concrete environment to make it less corrosive. One way would be to extract the chlorides from concrete such as electrochemical chloride removal.

Each state department of transport (DOT) has its own specifications and criteria for rehabilitation of deteriorated concrete bridge components. The following example illustrates the decision process for rehabilitation or deck replacement.

- If spalling of concrete is observed, the surface is checked for delamination by chain drag, and core samples are taken to determine the chloride concentration.
- If the chloride concentration is greater than 1.8 kg/m³, the coverage is removed to depths where the concentration of chloride is less than 1.2 kg/m³.
- If less than 75 mm is removed to reach an acceptable chloride level, the removed concrete is replaced by an overlay.
- If more than 75 mm is removed, the entire deck must be replaced.
- CP is applied only in the case of partial disbondment of the concrete and when there is no extensive spalling.

4.9.1.8 Surface Barriers The application of an overlay of low-slump concrete, latex-modified concrete (LMC), high-density concrete, polymer concrete, or bituminous concrete with membrane on the existing concrete provides a barrier that impedes continued intrusion of chloride anions, moisture, and oxygen that are necessary for continued corrosion. These barrier systems may be used only after decontamination because corrosive agents get trapped in the concrete, leading to loss in its capacity to function properly.

Traditionally, more than 90% of the rehabilitation jobs used low water-to-cement ratio concrete or LMC overlay as the preferred method. The estimated lifetime of the rehabilitation methods is about 15 years (16).

Many studies have reported performance and cost data for different overlay and patching systems (25–28). Table 4.17 gives the cost and life expectancy for overlay and patching options for concrete bridges (27).

4.9.1.9 Cathodic Protection This is a corrosion control method involving imposition of an external voltage on the steel surface such that the steel becomes cathodic and favoring cathodic reaction and decreasing the anodic reaction, that is, metal loss and hence mitigating corrosion. In other words, CP transfers the oxidation (anodic) reaction involving metal loss and hence corrosion of rebar, over to the anode of CP system. Thus, the selection of proper anode is critical.

The primary strength of CP is that it can mitigate corrosion after it has been initiated. Although CP is often placed on pipelines, underground storage tanks (USTs), and other structures during construction, it is generally installed on bridge members only after corrosion has initiated and some amount of deterioration has occurred. The primary reason for not installing a CP system on bridge components during construction is that corrosion often does not initiate for 10–20 years following construction. Therefore, the CP system maintenance and a large portion of CP system design life

TABLE 4.17   Cost and Life Expectancy for Overlay and Patching Options for Concrete Bridges

Type of Maintenance	Average Cost $1m²	Range of Cost $1m²	Average Expected Life (Years)	Range of Expected Life (Years)
Portland cement concrete overlay (includes latex-modified concrete)	170	151–187	18.5	14–23
Bituminous concrete with membrane	58	30–86	10	4.5–15
Polymer overlay/sealer	98	14–182	10	6–25
Bituminous concrete patch	90	39–141	1	1–3
Portland cement concrete patch	395	322–469	7	4–10

would be used on a structure that is not corroding. Furthermore, the use of CP on newly constructed bridge components is limited as materials such as epoxy-coated rebar provide economic, long-term corrosion prevention for these structures. The exception to this is the CP installed on newly constructed bridge pilings exposed to marine and brackish waters where corrosion is a serious problem.

Although in the early stages, CP systems cast a negative image that has improved over time, and the current technology for bridge decks has proven to be reliable, and improved technology for substructures is being developed. When properly applied and maintained CP mitigates corrosion of reinforcing steel and extends the useful life of a bridge. However, CP remains an underutilized technology for steel-reinforced concrete structures.

CP systems are characterized by the source of driving voltage that forces the rebar to become cathodic with respect to the anode. The two methods for applying CP are impressed current CP and sacrificial anode CP (galvanic). In an impressed-current CP, an external power source is used to apply the proper driving voltage between the rebar and anode. For impressed-current systems, the anode can be a wide range of materials as the driving voltage can be adjusted to suit the application and the anode material selected. For a sacrificial anode CP system, the driving voltage is created by the electrochemical potential difference between the anode and the rebar. Therefore, selection of the anode material is more limited.

4.9.1.9.1   Impressed-Current Cathodic Protection   The impressed-current system consists of the following:

1. An external power source.
2. Variation of voltage with a variable power source.
3. Variation of applied current.
4. Designing CP system for almost any current requirement.
5. Use of CP system in any level of resistivity.

So far, over 2.0 million m² of reinforced and prestressed concrete structures have been protected worldwide.

Anode selection and application have proven to be a challenge in designing CP systems for concrete structures with adequate life. The anode for concrete bridge deck should have the following features:

1. Capability to withstand traffic loads.
2. Resistance to environmental factors such as moisture and temperature fluctuations.
3. Design life equal to or greater than that of wearing surfaces.
4. Sufficient conductive surface area to minimize or completely prevent mature deterioration of the surrounding concrete.
5. Economical.

Over the past few decades, several anode configurations have been used in concrete bridge decks and substructures:

1. Coke–asphalt anode system used high silicon iron anode and required wear surface. The estimated cost for this anode system is $92/m² with service life of 20 years (29).
2. Nonoverlay slotted anode system used platinized-niobium–copper wire anode laid in regularly spaced slots designed to distribute CP current evenly to the rebar mat and was filled with a conductive polymer concrete. Estimated cost is $92/m² and 15 years of service life.
3. Conductive polymer mound anode system used platinized-niobium copper wire anode with the conductive polymer mounded on wire anode and rigid concrete overlay on top. Estimated cost is $137/m² with a service life of 20 years.
4. Activated titanium mesh secured to the concrete and covered with either conventional concrete or latex-modified concrete overlay at a cost of $137/m² and service life of 35 years.
5. Other anode systems specially developed are sprayable conductive polymer coatings, metallized zinc coatings, and conductive paints. Typical primary anode for the conductive polymer or paints is platinized niobium wire attached to the concrete prior to application. Estimated cost is $76/m² with a service life of five years.
6. Metallized zinc used either small stainless steel or copper plates epoxied to the concrete surface to make a connection to power service. Cost is $137/m² service life 15 years.

Some problems with the CP system are the following:

1. Debonding of the conductive coating in environments where concrete is constantly wet or when the materials applied before the concrete is dry.

2. Degradation of conductive coating after extended current passage.
3. Increase in electrical resistance between the anode and the steel because of insufficient moisture or accumulation of insulating by-products at the anode/concrete surface.

Of all the systems cited above, only the titanium mesh anode and metallized zinc are in extensive use at present. The titanium mesh on bridge decks is durable over long time. The thermal-sprayed zinc is free from debonding problems but suffers from an increase in resistance over time. However, the Oregon DOT has had significant success with thermal-sprayed zinc anode on substructure components. The thermal-sprayed titanium has shown promise as a new anode.

Some DOTs have investigated alternate energy sources for CP systems, such as solar power and long-life batteries for use on the substructure elements exposed to brackish waters.

In some cases, CP offers the only acceptable service life extension as an alternative to replacement of a critical ridge component. For example, a thermal-sprayed zinc CP system on historic bridges built in 1930s has been successfully implemented. Missouri DOT leads North America in the use of CP to extend the life of salt-contaminated and corroding concrete bridges. CP is primarily used for corrosion control of voided slab structures although it is used on steel frame and stringer type structures. Conventional methods of corrosion prevention on bridges built in the 1950s and 1960s were unsuccessful. Since 1975, Missouri has installed CP systems on more than 140 bridges. Many CP systems have been evaluated and used in Missouri. First introduced in 1986, the activated titanium mesh anode system with concrete overlay has become the exclusive CP system installed on Missouri DOT bridges. To date, this system has provided a high level of corrosion control to more than 30 bridges in the Kansas City and St. Louis areas.

4.9.1.9.2 Sacrificial Anode Cathodic Protection The basic characteristics of a sacrificial anode system are the following:

1. No external power source is required.
2. The driving voltage is fixed.
3. The applied current is dependent on the driving voltage and the resistance between the rebar and the anode.
4. The CP system is limited to relatively low current requirements.
5. The CP system is limited to low-resistivity concrete environments.

Sacrificial anode CP systems have been used for the corrosion control of bridge decks as long as impressed-current anode systems for corrosion control of bridge decks. Two of the earliest field trials (1977) for sacrificial anode systems were the following:

1. Perforated zinc sheets fastened on the deck with a bed of mortar, then covered with concrete overlay.

2. Zinc ribbons embedded in grooves cut into the concrete.

Both systems performed well for 14 years prior to removal because of failure of the asphalt overlay and the necessity of widening the structure. Although the sacrificial anode CP systems perform well, the majority of the CP systems on bridge decks are impressed-current systems.

Because of the relatively high resistivity of atmospherically exposed concrete substructures, most anodes utilize impressed current to achieve the necessary driving voltages to supply the current required for corrosion control. However, an exception to this is the use of sacrificial zinc anodes for CP of coastal bridges in Florida, which have a relatively low concrete resistance. However, studies continue to examine the use of sacrificial anodes because of the benefit of its low maintenance compared to impressed-current CP systems. Two of these studies are the following:

1. Hydrogen-gel anode system.
2. Thermal-sprayed alloy anode system

The zinc–hydrogen anode system uses 10–20 mm thick zinc sheet anodes attached to the concrete with ionically conductive hydrogel adhesive. Field trials have shown that this system is capable of supplying sufficient current for effective corrosion control. The thermal-sprayed alloy anode system utilizes a metallization (flame or arc spraying) process to form a metallized coating on the concrete surface. The two most promising anode materials were Al–Zn–In alloy and zinc (16).

The cost of CP systems varies depending on the type of system used. In this regard, Virginia DOT has published a report entitled, "Evaluation of Anodes for Galvanic Cathodic Prevention of Steel Corrosion in the Prestressed Concrete Piles in Marine Environments in Virginia" (30). The data in this report and the data in the literature published by Virmani (2) suggest that the sprayed Al–Zn–In alloy or the zinc-hydrogel systems with a life of 10–20 years cost \$108–129/m^2.

4.9.1.9.3 Cathodic Protection for Prestressed Concrete Bridge Members The primary concern for CP of prestressed concrete members is the possible hydrogen-induced cracking failure (also known as hydrogen embrittlement) of the tendons at operating loads. Hydrogen production at the steel surface is because of CP at potentials more negative than −0.9 V saturated calomel electrode. Because of this reason, CP for prestressed concrete has focused on the use of the sacrificial anode system and constant current or constant voltage rectifier impressed current systems. Another concern is the application on bridge members that have an uneven electrical resistivity across the concrete surface. This leads to uneven distribution of the CP current and the possibility of overprotection in the low-resistivity regions. It is generally agreed that CP of prestressed concrete members can be accomplished safely and reliably if proper care is taken to maintain minimum CP requirements and to prevent overprotection.

4.9.1.10 Electrochemical Chloride Extraction This method has been reviewed by Virmani and Clemena (16). When a direct current is passed through concrete, the mobile ions such as chloride, hydroxide, sodium, potassium, calcium will migrate, with each ion moving toward the electrode of opposite charge. The feasibility of removal of chloride from concrete without the excavation of contaminated concrete from a structure was explored in the 1970s by the Kansas DOT. Chloride ions could be expelled from concrete by passing direct current between the steel bars and anode as in CP at considerably greater current densities. However, the high levels of current used resulted in adverse effects on the concrete such as decrease in concrete-to-steel bond strength, increase in porosity, and increased cracking in concrete. These observed effects resulted in decreased interest in this method of chloride removal from concrete. Subsequent studies showed that in keeping the current level below 5 Amp/m^2 (0.5 A/ft^2) the adverse effects could be avoided.

Even at densities of the order of 5 Amp/m^2 the possible hydrogen-induced cracking does not favor electrochemical method of chloride removal from prestressed concrete structures. Pilot scale treatments showed it to be feasible and simple to treat full-sized reinforced-concrete bridge members, although difficult to conduct the treatment on concrete piers. One of the main difficulties is to predict the duration of treatment to reach the chloride levels to acceptable levels where corrosion is under control. Preliminary studies suggested a total charge of 600–1,500 A-h/m^2 with a total treatment time of 10–50 days.

It is impossible to remove all the chloride from concrete by the electrochemical method. But the level of chloride in contact with the steel is reduced by 45–95%. Field data have shown that the removal of chloride by electrochemical technique results in stopping corrosion for 8 years. It is predicted by FHWA that the electrochemical method of removal of chloride will extend the life of bridges by as much as 20 years (31). About 372,000 m^2 (4,000,000 ft^2) concrete has been treated by this method.

The costs of electrochemical removal of chloride are:

1. Treatment of bridge decks: \$53–129/m^2
2. Treatment of substructures: \$107–215/m^2
3. Treatment of very small substructures: \$269/m^2

4.9.1.11 Deicing Salts Calcium magnesium acetate (CMA) and potassium acetate (PA) have been found to be the most promising deicing agents. These salts contain 76% and 61% acetic acid, respectively. Annually, about 15.4 billion kg (17 million) tons of rock salt (sodium chloride) are used for deicing in the United States. A study conducted in 1987 showed that 910 kg (1 ton) of road salt costing only \$50 caused more than \$1450 in damages to vehicles, bridges, and the environment. CMA cost is \$1–10/kg versus \$0.04/kg of NaCl. Because of the large disparity in cost, CMA usage will be limited to critical structures sensitive to corrosion.

CMA is not only costly but also slower in action than rock salt. When applied as a solid, CMA exhibits marginal performance in light traffic freezing rain and dry and cold storm conditions. However, when CMA is applied as a solid or concentrated solution, the rate of action is similar to that of rock salt. The New York City DOT has used it on an experimental basis, a spray-on delivery of a liquid reagent for anti-icing of certain sections of the Brooklyn Bridge deck.

4.9.2 Steel Bridges

A discussion of various steel bridge coating installation and maintenance options along with the costs and expected life is as follows. In addition to the traditional coating methodologies used on steel bridges, research has identified several technologies and maintenance methods such as:

1. Zone painting.
2. The use of overcoating or maintenance repair painting techniques.
3. The selected use of metal spray coatings.

4.9.2.1 Traditional Coating System   A two to three coating system is traditionally applied over a clean, blasted system. These coating systems are the following (33):

1. Organic zinc primer, epoxy or polyurethane intermediate coat, and aliphatic polyurethane top coat.
2. Inorganic zinc silicate polymer, chemically curing epoxy or polyurethane intermediate coat, and aliphatic polyurethane top coat.
3. High-build high solids, good-wetting epoxy primer with aliphatic polyurethane topcoat
4. Three-coat waterborne acrylic
5. Three-coat, lead-free alkyd.

4.9.2.2 Zone Painting   Because of the high cost of repainting the whole bridge structure, it has become economically attractive to use zone painting. This approach is particularly attractive for larger structures and has been used on structures such as the Golden Gate and the Bay Bridges in California and others in the New York City area. These larger bridges have different exposure environments within the same structure because of their size and their location near saltwater. These bridges are toll bridges and have greater resources to focus on intermittent or periodic maintenance activities.

The vast majority of bridges in the United States are neither large nor maintained by toll authorities. The cost of full removal and repainting of even smaller structures is relatively high. Even on smaller structures, coating breakdown and corrosion is limited to areas with measurable levels of salt contamination and significant amounts of wetness.

For bridges in marine or semimarine environments, the entire structure is subject to corrosion. In the case of bridges in nonmarine environments, the corrosive areas are generally limited to expansion joints, drainage, traffic splash, and tidal areas. If these areas can be isolated and maintained by a better protection system, large expenses can be avoided in the maintenance of the bridges. Improved inspection procedures and standards will also be a positive contribution.

4.9.2.3 Overcoating Overcoating has become a more attractive option as the cost of full removal and repainting has increased. This approach limits the amount of surface preparation to the specific areas that have failed paint and corrosion. These areas are spot-primed, and one or two coats of paint are applied over the entire structure for uniformity of color. This approach can be effective in relatively less corrosive environments.

4.9.2.4 Metal Spray Coatings These are nontraditional bridge coating systems with potential long-term performance benefits. Metallized coatings appear to have excellent long-term corrosion resistance. Although the metallized coatings are expensive, the changing overall economics of bridge repainting operations has made their use more competitive in terms of life-cycle cost.

4.9.2.5 Coating Installation Maintenance Costs It is not easy to define the cost of coating system installation. The significant aspect is the dramatic increases in environmental and worker protection regulations that impact these operations. The use of containment structures to capture hazardous waste and pollutants generated during removal of old coatings and the gradual institutionalization of worker health and safety practices associated with hazardous materials removal have contributed to the high cost of these operations.

The problem of applying protective coatings to steel bridges to prevent corrosion involves the removal of the existing lead-based paint in compliance with environmental regulations. All wastes must be treated in accordance with congressional regulations.

According to the 1992 NCHRP data, nearly 80% of the bridges have been coated with lead-based paints. About 4100–$130 million is spent on painting annually. Lead paint removal generates an estimated 181 million kg (200,000 ton) of lead-contaminated abrasives.

The overall cost of coating consists of the cost of material, surface preparation, and application of the coating. The costs of some coating systems are given in Table 4.18.

The service life of the coating systems is significantly affected by service conditions. For example, a two-coat alkyd primer with the topcoat exposed to a rural or residential area would last only 3 years. The triple system of moist-cured urethane zinc-rich coat, a high-build acrylic urethane coat, and an acrylic urethane top coat will last 15 years.

TABLE 4.18   Cost of Alkyd, Epoxy, and Epoxy/Urethane Coatings

System	Cleaning cost $1m²	Material cost $1m2	Application cost $1m²	Total cost $1m²	System Life (Years)	Cost/Year $1m²
Two-coat alkyd	5.92–9.15	1.08	5.38	12.38–15.61	3–6	4.09–2.58
Two-coat epoxy	5.92–9.15	1.72	6.46	14.10–17.38	7.5–10.5	1.83–1.61
Two-coat epoxy urethane	9.15–10.76	2.26	7.00	18.41–20.02	10.5	1.94

The costs of total paint removal and repainting can range from $43/m^2 to $215.25/m^2, which may be explained by the fact that each job is unique in the sense of access for high structures or structures over water, the condition of bridge deterioration, and unusual traffic control.

An alternative to paint removal is overcoating that includes cleaning the structure, priming rusty portions, and applying intermediate coats and topcoats either on the repaired areas or over the full structure. The cost of overcoating bridges can range from $11 to $54/m^2 and the tighter standards may raise the cost to $86/m^2.

The effort to implement bridge corrosion control maintenance practices, which achieve regulatory requirements and cost efficiency, cannot be successful without the development of reliable task-based cost data for bridge painting tasks. These data depend on many factors such as local labor costs and structural factors such as accessibility to contractor costing rules.

It is estimated that nearly 50% of the cost of an average painting job is attributed to environmental protection and workers' health measures. This resulted in the increase in the cost of the coating removal job from $54.36/m^2 in 1992 to $114.10/m^2 in 1995. It is to be noted that the cost of the actual work such as surface preparation and coating materials remained relatively constant.

The estimated time to failure for several coating systems is given in Table 4.19. Table 4.20 gives the estimated costs for painting options (34, 35).

In Table 4.19, lifetime is defined as 10% degradation of the coatings. The data presented show that depending on the type of surface preparation (blasting versus surface coating) and the type of coating, the service life can vary considerably from 3 to 30 years. Similarly, Table 4.20 suggests that the longevity of coating is related to the costs of surface preparation and coating application. For example, overcoating costs $3.22/m^2 and lasts only a few years while near-white metal blasting followed by metallizing costs 10 times more and lasts for 30 years.

The data given in Table 4.20 also contains extra costs such as containment and waste-disposal-related costs and worker health and safety costs. The numbers show that these types of costs are equal to or exceed the costs of surface preparation, coating material, and coating application.

A sample cost distribution shown in Table 4.21 for a typical heavy-duty maintenance job on a steel bridge structure shows that only a small portion of the total job cost is attributed to paint and its application (34).

TABLE 4.19   Coating System Time-to-Failure Estimates in Marine Environment

Coating System	Estimated Life
Ethyl silicate inorganic zinc/epoxy polyamide Aliphatic urethane over SP-10 near-white metal blast	15 years
Epoxy mastic/aliphatic urethane over Sp-10 near-white metal blast	10 years
Epoxymastic/aliphatic urethane overcoat over existing paint and SP-3	4 years
85% zinc and 15% aluminum metallizing over SP-10, near-white metal blast	30 years
Low-VOC alkyd three-coat system overcoat over existing paint and SP-3	3 years

TABLE 4.20　Estimated Costs for Painting Options

Category	Type	Estimated Cost ($/m²)
Surface preparation	SP-10 near-white metal blast	13.45
	SP-3 power-tool cleaning	6.46
Coating application	Three-coat full painting	13.45
	Overcoating	3.23
	Metallizing	26.91
Coating material	IOZ/epoxy/urethane	5.27
	Epoxymastic/urethane	4.52
	Metallizing	16.15
	Moisture-cured urethane	2.69
	Three-coat alkyd	2.05
Other job costs	Containment and air filtration system SP-3	5.38
	Containment and air filtration SP-10	21.53
	Inspection SP-3 only	5.38
	Inspection SP-10 only	10.76
	Rigging	5.38
	Mobilization	5.38
	Hazardous waste storage and disposal	10.76
	Hazardous waste storage and disposal SP-10	26.91
	Workers health and safety SP-3	10.76
	Workers health and safety SP-10	21.53

TABLE 4.21　Cost Distribution of Coating Application on Steel Bridge Structure (34)

Description of Job	Percent Cost
Coating material	4.0
Access costs	28.0
Containment	19.0
Worker health	15.0
Coating application	10.0
Surface preparation	10.0
Environment monitoring	9.0
Waste disposal	5.0

Examples of severe corrosion resulting in deficient bridges are shown in Figure 4.4. Examples of bridge deck corrosion are shown in Figure 4.5. Corrosion in the free length of tendon is shown in Figure 4.6a and that of failed strands in Figure 4.6b.

The annual direct cost of corrosion may be divided into: (i) cost to replace structurally deficient bridges, and (ii) corrosion-associated life-cycle cost for remaining (nondeficient) bridges, including the cost of construction, routine maintenance, patching, and rehabilitation.

The annual cost of structurally deficient bridges is estimated as the cost to replace these bridges over a 10-year period. It is calculated using 23.9 billion as a present

Figure 4.4   Severe corrosion resulting in deficient bridges (6).

value of the cost at 5% annual percentage rate, assuming annual payments for the replacement cost, that is, the annual cost to replace structurally deficient bridges (both steel and reinforced concrete) over next 10 years is 3.79 billion/year. This cost is for the current number of deficient bridges. The total cost may be greater than the value given here.

There are 543,019 concrete and steel bridges of which 78,448 are structurally deficient, leaving 464,571 bridges to be maintained for estimating purposes; it is assumed that all these bridges have a conventionally reinforced concrete deck. The annualized life-cycle direct cost of original construction, routine maintenance, patching and rehabilitation for a black steel rebar deck costs between $18,000 and $22,000. These costs are both corrosion- and non-corrosion related.

The cost of a corrosion-free scenario used the same as the scenario of a bridge deck with corrosion with the assumptions:

Figure 4.5   Examples of bridge deck corrosion (6).

(a)   (b)

Figure 4.6   (a) Corrosion in the free length of tendon (6). (b) Failed strands.

1. Annual routine maintenance is the same as for the deck with corrosion.
2. No patching is required.
3. An overlay is required for improved skid resistance at 50 and 85 years (overlay life of 35 years) giving a bridge life of 120 years.
4. Deck is removed after 120 years

The annual cost of a "corrosion-free" bridge deck amounts to $15,700. The annual cost of corrosion for an average bridge deck is the difference between the annual cost of corrosion ($22,000) and that of a corrosion-free deck ($15,700) or $6300–2300 ($18,000–15,700). The total estimated cost of corrosion for bridge decks is $2.93–$1.07 billion.

The differences in the two maintenance scenarios that resulted in the range of corrosion-related costs were experience-based maintenance and information-based maintenance with crack repair. This difference represents the range of maintenance from minimal practice to best practice. The cost analysis estimated the cost of corrosion from $6300 (minimal practice) to $2300 per deck per year (best practice). These values show that a saving of 63% of the corrosion is possible by adopting the best practice although the actual bridge maintenance is between the minimal and best practice.

Assuming an average of the maintenance of 46% savings or $2000 per bridge per year can be achieved by improving maintenance practice. These savings were calculated for black steel rebar decks for which improved maintenance can still provide savings. However, corrosion of many black steel rebar decks has progressed to the extent that improved maintenance will not make a significant difference. For these decks, other rehabilitation options must be considered such as CP, overlays, or electrochemical chloride removal. The savings of $2000 per bridge per year translates into $0.93 billion for all the bridges per year.

The area of the substructure and superstructure (minus deck) was estimated to be similar to the deck surface area of an average bridge. Estimation of the cost of substructures and superstructures without deck consisted of:

1. Repair and maintenance for the substructure/superstructure cost significantly more per surface area than the deck.
2. In nonmarine applications, the percent of surface area deteriorated because of corrosion of reinforcing steel is much less and often is limited to area beneath expansion joints and drains, which are exposed to deicing salt runoff.
3. Corrosion problems are hence more prevalent on substructures than decks in severe marine environments.

On the basis of these considerations, the cost of corrosion of substructures and superstructures without deck is estimated to be 2.93–1.07 billion dollars.

In the case of steel bridges, there is an additional cost for maintenance painting, and expenditure for painting is estimated at $0.5 billion per year. The total annual direct cost of corrosion of bridges: estimated to be $10.15 billion to $6.43 billion itemized as

$3.79 billion to replace structurally deficient bridges over 10 years; $2.93–1.07 billion for maintenance and cost of capital for concrete bridge decks and $2.93–1.07 billion for maintenance and cost of capital for concrete substructures and superstructures; $0.5 billion for maintenance painting cost for steel bridge. The average annual cost of corrosion of bridges amounts to $8.29 billion.

The cost of corrosion can be greater in the case of historically significant bridges and bridges that are critical to traffic flows. In addition, problems in posttensioned bridges or cable and suspension bridges can be very costly to repair. Although the direct costs presented are estimated costs based on broad assumptions, the calculated cost represents the relative cost of corrosion for the highway bridge industry sector. Life-cycle analysis estimates indirect costs such as traffic delays and lost productivity at ten times the direct cost of corrosion.

4.10 MITIGATING CORROSION OF REINFORCING STEEL IN UNDERWATER TUNNELS (36)

Corrosion of the outermost rebar may be mitigated by installing CP; however, a conventional CP system with anodes installed on the concrete surfaces likely will not be effective because of a dry environment at the anode-concrete interface that does not facilitate current flow caused by electrochemical osmosis. Because the interior of the tunnel is not exposed to rain or splashing sea water, the moisture available would come from the air. Water from inside the concrete would not reach the surface anodes because the surface cracks were sealed.

To address this type of corrosion mitigation problem, a specialized CP system was developed that can minimize the drying effect at the anode/concrete interface without direct water exposure. By creating a low-resistance anode interface using a semiconductive layer the effect of electrochemical osmosis at the anode/concrete interface can be minimized, so that the CP system is not dependent on water or moisture from the outside environment and can also operate at low voltage. In addition, the semiconductive layer would minimize acid generation if the anode comes into direct contact with saltwater present because of new water leakages.

4.11 CORROSION OF UNDERGROUND GAS AND LIQUID TRANSMISSION PIPELINES

Corrosion of the pipe wall can occur either internally or externally. Internal corrosion occurs when corrosive liquids or condensates are transported through the pipelines. Depending on the nature of the corrosive liquid and the transport velocity, different forms of corrosion may occur, including uniform corrosion, pitting/crevice corrosion, and erosion–corrosion.

There are several different modes of external corrosion identified in buried pipelines. The primary mode of corrosion is a macrocell form of localized corrosion because of the heterogeneous nature of soils, local damage of the external coatings

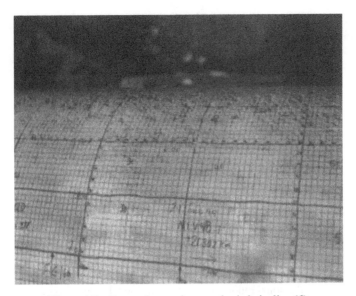

Figure 4.7   External corrosion on a buried pipeline (6).

(holidays), and/or the disbondment of external coatings. Figure 4.7 shows typical external corrosion on a buried pipeline. The 25 mm (1 in.) grid pattern was placed on the pipe surface to permit sizing of the corrosion and nondestructive evaluation (NDE) wall thickness measurements.

4.11.1   Stray Current Corrosion

Corrosion can be accelerated through ground currents from dc sources. Electrified railroads, mining operations, and other similar industries that utilize large amounts of dc current sometimes allow a significant portion of current to use a ground path return to their power sources. These currents often utilize metallic structures such as pipelines in close proximity as part of the return path. This "stray" current can be picked up by the pipeline and discharged back into the soil at some distance down the pipeline close to the current return. Current pick-up on the pipe is the same phenomenon as CP, which tends to mitigate corrosion. The process of current discharge in the pipe and through the soil of a dc current accelerates corrosion of the pipe wall at the discharge point. This type of corrosion is termed as stray current corrosion (Fig. 4.8).

4.11.2   Microbiologically Influenced Corrosion (MIC)

This form of corrosion is defined as corrosion influenced by the presence and activities of microorganisms including bacteria and fungi. About 20–30% of all corrosion on pipelines is MIC related. MIC can affect either the external or internal surfaces of a pipeline. Microorganisms located on the metal surface do not directly attack

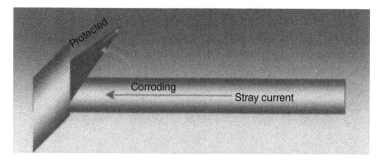

Figure 4.8   Schematic of stray current corrosion (6).

the metal or cause a unique form of corrosion. The by-products from the organisms promote several forms of corrosion such as pitting, crevice corrosion, and under-deposit corrosion. The products of a growing microbiological colony accelerate the corrosion process by either (i) interacting with corrosion products to prevent natural film-forming properties of the corrosion products that would inhibit further corrosion, or (ii) provide an additional reduction reaction that accelerates the corrosion process.

Both aerobic and anaerobic bacteria are implicated in exacerbating corrosion of underground pipelines. Obligate aerobic bacteria can survive only in the presence of oxygen while obligate anaerobic bacteria can survive only in the absence of oxygen. Facultative aerobic bacteria prefer aerobic conditions but can live under anaerobic conditions. Aerobic bacteria include metal-oxidizing bacteria, while acid-producing bacteria are facultative aerobes.

The most aggressive corrosive attacks occur in the presence of microbial com-munities that contain a variety of bacteria. In these communities, the bacteria act cooperatively to produce favorable conditions for the growth of each species. For example, obligate anaerobic bacteria can thrive in aerobic environments when present beneath biofilms/deposits in which aerobic bacteria consume the oxygen. In the case of underground pipelines, the severe attack has been associated with acid-producing bacteria in such bacterial communities (Fig. 4.9).

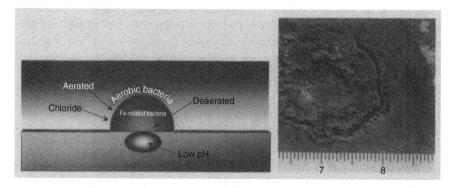

Figure 4.9   Iron-related bacteria reacting with chloride ions to create acidic environment (6).

4.11.3 Mitigation of External Corrosion

Corrosion is an electrochemical phenomenon and hence can be controlled by altering the electrochemical condition of the corroding interface. For external wall surfaces altering the nature of the corroding surface is simply done by altering the voltage field around the pipe. By applying a negative potential and making the pipe a cathode, the rate of corrosion (oxidation) is reduced and the reduction process accelerated. This means of mitigating corrosion is known as CP. CP may be achieved by one of two primary systems such as sacrificial anode (galvanic anode) CP and impressed-current CP. Sacrificial anode CP uses an anode material that is electronegative to the pipe steel. When connected to the pipe, the pipe becomes the cathode in the circuit and corrosion is mitigated. Typical anode materials for underground pipelines are zinc and magnesium.

Impressed-current CP uses an outside power supply such as a rectifier to control the voltage between the pipe and an anode (cast iron, graphite, platinum clad, mixed metal oxide) in such a manner that the pipe becomes the cathode in the circuit and corrosion is mitigated.

CP is often used in conjunction with a coating. There are always flaws in the coatings, because of application inconsistencies, construction damage, or a combination of natural aging and soil stresses. If left unprotected, corrosion will occur at these coating flaws (holidays). Often, the rate of attack through the wall is much higher at the holiday than the general attack of a bare metal surface. The use of a coating greatly reduces the total amount of current required to achieve protection of the pipeline system; therefore, CP and external coatings are used together wherever possible.

CP can be used to mitigate all forms of corrosion such as general, stray current, MIC, and SCC. Sometimes it is difficult to determine the level of CP necessary to mitigate the different corrosion mechanisms and to identify which type of corrosion mechanism is present. Stress corrosion cracking (SCC) presents additional problems. The high pH form of SCC is found only on pipelines protected with CP. The products that result from cathodic reactions occurring on the pipe surface during CP in conjunction with soil chemistry produce the environment necessary for high-pH SCC. As high-pH SCC propagates only in a very limited potential range, maintaining the potential of the pipe surface outside of this range by proper CP control will prevent the growth of the high-pH SCC cracks. In addition, it has been established that proper CP control can inhibit the growth of near-neutral SCC cracks.

Internal corrosion is also an electrochemical process; however, CP is not a viable option for mitigating internal corrosion in a pipeline. One of the first defense systems against corrosion of transmission pipelines is to ensure that the product being transported is free of moisture. Dry, deaerated natural gas and moisture-free oil and petroleum products are not corrosive. In order for corrosion to occur, there must be moisture, CO_2, oxygen, or some other reductant, such as the one produced by microbes. Operators generally control moisture, oxygen, and CO_2 contents of the transported product, but these constituents can enter the pipeline through compressor or pump stations, metering stations, storage facilities, or other means. Gathering lines in production fields have a much more significant problem with internal corrosion than the typical transmission pipeline.

One of the options available for mitigating internal corrosion is chemical treatment of the product being transported. Chemical corrosion inhibitors for mitigation of corrosion and biocides to prevent microbiological activity and hence microbial enhanced corrosion may be used. Both of these methods have been found to be effective in either natural gas or liquid pipelines. The cost of either the chemical corrosion inhibitor or biocide is significant. It should be noted that very large volumes of products are continuously flowing through the pipeline. To mitigate corrosion through the chemical treatment requires continuous injection or regular batch treatment of the inhibitor or biocide.

4.11.4 Operations and Maintenance

Significant maintenance costs for pipeline operation are associated with corrosion control and integrity management. The driving forces for the expenditure of maintenance dollars are to preserve the asset of pipeline, which is equal to $93.3 billion in book value and $541 billion in replacement value and to ensure safe operation without failures that jeopardize public safety, result in lost product and throughput, and cause property and environmental damage that is estimated at $470–870 million/year.

A survey of major pipeline companies indicated that the primary cause for the loss of corrosion protection was coating deterioration (30%) and inadequate CP current (20%). Other contributing causes were shorts or contacts of about 12% and about 7% because of stray current. Most of the general maintenance consisted of monitoring and fixing these problems. Integrity management concerns are focused on conditions assessment, mitigation of corrosion, life assessment, and risk modeling.

The use of coatings in conjunction with CP is the most popular form of corrosion protection of pipelines. Some of the coatings used are fusion-bonded epoxy, extruded polyethylene, coal tar enamel, liquid epoxy, tape, polyurethane, mastic, and wax. Pipelines with each of these coatings remain in service at the present. The most widely used coating on pipelines is fusion-bonded epoxy. New multilayered coatings are now on the market.

Coatings have been specified for all new pipelines since the 1960s. The average cost of coating pipe for new construction is estimated at $24,000/km and for a total length of 778,900 km the total coating corrosion prevention can be estimated at $18.4 billion in replacement costs.

Nearly 30% of the operational pipeline corrosion problems are attributed to coating deterioration. A large portion of corrosion control costs is because of monitoring, identifying, and repairing coating anomalies. Coating deterioration also affects the ability and effectiveness of CP. Thus, to extend the operating life of a pipeline, pipeline rehabilitation (recoating the pipeline) is recommended.

CP is the required method of corrosion control of buried pipelines. The two forms of CP are impressed-current and sacrificial anode systems. Both forms of protection have been in use in industry for quite some time and the industrial personnel are familiar with their installation and operation (NACE Standard RPO169-96).

Impressed-current CP systems are extensively used in transmission pipelines. Impressed-current systems can be readily adjusted to compensate for changes in the amount of current required to adequately protect the structures; however, they may also contribute to the interference of other structures in the vicinity.

Depending on soil type pipe coating properties and pipe size, impressed-current CP systems can be used to protect long pipelines. However, impressed-current CP systems require more expensive installation and equipment, increased monthly monitoring, and greater power consumption charges than that of sacrificial anode systems. It is estimated that there are between 48,000 and 97,000 CP rectifiers in operation at present:, the corresponding pipe length of 778,900 km (484,000 miles) with rectifiers every 8–16 km (5–10 miles) with an average installation cost of $12,000 per rectifier/ground bed. The total investment in impressed-current CP system is between $0.6 billion and $1.2 billion. It is estimated that the annual investment by pipeline companies in impressed-current CP systems consisting of new installations and replacement of existing system is $40 million.

Sacrificial anode systems are widely used to protect gas distribution pipelines but are applied as a remedial measure for problem areas on transmission pipelines. Sacrificial anodes are relatively inexpensive, do not require external power supply and require no regular monitoring of the anodes because of their low driving voltages, sacrificial anodes are not applicable in all environments and do not have the power to protect long pipelines. Sacrificial anodes are often used to complement impressed-current CP systems by providing protection to local areas where additional protection is required because of inadequate coating quality. It is estimated that $30 million of sacrificial anode material (zinc and magnesium) are purchased by the pipeline industry each year. Assuming 30% of the sales is for transmission pipelines, the annual cost is $9 million. The major portion of $21 million may be assigned to distribution pipelines.

Internal pipeline corrosion can be mitigated through various measures such as dewatering, inhibition, cleaning (pigging), and internal pipeline coatings. Dewatering consists of removal of corrosive fluids prior to their entry into the pipeline. Dewatering components are generally located at pipeline compressor and pump stations. In other cases, specific low points are selected along the pipeline right-of-way for the installation of "drips" that allow the collection of corrosive fluids periodically and removed from the line to prevent corrosion of the pipe downstream from the site.

Another method of mitigating internal corrosion of pipelines is by periodic cleaning of the line. This is known as "pigging" and consists of cleaning and scraping pigs. Pigging involves insertion of different "pigs" into the line and propelling it through the line with gas or another product. As the pig passes through the line, it pushes and/or scrapes fluids, waxes, and debris from the line. These cleaning operations can also make use of various cleaning media, including solvents, biocides, acids, and detergents to aid in cleaning effectiveness. Corrosion is reduced by the elimination of the corrosive environment from the line. Costs of pigging process include the cost of preparing the line for pigging (installation of pig launchers and receivers, removal of appurtenances that could cause the pig to become lodged), possible reduced throughputs during the pigging operation, cost of pigs, solvents, and the cost of disposal of the material from the pipe.

Rehabilitation of internally corroded pipelines is somewhat more difficult to manage than external corrosion. Internally corroded pipes require cutting out and replacing the affected sections of the pipeline. Other methods of internal rehabilitation consist of pulled liners and epoxy flood coating. Cost estimates for these options can vary greatly and predominantly dependent on the extent of cleaning required to prepare the internal surface for coating.

4.11.5 Cost of Operation and Maintenance (Corrosion Control)

The most effective way to account for all of the related operating and maintenance costs associated with corrosion is to examine the total operating and maintenance budgets for representative companies (Table 4.22).

The three specific conditions that make replacement/rehabilitation necessary are: (i) severe corrosion damage of a pipeline not cathodically protected; (ii) severe coating deterioration leading to increased CP requirements, and (iii) SCC along a large area of pipeline. Pipeline integrity management programs are used by pipeline operators to determine the locations in which corrosion defects pose a threat to safe operation. Repairs at these locations can vary from the installation of a reinforcing sleeve to the implementation of a large-scale pipe rehabilitation or replacement program. For localized corrosion flaws, the repair process can include composite sleeves, full-encirclement steel sleeves, or replacement of a pipe segment. For local flaws, the repair process can be handled by company procedures and criteria. For large-scale corrosion and/or coating deterioration issues, the replacement/rehabilitation decisions must consider both operational and economic factors.

In-line inspections (ILI) are useful in obtaining a profile of defects in a pipeline. High-resolution UT and MFL ILI tools can be used to determine the geometry and orientation of corrosion defects.

These inspections can be used to determine the number and locations of near-critical flaws that should be examined by a dig program to verify and repair the flaw. Using appropriate growth models, predictions can be made on future dig/repair and/or reinspection requirements for the ULI inspected line. If the density

TABLE 4.22 Estimated Costs for Operation and Maintenance Associated with Corrosion and Corrosion Control

Company	Miles of Pipe	Total Costs	Corrosion Costs	Corrosion Cost Per Mile
A	11×10^6	359×10^6	15%	$4894
B	10×10^6	707×10^6	15%	$10605
C	5×10^6	192×10^6	15%	$5760
Average cost				$7086

of corrosion defects is high or the potential exists for continued increase in dig/repair frequency, the affected pipe section may require repair or replacement.

4.11.6   Aging Coating

Maintaining the required level of CP is of concern because of the cost. An increased number of coating defects requires an increased amount of CP current. The increase in CP current is accomplished by increasing the current output of the impressed-current rectifiers, installing impressed-current rectifiers at more locations along the pipeline or installing additional sacrificial anodes. Coating defects can be identified by conventional potential surveys or by specific coating defect surveys and verified by direct visual inspection also known as a dig program. Under certain conditions, coatings fail in a manner that makes assessment of the corrosion condition of the pipe through conventional surveying methods difficult. Aging coating and the associated increase in coating defects can make the continuous need for CP upgrading uneconomical.

4.11.7   Stress Corrosion Cracking (SCC)

This form is defined as the brittle fracture of a normally ductile metal by the conjoint action of a specific corrosive environment and a tensile stress. On underground pipelines, SCC affects only the external surface of the pipe that is exposed to soil/ground water at locations where the coating is disbonded. The primary component of the tensile stress on an underground pipeline is in the hoop direction and results from the operating pressure. Residual stresses from fabrication, installation, and damage in service contribute to the total stress. Individual cracks initiate in the longitudinal direction on the outside surface of the pipe. The cracks typically occur in colonies that may contain hundreds or thousands of individual cracks. Over a period of time, the cracks in the colonies interlink and may cause leaks or rupture once a critical-size flaw is achieved. Figure 4.10 shows an SCC hydrostatic test failure on a high-pressure gas pipeline.

The two basic types of SCC identified on underground pipelines are: high pH cracking (pH of 9–10), which propagates intergranularly, and "near-neutral pH" cracking, which propagates intergranularly. Near-neutral pH SCC (<pH 8) is most commonly found on pipelines with polyethylene tape coatings that shield the CP current. The environment that develops under the tape coating and causes this form of cracking is dilute carbonic acid. Carbon dioxide from the decay of organic material in the soil dissolves in the electrolyte beneath the disbonded coating to form the carbonic acid solution. High pH SCC is most commonly found on pipelines with asphalt or coal tar coatings. The high pH environment is a carbonate–bicarbonate solution that develops because of the presence of carbon dioxide in the groundwater and the CP system.

The presence of extensive SCC may necessitate replacement or rehabilitation of a pipeline. Because SCC is dependent on unique environmental conditions, a large-scale recoating program may protect against these environmental conditions

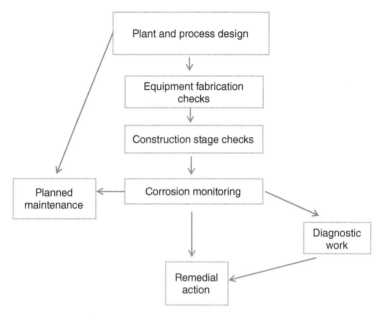

Figure 4.10   The phases of corrosion control.

and permit continued operation of the pipeline. Severe and high density of the stress corrosion cracks may necessitate pipe replacement from the economic point of view.

Some of the considerations that play a role in deciding on rehabilitation or replacement of a pipeline are terrain conditions, expected or required life, excess capacity, throughput requirements, and internal versus external corrosion (Table 4.23).

The location of a pipeline is critical to repair considerations. For instance, a pipe in swampy clay would exclude recoating as a repair option. On the contrary, pipes in prairies are conductive to recoating, with firm footing for the equipment and good accessibility.

If the expected life of a section of pipe is short, the operator must decide whether recoating and repair would extend the life of a pipe section to match the rest of the pipeline. Otherwise, recoating is not a recommended solution. Replacing the pipeline may be the best solution.

Alternatives other than rehabilitation and replacement such as abandonment of the pipe section with a bypass loop increasing the frequency of ILI enables greater accuracy in determining the point of failure for existing defects. Derating the pipe may extend the life of the pipe provided throughput requirements are met. For the replacement or repair option, it is generally necessary to have a loop system and allow for pressure restrictions or interruptible service to facilitate repairs.

It is more difficult to combat internal corrosion than external corrosion. Internal corrosion can be mitigated by having an internal coating or lining. For internal coating repair, the pipeline will have to be completely out of service. The pipeline need not be out of service for external recoating operations.

TABLE 4.23 Factors to be Considered for Pipeline Rehabilitation

Pipe	Anomalies	Burial	Operating	Labor	Material	Equipment
Size	Position	Depth	Interruptibility	Contractor versus internal	Pipe	Digging equipment
Span	Size	Location	Ability to lower pressure	Bidding versus time and material	Coating	Nondigging equipment
Grade	Quantity	Soil condition	Cathodic protection	Employee skill level	Sleeves	Specialty items
Wall thickness	Profile	Drainage conditions	Welding issues	Job limitations	Fittings	Transportation
Operating pressure	Concentration	Season	Regulations	Availability	Cathodic protection	
Availability	Wall loss	Other facilities	Company standards	Location	Specialty repair items	
Company specifications	Cause	Environmental issues Legal issues	Union requirements Benefits	Site restoration Availability		

4.12   GAS DISTRIBUTION

The gas distribution pipeline sector is part of the oil and gas industry. The different components of natural gas production, transmission, storage, and distribution systems are production wells, gathering lines within the production fields, processing plants, transmission pipelines, compressor stations (periodically along the transmission pipelines), storage wells and associated gathering pipelines, metering stations and city gate at distribution centers, distribution piping, and meters at residential or industrial sites. A schematic representation of natural gas production transmission and distribution is as follows:

Several different materials have been used for main and service distribution piping. Summary of miles of gas distribution main and number of services by material are noted in Table 4.24.

A large percentage (57%) of mains and services (46%) is metal (steel, cast iron or copper), and corrosion is a major issue. For distribution pipe, external corrosion is of primary importance, although internal corrosion has been noted in some cases. The methods of monitoring corrosion on cathodically protected pipe are similar to those in the transmission pipeline sector, including pipe-to-soil potential and coating surveys. One difference is that in distribution systems, leak detection is an acceptable method of monitoring for these pipelines without CP (nearly 15% of the steel mains).

For gas distribution piping, corrosion mitigation is primarily sacrificial CP. Techniques such as in-line inspection are typically not an option for the relatively complex network of distribution mains and services, which makes integrity assessment of the piping difficult if not impossible.

There are two different costs, namely, funding for corrosion control in maintaining the existing piping system and the cost for replacing the infrastructure. The average cost of main replacement (1993 dollars) ranged from $328/m in urban areas to $482/m in developed areas. The average cost of a service replacement was $950 per service.

TABLE 4.24   Summary of Miles of Gas Distribution Main and Number of Services by Material

Material	Total Miles of Main by Material	Total Number of Services
Steel	583,711	23,814,222
Cast iron	46,023	–
Copper	–	1,497,638
Plastic PVC	21,526	1,198,017
Plastic polyethylene	439,907	27,308,110
Other	8,041	1,151,019
Total by size	1,099,208	

4.12.1 Pipe Failures

Low-pressure gas distribution pipeline failures result in leaks rather than catastrophic ruptures. The main concern is that a leak goes undetected and the gas collects in a confined space, eventually igniting and causing an explosion.

4.12.2 Plastic Pipe

It is sometimes suggested that plastic pipe is safer than steel pipe because of corrosion of steel pipe. The aging or degradation of plastics may play an important role in plastic pipe failures. The degradation processes that lead to plastic pipe failures in operation are not well documented. The vulnerability of older plastic gas distribution pipe to brittle-like cracking has brought to light that plastic pipe is susceptible to certain aging and degradation processes. The brittle-like cracking in plastic pipe relates to a part-through crack initiation in the pipe wall followed by stable crack growth at stress levels much lower than yield stress, resulting in a very tight slit-like opening and gas leak. Although significant cracking may occur at points of stress concentration and near improperly designed or installed fittings, small brittle-like cracks may be difficult to detect until a significant amount of gas leaks out of the pipe and potentially migrates into enclosed spaces such as a basement. Premature brittle-like cracking requires relatively high localized stress intensification that may result from geometric discontinuities, excessive bending, improper fitting assemblies, and/or dents and gouges. The older polyethylene piping manufactured from the 1960s through the early 1980s may fail at lower stresses and after less time than was originally projected.

4.13 WATERWAYS AND PORTS

The United States has more than 7750 commercial water terminals, 192 commercially active lock sites with 238 chambers, and 40,000 km of inland, intracoastal, and coastal waterways and canals; 41 states, 16 state capitals, and all states east of the Mississippi River are served by commercially navigable waterways. Public and private works associated with waterways and ports have corrosion-related problems in both freshwater and seawater environments.

The reinforced concrete structures exposed to marine environment suffer premature corrosion-induced deterioration by chloride ions in seawater. Corrosion is typically found in piers and docks, bulkheads and retaining walls, mooring structures, and navigation aids.

The marine environment can have varying effects on different materials depending on the specific zones of exposure. Atmosphere, splash, tide, immersion, and subsoil have very different characteristics and therefore have different influences on corrosion.

Atmospherically exposed submerged zones typically experience the greatest corrosion. These zones are found on piers and docks (ladders, railings, cranes, and

TABLE 4.25 Total Annual Corrosion-Related Costs

Agency	Cost (Millions)
US Army Corps of Engineers (maintenance @ 5%)	70.0
US Public Ports – Corrosion-related maintenance	87.3
Corrosion-related replacements	95.0
US Coast Guard – Lighthouse maintenance	23.5
Replace steel ocean buoys	2.0
Paint buoys	5.0
Replace river buoys	2.0
Corrosion-related maintenance	8.6
Total	$293.4

steel support piles), bulkheads and retaining walls (steel sheet piling, steel-reinforced concrete elements, backside, and anchors on structures retaining dredged fill), and mooring structures and dams (steel gates, hinges, intake/discharge culverts, grates, and debris booms). Stationary navigational aids suffer from corrosion of support piles and steel-reinforced concrete pile caps. Floating steel buoys are also prone to corrosion (Table 4.25).

Typical corrosion control methods for freshwater structures include coatings for atmospherically exposed steel and corrosion allowances for submerged and splash zone steel. Dielectric coatings are normally used for structural steel above water, while galvanizing is often used for railings, ladders, gates, and gratings. Copper containing steel alloys are sometimes utilized for structural elements and sheet pile walls. These alloys, which form a tenuous oxide film in the atmosphere, provide little protection when buried or submerged. CP is occasionally used on submerged steel elements.

Marine corrosion control methods also include coatings for atmospherically exposed steel elements and a corrosion allowance for submerged and splash zone steel structures. Specialty marine dielectric coatings are normally used for structural steel above and often below water.

Although galvanizing is used for railings, ladders, gates, and gratings non-ferrous alloys provide better service in the aggressive saltwater marine conditions. Marine structures commonly use CP to control corrosion on submerged steel. CP is occasionally used on atmospherically exposed steel-reinforced concrete, in particular, in warm climates. The most cost-effective corrosion control on submerged and splash zone steel is achieved by using CP in conjunction with a heavy dielectric coating. Although corrosion allowances are often used for saltwater marine structures, they are not as helpful as in freshwater because the corrosion damage tends to be localized in the tidal zone (wet/dry cycling) and at the mud interface zone.

Although corrosion is a significant issue within waterways and ports, funding for protection against corrosion is in short supply. Out of 276 lock chambers at 230 sites, only 191 sites with 237 lock chambers were funded in 1998 toward maintenance work. Neglected structures include single-pile navigational aids left in service until failure occurs. It is estimated that a new $15,000 navigational aid is necessary and the underwater pole needs to be removed.

Structures with higher initial capital costs are more likely to be protected with coatings and CP.

In the past 50 years, waterways and ports have benefited from advances in coating systems such as metallizing, epoxies, and solids coating. The choice and development of coatings are governed by environmental regulations specifying which coatings can be exposed to water streams, such as the amount of VOC that can be used in coatings. Coatings with 100% solids have been developed, which contain no VOC. In addition to epoxy coatings, antifoulants are applied to submerged sections of the structure to prevent microbiologically induced corrosion (MIC).

Epoxy coatings cost approximately $4.7–5.3/l while antifoulants are more expensive at $11.8–21.1/l. Environmental regulations have led to decreased amount of chemicals released from industrial installations along waterways, especially corrosives such as chlorine. The materials of construction for some water structures have also changed. Piers and docks are no longer constructed with wood, but instead are constructed with steel-reinforced concrete. To improve the lifespan of the structure and prevent corrosion of reinforcing steel, fusion-bonded epoxy-coated reinforcement or corrosion-inhibiting admixtures are sometimes utilized in the concrete mix.

4.14 HAZARDOUS MATERIALS STORAGE

Hazardous materials consist of the following:

1. Explosives
2. Flammable and compressed gases
3. Flammable liquids
4. Flammable solids
5. Oxidizers
6. Poisonous materials
7. Radioactive materials
8. Corrosive materials
9. Miscellaneous materials.

A significant portion of hazardous materials (HAZMAT) concerns petroleum and petroleum products. The petroleum industry processes 65% of the energy that Americans consume. These include transportation fuels, home heating oil, industrial fuels, petrochemicals used in the manufacture of countless consumer products. Small quantities of corrosive materials are stored in corrosion-resistant drums or containers.

4.14.1 Nuclear Waste Storage

Nuclear wastes are generated from spent nuclear fuel from electric power plants, dismantled weapons, and products such as radio pharmaceuticals. The most important design item for the safe storage of nuclear waste is the effective shielding of radiation. To reduce the probability of nuclear radiation exposure, special packaging is designed to meet the protection standards for temporary dry or wet storage or for permanent underground storage. The most common materials of construction of storage containers are steel and concrete. The wall thickness of the packaging is generally thick in comparison to the contained volume.

Corrosion is a form of material degradation that results when moisture or water comes into contact with the packaging materials. A corrosion failure may not result in a large release of nuclear waste and radiation; however, a leak would be considered potentially hazardous and therefore unacceptable. At present, nuclear waste is stored at temporary locations, including water basins in nuclear power plants and at dry locations aboveground. Deep underground storage in Yucca Mountain, Nevada, has been proposed as a permanent storage solution.

The vast majority of nuclear shipments are very small in size (0.45 kg or 1 lb) and total 2.8 million shipments per year. Spent fuel shipments weigh 0.5–1.0 metric tons for truck shipments and up to 10 metric tons for rail shipments. In addition, protective lead shipping casks for containment of the spent fuel weigh many more additional tons. Corrosion is not a problem in the transportation of nuclear waste because of the stringent package requirements and short duration of transport. However, corrosion is an important problem in the design of casks used for permanent storage.

Table 4.26 indicates the volume of low-level waste received in US disposal facilities.

In 2000, interim storage facilities for nuclear waste were numerous. Low-level waste can be solid or liquid. It is stored "dry" aboveground or relatively shallow underground. At present, there are a total of 250,000 m³ of buried low-level waste and 106,000 m³ stored aboveground in the US Department of Energy (DOE) facilities. The cost of dry storage is reported to be $1.2 million per cask.

High-level waste consisting of spent nuclear fuel from nuclear power plants is generally stored in water basins at the location of the nuclear plants. Nearly 30,000 metric tons of nuclear waste is stored at the reactor sites. Both dry and wet basin storage are meant to be temporary solutions. A long-term storage repository is being considered.

Two concrete basins were built in 1951 for the temporary storage of nuclear fuel produced at DOE's Hanford site. Storage of the nuclear waste at this site was planned

TABLE 4.26 Volume of Low-Level Waste Received in the US Disposal Facilities

Year	Volume of Waste (Cubic Meters) × 1000
1985	75.4
1986	51.1
1987	52.2
1988	40.4
1989	46.1
1990	32.4
1991	38.8
1992	49.4
1993	22.1
1994	24.3

to terminate in 20 years. The two basins are still being used for storage of nuclear waste. It is reported that the fuel rods in open canisters corroded with the subsequent release of isotopes into the water. Basin cleanup waste removal and groundwater contamination were reported.

4.14.1.1 Cask Design for Permanent Storage of Nuclear Waste Aging because of exposure to radiation from radioactive elements and aging because of exposure to corrosion are of concern in long-term storage of nuclear waste. The heat generated by radiation can also increase the corrosion rate. Several cask designs with different materials of construction such as carbon steel, stainless steel, and concrete construction have been advanced.

At present, all the nuclear waste generated is solid waste. This waste is noncorrosive, and the risk for internal corrosion damage is minimal. The considerable amounts of liquid nuclear waste can cause internal corrosion problems. It is also possible that any moisture in solid waste can also lead to corrosion. The tanks, although buried, are exposed to groundwater and are prone to external corrosion. The consequences of leaks in nuclear waste storage tanks or container are numerous: damage to environment, loss of public trust, and exposure to radiation. Thus it is fair to state that long-term extrapolations must be made to ensure the structural integrity of the nuclear waste storage containers for centuries to come.

The scientific activity with respect to corrosion of permanent storage containers continues as evidenced by: a literature review and a summary of plutonium oxide and metal storage packing failures (37). Metal oxidation in nonairtight packages with gas pressurization was identified as the most common mechanism of packaging failure. Another corrosion problem was observed in hydrogen/oxygen recombination and generation of plutonium storage environment. The service life of steel in concrete for the storage of low-level waste has been predicted (38).

The US Nuclear Regulatory Commission and the Department of Environment (DOE) collaborated on the needed research into the effects of corrosion processes on the lifetime of the containers. The designers are facing a wide range of problems with respect to material issues. In nuclear waste containers, corrosion can occur in both inside and outside the containers. The issues that require attention are general and localized corrosion of the waste package outer barrier; methods for corrosion rate measurements; documentation on materials such as alloy 22 and titanium; the influence of silica deposition on the corrosion of metal surfaces; stability of passive film on welded and aged material; electrochemical potentials; microbiologically influenced corrosion (MIC); stress distribution because of laser peening and induction annealing; SCC and its influence on rock fall impact strength; dead load stressing and the effects of fabrication sequence and of welding.

4.14.1.2 Effect of Location Site on Corrosion of Nuclear Waste Storage Containers The plan consists of a permanent nuclear depository to build at a relatively dry site with a depth of several hundred meters below the earth's surface. It is envisaged that the presence of water will eventually corrode the storage containers. In the United States, the Yucca Mountain site is reported to be a good location because of its low water content. The proposed design for waste disposal is for steel canisters containing the nuclear waste to be stored within other steel canisters and buried horizontally in chambers 300 m below the earth's surface. The canisters are designed to last 1000 years and will depend on the mountain itself to provide a natural barrier to survive the minimum 10,000 years required by the government. However, there is no guarantee that the canisters at Yucca Mountain will be free from water flow for 10,000 years.

The total repository costs for radioactive waste in Yucca Mountain by construction phase and the cost of nuclear waste packaging fabrication for permanent storage have been documented (39) by the US DOE.

4.15 CORROSION CONTROL OF STORAGE TANKS

Corrosion of tank bottoms, walls, roofs and roof structures can pose dangers to their structural integrity. Corrosion may cause leaks that result in loss of product or pollution of the soil and water around a tank. Leaks can result in water penetrating into the tank and contaminating the product.

Corrosion control and prevention can take many forms. It may take the form of a design detail, such as the application of corrosion allowance to sophisticated lining systems and CP devices. The following most common methods of corrosion control and prevention have been listed by Myers (40).

1. Linings ("bladder") coatings (Paints)
2. Corrosion allowances
3. Design (avoidance of dissimilar metals, galvanic couples, improper materials, high fluid velocities in inappropriate places, caulking or seal welding of areas prone to crevice corrosion, roof design, etc.)
4. Sacrificial anodic systems
5. Impressed-current CP
6. Use of high-alloy (corrosion-resistant) materials. Tanks designed for materials that produce corrosive vapors often include roof and roof support structures (pontoons for floating roofs) that are made of corrosion-resistant materials. Petroleum tanks that are subject to a contaminated water layer are internally coated and cathodically protected on the bottom and partially along the wall. The external bottom corrosion of the site-fabricated tanks (4 m in diameter) can be controlled with a combination of select sand/concrete foundation pads, impervious liners, and CP. The following is a list of corrosion control methods for aboveground storage tanks. A list of corrosion control methods for aboveground storage tanks (AST) is given in Table 4.27.

4.15.1 Aboveground Storage Tanks – Internal Coatings

Internal coatings protect the structural integrity of the tank by preventing internal corrosion. These coatings have a design life of 10 years or more for larger tanks.

TABLE 4.27 Corrosion Control Methods for Aboveground Storage Tanks

Corrosion Mode	Control Method
Uniform corrosion	Inhibitors; protective coating; cathodic protection
Intergranular corrosion	Avoiding temperatures that cause contaminant precipitation during heat treatment or welding
Pitting corrosion	Protective coating; allowance for corrosion in wall thickness
Stress corrosion	Reducing residual or applied stress; redistributing stresses; avoiding misalignment of sections joined by bolts, rivets, or welds; using materials of similar expansion coefficients in one structure; protective coating; cathodic protection.
Corrosion fatigue	Minimizing cyclic stresses and vibrations; reinforcing critical areas; redistributing stresses; avoiding rapid changes in load, temperature, or pressure. Inducing compressive stresses through peening, swagging, rolling, vapor blasting, chain tumbling
Galvanic corrosion	Avoiding galvanic couples; completely insulating dissimilar metals; using filer rods of same chemical composition as metal surface during welding; avoiding unfavorable area relationships; using favorable area relationships; cathodic protection; inhibitors
Thermogalvanic corrosion	Avoiding nonuniform heating and cooling; maintaining uniform coating or insulation thickness
Crevice corrosion concentration cells	Avoiding sharp corners and stagnant areas; minimizing crevices in heat transfer areas and in aqueous media containing dissolved oxygen or inorganic solutions; enveloping or sealing crevices; protective coating; removing dirt and mill scale during clearing surface
Erosion impingement attack	Decreasing flow velocity to laminar flow; minimizing abrupt changes in flow direction; streamline flowing; installing replaceable impingement plates at critical points in flow line; using filters and steam traps to remove solids and steam protective coatings; cathodic protection
Cavitation damage	Maintaining pressure above liquid vapor pressure; minimizing hydrodynamic pressure differences; protective coating; cathodic protection; injecting or generating larger bubbles
Fretting corrosion	Installing barriers that allow for slip between metals; increasing load to stop motion, but not above load capacity; porous protective coating; lubricant
Hydrogen embrittlement	Low-hydrogen welding electrodes; avoiding incorrect pickling, surface preparation, and treatment methods; inducing compressive stresses; baking metal at 93–148 °C to remove hydrogen; impervious coating such as rubber or plastic
Stray-current corrosion	Providing good lubrication on electric cables and components; grounding exposed components or electrical equipment; draining off stray currents with another conductor; electrically bonding metallic structures; cathodic protection
Different environment cells	Protective coating; cathodic protection; avoiding partially buried structures; underlaying and backfilling underground pipelines and tanks with the same material

A coating is selected on the basis of the location within the tank: bottom, water layer, product exposed, vapor space and roof structure. In addition, coatings are sometimes used to maintain product purity. Often the internal bottom surface must be able to withstand the abrasive effects of slurry movement caused by internal flow patterns, mixers, or inlet and outlet flows or by mechanical action, such as the movement of roof drain hoses lying on the tank bottom. A benefit of a bottom liner is that it reduces the cleaning effort when the tank is removed from service for repairs or for inspection.

4.15.2 Aboveground Storage Tanks – External Coatings

Painting the exposed external surfaces of an AST provides corrosion protection, improved appearance, and reduced evaporation loss. Selection of coating type depends on the tank operating temperature and the presence of insulation that contains minerals and salts that may cause corrosion. External coatings must be able to withstand the effects of weather, ultraviolet light, and industrial or marine atmospheres.

4.15.3 Aboveground Storage Tanks – Cathodic Protection

AST forms have a network of CP rectifiers and anodes to protect the tank bottoms. The design of CP for new or existing ASTs can be done according to the API Recommended Practice for CP of ASTs (41).

Design considerations include the proximity to other metallic structures and existing CP systems, the type of grounding, the estimated remaining service life of the tank, the type and temperature of the stored product, the amount of product stored, the cycling rates, the method of tank bottom plate construction, the type of tank foundation, the type of secondary containment, if any, and the backfill soil characteristics.

There are two types of CP: (i) sacrificial anode CP, by zinc or magnesium ribbon or ingot anodes; (ii) impressed-current CP, using perimeter, deep-buried, angle-drilled anodes or vertical, loop or string undertank anodes. Depending on the above parameters, the CP type, and the diameter of the tank, CP installation costs for an AST tank bottom may range from $10,000 to $25,000 per tank and the average cost may be $15,000.

4.15.4 Underground Storage Tanks – Corrosion Control

Corrosion control of the external surfaces of USTs (42) can be achieved with a combination of CP and dielectric coating. However, the external coating should be applied when the tank is new. A buried tank cannot be retrofitted with an external coating unless it is removed from the ground. Internal corrosion protection, when required because of contamination or corrosive products, is commonly protected with an internal liner and at times in combination with galvanic CP.

4.15.5   Underground Storage Tanks – Cathodic Protection

CP depends on the reversal of electrochemical current that occurs during corrosion process. The two CP systems used are: (i) sacrificial anode systems and (ii) impressed-current systems.

The sacrificial anode system consists of burial of anodes in the electrical proximity of the tank. The anodes are made of magnesium or aluminum, which are less noble than the steel tank. This enables the flow of current from the sacrificial anode (Al or Mg) to the cathodic steel tank. Over a period of time, the anodes are consumed and hence replaced with new anodes in order for continued corrosion protection of the tank.

The second system is based on the application of impressed current that is forced through anodes to the protected structure such as the tank by a current source of sufficient potential. Properly designed CP systems that are well maintained and operate at the correct current density are a proven method of protecting tanks from the corrosive effects of contact with corrosive soils. In addition to protection of underground tanks, CP is also useful for aboveground double-bottom tanks and for internal corrosion protection.

4.15.6   Polymer Tanks

These are used when avoiding maintenance problems. This involves use of a corrosion-resistant material. High-density polyethylene (HDPE) is commonly used for chemical storage. For storage at higher temperatures, fiberglass reinforced tanks made with vinylester or epoxy can be used. For temperatures greater than 200 °C metal storage tanks are used.

4.16   AIRPORTS

A typical airport infrastructure is relatively complex, and the components that might be subject to corrosion are the following:

1. Natural gas distribution system
2. Jet fuel storage and distribution system
3. Deicing storage and distribution system
4. Water distribution systems
5. Vehicle fueling systems
6. Natural gas feeders
7. Dry fire lines
8. Parking garages
9. Runways and runway lighting

In general, each of these infrastructure components is owned and/or operated by different organizations and companies. Given the above, airports do not have any specific corrosion-related problems that cannot be found in other sectors of the national economy, for example, corrosion of heat, ventilation and air-conditioning systems; corrosion of reinforced concrete floors in parking garages, or corrosion of buried metallic structures. Corrosion of buried metallic structures is primarily manifested in USTs or buried fuel lines transporting fuel from tank farms. Larger airports generate considerable volumes of wastewater during the deicing season and may have wastewater treatment facilities.

Some relevant references are given below:

1. US Bureau of Transportation statistics data, 1999.
2. Chapter 7, Airports, Airline Handbook, www.air-transport.org, June 2000.
3. FAA DOT/TSC ACAIS database, 1999.

4.17 RAILROADS

Published information with respect to corrosion-related issues is scarce. Corrosion-prone items are rail and steel plates for wooden ties. Barlo et al. identified the corrosion of electrified trains that covered a number of transit systems. Stray current that occurs on the electrified rails system was found to cause corrosion.

Accelerated corrosion of the insulators of the rail fasteners and wood tie spikes has been observed, and the wood tie spikes had to be replaced after 6 months instead of 25 years. Corrosion damage to other railroad-owned property such as bridges, rail yard structures occurs.

4.17.1 Corrosion of Railroad Cars

The largest costs to the railroad cars are because of corrosion of the exterior and interior of the railroad cars. External corrosion of the cars is primarily because of atmospheric exposure. Although corrosion damage is of concern, it is the appearance that takes precedence. Therefore, the car manufacturers or lessees often choose to apply an exterior paint coating. The paint systems used are direct-to-metal (DTM) epoxy or epoxy with a urethane coat. The epoxy coating also protects the railroad cars against ultraviolet radiation. Certain goods such as chemicals can be corrosive. Thus, internal corrosion of the railroad cars can be prevented by using coating systems and rubber linings for internal surfaces. Coatings and linings not only prolong the service life of the fleet, but also preclude the contamination of the transported commodity by a corroding metal surface. Corrosion prevention measures are of absolute necessity particularly when transporting corrosive goods such as chemicals. The largest segment of the freight has been coal, chemicals and allied products transported, while food and kindred products make up only 5% of the transported goods. These two groups of commodities are either corrosive or sensitive to corrosion. Nearly 130,000 of covered hopper cars are used in transporting plastic pellets, which require liners to preserve product parity. The life of a liner is 8–10 years.

Transportation of coal presents a problem because when mixed with moisture, it becomes highly acidic and corrosive to the carbon steel. A large number of cars are prone to this corrosion problem. Corrosion is likely to advance further because of the thawing sheds during the winter months in cold climates in which the cars are heated to thaw the coal. It is estimated that about 100,000 cars are used in coal service leading to severe corrosion problems.

Another corrosive is rock salt, which, when transported, causes severe corrosion that can last for 3 years. The high cost of rehabilitation of salt cars created a trend toward using unlined, covered hopper cars previously used for transporting grain for rock salt service. The cars were scrapped when the corrosion was excessive.

To tolerate the properties of cargo, in addition to using linings and coatings, certain components of the cars, such as valves, undergo an upgrade from the lower corrosion-resistant carbon steel to the higher-resistant steel grades, such as stainless steel.

Rubber linings are often used in handling strong acids such as hydrochloric and phosphoric acids. In the case of extremely aggressive nitric acid, the entire tank body is made of type 316L stainless steel.

4.18 DRINKING WATER AND SEWER SYSTEMS

The most commonly used corrosion control methods for water systems are shown in Table 4.28.

4.18.1 Corrosion Control in the Water Supply

Each water utility tries to have a sufficiently large supply of water to fulfill the needs of its customers. Rain water is the main source of ground water, while river water and lakes are the main source for surface water. Lakes and underground reservoirs are used to store large amounts of raw water for times when the water level in a river is too low.

Infrastructure in and connected to the reservoirs includes dams, water intake structures, and piping. Corrosion is generally not a very significant issue here. For example, metal dams are given a corrosion tolerance with regard to the thickness of the steel walls, allowing for metal loss because of general corrosion during the expected service life.

4.18.2 Corrosion Control in Water Treatment Facilities

The infrastructure of water treatment facilities is designed to remove contaminants from water.

A series of filtration procedures and several chemical treatments are used in cleaning the raw water to prepare it for consumption. Mixing of waters from different sources is often used as an option to change quality and reduce corrosivity. In some cases, aeration may be used in drinking water treatment. In addition to removing

TABLE 4.28   The Most commonly Used Corrosion Control Methods for Water
Systems

Components in Water System	Corrosion Control Method
Steel drums	Increased wall thickness
General water infrastructure	pH adjusters, corrosion inhibitors, alkalinity controllers, hardness controllers
Storage tanks	Cathodic protection (CP); internal coatings; external paint coatings
Ductile iron, cast iron, and steel pipes-internal corrosion	Internal linings, internal inspection
Cement-based pipe	Internal lining-cement mortar; cathodic protection (CP)
Ductile iron, cast iron, and steel pipes	External coatings
External corrosion	Corrosion coupons, test stations, corrosion data loggers
Lead pipe	Replacement with copper pipe
Copper pipe	Prevention by improved tube production
Nonferrous alloys – Fittings, fixtures, joints	Replacement with corrosion-resistant components
Sewage pipes	Increased wall thickness

hydrogen sulfide, methane, radon, iron, manganese, and VOC aeration is effective
for the removal of carbon dioxide. Carbon dioxide in turn directly affects pH and dissolved inorganic carbon, the two parameters that significantly influence the solubility
of lead and copper. Under the right water quality conditions, aeration can serve as a
potential corrosion control treatment by removing CO_2 and subsequently increasing
pH and decreasing dissolved inorganic carbon. The degree to which aeration affects
corrosion depends on such raw water quality parameters as pH, dissolved inorganic
carbon dissolved oxygen, as well as the efficiency of removal of CO_2 (43).

Chemicals used to treat raw water and improve its quality include corrosion
inhibitors, pH adjusters and alkalinity, and hardness-controlling agents. The
commonly used water treatment chemicals are soda ash, sodium bicarbonate,
sodium hydroxide (caustic soda) plus carbon dioxide, lime, alkaline media filters,
combinations such as limestone slurry, carbon dioxide, sodium hydroxide. All US
water utilities are required to always monitor the water quality by an analysis of
treated water. The samples for analysis are taken at regular time intervals and at
different locations spread out over the system.

4.18.3   Corrosion Inhibitors, pH Control, and Alkalinity Adjusters

In addition to water quality control as per the Safe Drinking Water Act (SDWA),
the application of chemicals for adjusting pH is one of the main options of internal
corrosion control. It should be noted that pH control alone is not sufficient to counter
corrosion problems. In such cases, corrosion inhibitors may be used for internal corrosion control. Corrosion inhibitors consist of chemicals that are used in small quantities
to counter corrosion. The impact of inhibitors on water quality and their effectiveness

on different materials is complex (44). The rather stringent limits concerning lead and copper and other ions in drinking water limit the use of inhibitors for corrosion control.

Corrosion inhibitors for water treatment consist of naturally occurring inhibitors. Natural inhibitors consist of naturally occurring organic compounds, dissolved silica, and phosphate. Corrosion protection of iron, zinc coatings, lead, and copper can be achieved by using naturally occurring inhibitors. The added inhibitors in small quantities produce a passivating film at anodic sites to suppress the anodic corrosion reaction or inhibit the cathodic reaction and leading to a decrease in the corrosion rate. Some of the added inhibitors are orthophosphates, molecularly dehydrated polyphosphates, bimetallic (zinc-containing) phosphates, silicates, and phosphate-silicate mixtures.

Selection of corrosion inhibitors is a complex task that depends on many factors. The cost effectiveness of the inhibitor may be obtained from the relationship:

Cost effectiveness = Relative effectiveness × dosage × price per weight. The inhibitor dosage rate depends on the local water conditions and temporal factors, such as the time of the year. It should be quantified in terms of percent corrosion inhibition and extension of useful life. Table 4.29 lists inhibitors used in potable water systems.

Table 4.30 gives the costs of chemicals used for corrosion control (47).

TABLE 4.29 Commonly Used Inhibitors in Potable Water systems (45, 46)

Inhibitor	Dosage (mg/l)	Cost ($1 kg)
Lime	10–30	0.04
Caustic soda	10–30	0.44
Soda ash	10–30	0.27
Sodium hexametaphosphate	1–4 (PO_4)	2.00 (PO_4)
Bimetallic phosphate	0.5–2 (PO_4)	3.33
Zinc orthophosphate	0.1–0.5 (Zn)	4.99 (PO_4)
Sodium silicate	4–10 (SiO_2)	0.67 (SiO_2)
Carbon dioxide	5–10	0.11
Phosphoric acid	0.5–3 (P)	1.33 ((PO_4))
Monosodium phosphate	0.5–3 (P)	2.66 (PO_4)
Orthopolyphosphate blend	0.2–1 (PO_4)	5.54 (PO_4)

TABLE 4.30 Costs of Chemicals Used for Corrosion Control (47)

Chemical	Use	Feed Rate (mg/l)	Cost Per Unit ($)
Quicklime, CaO	pH control	1–20	95/ton bulk
Hydrate lime, $Ca(OH)_2$	pH control	1–20	117/ton bulk
Caustic soda, NaOH 50% solution	pH control	1–20	300/ton bulk
Soda ash, Na_2CO_3	pH control	1040	228/ton bulk
Inorganic phosphates	Inhibitor	3	98 cwt bag
Sodium silicate	Inhibitor	2–8	8/cwt tank

Phosphates and silicate corrosion inhibitors have been used with or without pH control, to reduce the metal release and to prolong the service life of distribution systems or domestic installations. When the concentration is limited, the inhibitors may not avert localized corrosion such as pitting or the corrosion of galvanized steel, steel, cast iron, copper, or lead, sufficiently to extend the life of the system beyond 75–100 years. Corrosion inhibitors are useful when concerns about water quality deterioration have to be resolved. Unfortunately, there is no simple solution for balancing water quality, health risks, system reliability, and environmental impact.

4.18.4   Corrosion Control in Water Storage Systems

After treatment of raw water in treatment facilities, clean water may be stored in aboveground or underground tanks or underground clear wells. If left unattended, both internal and external corrosion may pose a structural risk because of loss of wall thickness. Hence the need for a periodical inspection of water tanks and towers. With periodic maintenance, water tanks may have a life of over 100 years.

4.18.4.1 Internal Corrosion of Storage Towers and Tanks   The predominant forms of internal corrosion are general corrosion, galvanic corrosion and MIC in standing water. The microbiological contaminants are regulated under the surface water treatment rule (SWTR) and the total coliform rule (TCR). Corrosion control for these types of corrosion is CP and lining or painting of the interior of the tanks. CP is usually performed on a project basis, while painting generally is performed as part of a long-term maintenance program.

4.18.4.2 External Corrosion Because of Weather Conditions   External corrosion originates from moisture, rain, and changes in weather. Generally, water tanks and towers are designed with the so-called corrosion allowance, which is an allowable rate of corrosion. The thickness of walls of the tanks is measured from time to time and the corrosion rate determined. If the tanks are originally designed to withstand the loss in thickness because of corrosion with time the tank is considered to be structurally fit for service. The common corrosion control method consists of painting the tower or storage tank. Deterioration of the appearance of water towers by external corrosion is another consideration for painting the towers.

The costs of corrosion control for water storage tanks are determined by the type of CP and the type of coatings utilized. Comparative case studies of the economics of corrosion protections systems showed that it costs large sums of money to recoat and repair interior coatings while CP would mitigate corrosion activity and prolong the necessity of coating maintenance. The corrosion control method consists of painting the towers to prevent its deterioration because of corrosion. Using economic models, Robinson determined that long-term cost benefits can be realized by the application of CP to water storage tanks (48).

TABLE 4.31 Materials Used in Transmission Water Pipes

Material	Percent
Cast iron	48
Ductile iron	19
Concrete and asbestos concrete	17
PVC	9
Steel	4
Other	2

4.18.5 Corrosion Control in Water Transmission Systems

Water is pumped from temporary storage tanks or from the treatment facilities through large-diameter transmission water pipes. The transmission water piping system contains large valves where the water amounts are measured using water meters. The materials of construction of transmission pipe are cast iron, ductile iron, prestressed concrete, asbestos concrete, PVC and welded steel piping. All of these materials except PVC contain ferrous metal components, which need corrosion protection. The different materials used in transmission are as shown in Table 4.31.

Ductile iron pipe is used extensively for drinking water and wastewater systems. The pipes are made from the manufactured sections of pipe, with a bell-and-spigot connection sealed with O-rings.

The most common failure modes of these pipes are uniform corrosion (both external and internal), graphitization, and pitting under unprotected corrosion scales. Loose tubercles may cause blockage of pipes. The corrosion control of loose particles is by the addition of corrosion inhibitors, which protect the inside pipe walls or internal lining of the pipe. Other protective linings are specialty cement mortars, epoxies, polyethylene, and polyurethane.

In some cases, coal tar has been used, but these pose health hazards and oily organic residues given off by coal tar coatings limit their use. Steel pipes are used for about 4% of the US transmission water lines. The most common corrosion control methods for external corrosion of steel pipes are coatings or coatings and CP.

4.18.5.1 Effect of Reduced Pipe Wall Thickness Significant problems occur in older transmission pipes made from cast iron and ductile iron, as the wall thickness is reduced by corrosion until a leak occurs. Problems in newer iron pipes are similar to those found in older iron pipes, but occur after shorter time periods because of decreased wall thickness. Therefore, an effective corrosion control method is the selection of thicker wall pipe to provide a larger corrosion tolerance to wall thinning. Expensive thicker wall pipe will be cost-effective because of its long life and low need for maintenance.

4.18.5.2 Corrosion of Cement-Based Pipes Nearly 17% of US transmission water lines are built from concrete and asbestos concrete materials. Internal steel reinforcement wires and bars (rebar) steel mesh and steel plates are used to provide tensile strength. Chloride ions migrate to the steel surface and cause corrosion. The corrosion products take up more volume than the original steel, causing cracking of the concrete, further accelerating corrosion.

In asbestos cement pipes, asbestos fibers are used as reinforcement for tensile strength. The main concern with these pipes is the release of asbestos fibers into the drinking water. Other effects of cement-based material degradation are increased hardness because of calcium dissolution (increased water hardness), increased pH values, increased alkalinity, and migration of aluminum into the drinking water. A common corrosion control method for concrete pipe is the application of internal protection using a cement mortar lining. The quality of the lining is measured in terms of calcium oxide leaching resistance, which is a function of the mortar density.

4.18.5.3 Cement Mortar Linings New iron and steel pipelines are commonly lined with cement mortar. The cement mortar linings are also used in rehabilitation of older ductile iron, cast-iron, and steel water pipeline networks. The linings can eliminate limited leaks of pipes and pipe connections as a result of the high resistance of cement mortar to pressure, enhance the hydraulic characteristics of the mains, and prevent further internal corrosion damage. The estimated costs for water pipe rehabilitation by cement mortar lining as a percentage of pipe replacement costs are as follows: (i) cleaning and cement mortar lining; (ii) excavation, pipe fitting, and restoration of the road surface; (iii) materials; and (iv) labor costs (Table 4.32).

External corrosion modes of transmission water piping are general or localized corrosion because of corrosive soils, galvanic corrosion through connections to other utilities and structures, microbiological corrosion, ac stray current from power lines, and dc stray current corrosion from CP systems on nearby structures. Corrosion control to mitigate these forms of corrosion consists of application of coatings and CP by installation of impressed current or sacrificial anode systems. External coatings on older water pipes are asphalt coatings and coal tar enamel coatings, while external coatings on new pipes are coal tar enamel coatings, polyethylene base coatings and fusion-bonded epoxy coatings.

4.18.5.4 CP and Coatings CP design, installation, and regular inspection of the system should be done by specialists. The pipe must be electrically continuous for the application of CP. CP can be applied on welded pipes with ease as these pipes are electrically continuous. CP is more effective when used supplementally to the

TABLE 4.32 Estimated Costs for Water Pipe Rehabilitation by Cement Mortar Lining

Pipe Diameter (cm)	Cost for Rehabilitation in Relation to Pipe Replacement (%)
8–15	39.5
20–30	41.0
50	33.2
60–80	19.4
100–120	13.0

coating system. CP system in the absence of a coating is economically unfavorable for water pipes. For prestressed concrete pipe, CP can be used to augment the protection provided by standard mortar coating in aggressive soil environments (50). Care must be taken not to overprotect the prestressing steel.

The two types of CP systems are: (i) impressed-current systems that require rectifiers necessitating periodic inspection, monitoring, and adjustment by trained operators, and (ii) sacrificial systems that require less attention. Thus a sacrificial anode system consisting of buried zinc or magnesium is generally preferred for welded steel pipe.

4.18.6 Corrosion Control in Water Distribution Systems

Distribution water pipes are of smaller diameter than larger transmission pipes and are made of ductile iron, PVC, and copper. Corrosion problems and corrosion control methods for ductile iron and the deterioration of PVC are similar in pipes of both small and large diameters.

Use of chloramines for disinfection, as a means to reduce trihalomethane, accelerates corrosion and degradation of metals and elastomers common to distribution plumbing. The oxidation effects of free and combined chlorine species on mild steel, copper, bronze, Pb/Sn solder, Sn–Sb solder, Sn–Ag solder, elastomers such as natural rubber, acrylonitrilebutadiene, styrene–butadiene, chloroprene, silicone, ethylene–propylene, fluorocarbon, and three thermoplastics were studied. The results showed with few exceptions that solutions of chloroamines produced material swelling, deeper and denser surface cracking, rapid loss in elasticity, greater loss of tensile strength than with an equivalent amount of chlorine. Only newly engineered synthetic polymers performed well in chloroamine exposure. All the tested chlorine disinfectants accelerated corrosion of copper and its alloys. Unlike elastomers, free chlorine exerted greater oxidizing effect than chloroamines. There was no significant effect on the galvanic corrosion of solder while lead-free and tin-based solders were immune to chlorine attack.

In addition to ductile iron and PVC, copper and lead are used in pipes, and brass in fixtures and connections. Lead is released because of uniform corrosion. Copper is also released because of uniform corrosion, localized-attack cold water pitting, hot water pitting, MIC, corrosion fatigue, and erosion–corrosion. Lead pipes and lead-tin solder exhibit uniform corrosion. Brass corrosion includes erosion–corrosion, impingement corrosion, dezincification, and SCC. The direct health impacts are because of increased copper, lead, and zinc concentrations in the drinking water. Mechanical problems because of corrosion include leaks from perforated pipes, rupture of pipes, and the loss of water pressure because of blockage of pipes by corrosion products.

4.18.6.1 Corrosion of Lead Pipes and Solders   Lead is generally not present in domestic water supplies such as rivers and lakes. The lead content of drinking water is generally below detection limits; however, lead can enter water because of the corrosion or wear of brass fixtures, lead pipes, or solders. Stagnant water in pipes can

TABLE 4.33   Types and Frequency of Failure of Copper Plumbing in the United States in 1983

Cause of Failure	Frequency (%)
Pitting corrosion	58
Erosion-corrosion	24
Corrosion of outer surface of tubes	7
Faulty workmanship	5
Fatigue	2
Other	4
Total	100

result in lead levels greater than an EPA level of 15 ppb. Cold water lines generally have less lead than hot water lines. Orthophosphate may be used to control the release of lead in water (51). When the lead content of the water is too high, the alternate control method is to replace the home plumbing with new (copper) pipes.

Copper has been the most common material for consumer plumbing because of its excellent characteristics such as ease of installation, low cost, and corrosion resistance. Cooper accounts for 50–90% of all tubes installed in drinking water services (52). Corrosion problems, although infrequent, can be severe for the affected consumers and systems. Failure of copper tubing by pitting, blue or green water problems, and failure to meet the US EPA levels for copper in tap water are major problems when they occur. The different causes of copper tubing failure are noted in Table 4.33 (45).

Corrosion control methods for copper corrosion are: (i) improved production techniques that yield a clean inner bore and free from carbon films, which initiate pitting. The main practice is to use iron grit for blasting to remove carbon films; this process reduces the frequency of pitting by 90%. Another technique is to preoxidize the inner bore that removes the carbon film and produces an oxide film that improves the resistance to corrosion. Chemical treatment of the water supply is also a method used to reduce the corrosive attack on copper pipes.

Nonferrous alloys are used in domestic plumbing systems such as fixtures, fittings, or joints. These alloys act as a source of lead contamination in drinking water. Other elements of concern are copper, tin, zinc, antimony, and bismuth. The corrosion mechanisms vary greatly for each different alloy system, and the local water composition has profound influence on the corrosion susceptibility of different alloys. Corrosion control methods for nonferrous alloys include preventive measures such as replacement of fixtures or a complete change of material design. Corrosion of nonferrous alloys is minimized by using industry-standard materials and workmanship in the installation of copper tubing systems.

4.18.6.2 Requirements to Conduct Corrosion Control   Chemical analysis of the water and adoption of a corrosion control strategy to reduce the lead release rates are required. Drinking water is analyzed for corrosion by: (i) weight loss, (ii) total concentration of metals, cation and anion concentrations, (iii) alkalinity, (iv) hardness, (v) pH, and (vi) chlorine concentration.

Internal condition of pipes can be assessed through visual inspection photomicrographs, weight loss, pitting potential measurements, scale analysis, and corrosion probe data. From the corrosion data, service life of the pipes can be estimated.

Metal release tests may be used to measure the corrosion rates and the accumulation of corrosion products in water flowing through a plumbing system or distribution network. It is desirable to use large loops to evaluate metal release rates to select the suitable control strategy. The two systems in use are: (i) closed-loop system and (ii) recirculation loop system. The pipe loop corrosion control system simulates closely the consumer's tap under different corrosion control strategies. When determining the lead concentrations, the protocol for sampling the water is important. Stagnation time, flushing, and the specific conditions of the installation under study influence the results to a great extent.

4.18.7 Corrosion Control in Sewage Water Systems

Sewage water is pumped through sewer water piping to a treatment plant. The US investment in sewage lines in the United States is nearly $1.8 trillion. Wastewater is collected through relatively small diameter pipes and transmitted to treatment plants through larger diameter pipes. Common materials of construction for sewage water systems are concrete piping, steel, and ductile iron piping.

The mechanism of material degradation in sewer pipes is similar to potable water systems. The internal corrosion may be more severe than in potable water because the wastewater is not clean. The winterizing treatments of roads are a source of chloride, which comes into contact with the pipe. Cement-based pipe experiences corrosion of reinforced steel. The corrosion control method consists of using thicker pipe walls, which provide for larger corrosion tolerance and a longer design life.

Table 4.34 represents a corrosion control program implementation flowchart for internal corrosion.

4.18.8 Optimized Management by Combining Corrosion Control Methods

Increased wall thickness is one way of decreasing corrosion impact. The most commonly used repair methods for water systems with corrosion damage are given below. The repair methods consist of addition of corrosion inhibitors, pH adjusters, alkalinity controllers, hardness controllers to the water, application of CP, internal coatings and linings, internal inspection, external coatings, monitoring with coupons, test stations, and corrosion data loggers.

To prevent any further problems in the cases where lead and copper release is the problem, it is advisable to consider the complete replacement of the tubes, fittings, fixtures and joints by corrosion-resistant components. The most commonly used repair methods for water systems with corrosion damage are given in Table 4.35.

TABLE 4.34   Corrosion Control Program Implementation Flowchart for Internal Corrosion

Step 1	Document extent and magnitude of corrosion	(a) Water quality impact
		(b) Piping deterioration
		(c) Environmental issues
Step 2	Determine possible causes of corrosion	(a) Water quality characteristics
		(b) Susceptible piping
		(c) Workmanship and materials
Step 3	Develop and assess corrosion control alternatives	(a) pH and alkalinity
		(b) $CaCO_3$ saturation
		(c) Inhibitors
		(d) Material replacement
Step 4	Evaluate alternatives and select corrosion control strategy	(a) Performance
		(b) Cost
		(c) Side effects
Step 5	Document findings in a report	(a) Formal documentation
		(b) Regulatory requirements
Step 6	Implement corrosion control and monitoring effectiveness	(a) Design, construction
		(b) Treatment monitoring
		(c) Distribution system monitoring

TABLE 4.35   Repair Methods for Water Systems with Corrosion Damage

Water System	Damage	Repair Method
Any system	Small corrosion area or general corrosion over large area	Evaluate structural integrity. If fit for service, then apply coating to protect metal and inspect according to schedule
	Localized corrosion	Identify root cause of corrosion. Remove materials causing corrosion. Replace damaged material
Wall of dam or storage tank	Wall thinning	Evaluate structural integrity. If necessary, reinforce wall with extra steel
Metal pipe	Small leak	Clap or sleeve around pipe or replace small pipe section
	Multiple leaks	Replace pipe section
	Large leak/rupture	Replace pipe section
	Internal corrosion	Apply cement lining. Insert PVC tubing in pipe
	External corrosion	Evaluate structural integrity. If fit for service, then apply coating to protect metal and/or apply cathodic protection to reduce corrosion rate
Cement-based pipe	Reinforcement corrosion	If localized, remove loose concrete. Reapply concrete

4.19 ELECTRIC UTILITIES

The total cost of electricity sold in the United States in 1998 was 3.24 million gigawatt hours at a cost to consumers of $218.4 billion. The electricity generation plants use fossil fuel, nuclear, hydroelectric, cogeneration, geothermal, solar, and wind energies. The major players are fossil and nuclear steam supply systems. The two types of nuclear reactors are boiling water and pressurized water reactors. Some relevant data on the costs of corrosion estimated in 1998 are as follows: nuclear facilities $1.546 billion; fossil fuel sector $1.214 billion; transmission and distribution $607 million; hydraulic and other power $66 million. The total cost of corrosion in the electrical utilities industry in 1998 is estimated at $6.889 billion/year.

Because of the complex and often corrosive environments in which the power plants operate, corrosion has been a serious problem. Corrosion continues to be a problem with electrical generators and turbines. Specifically, SCC in steam generators in PWR plants and boiler tube failures in fossil fuel plants continue to be problems.

If corrosion problems are not addressed in a timely manner, the materials will corrode to an extent that major repair and rehabilitation will be required with attendant increased costs.

It is recommended that economic corrosion control programs be developed to provide a strategic cost-effective approach. The programs should focus on: (i) implementation of corrosion control in equipment design and use of corrosion-resistant alloys; (ii) selection of proper on-line corrosion monitoring techniques; (iii) implementing corrosion maintenance programs and (iv) development of educational and training programs for corrosion control and prevention. In the United States, the Electric Power Research Institute (EPRI) could be of assistance in this regard.

By providing education and training on corrosion control and prevention to plant personnel at all levels, it should be possible to control and prevent corrosion. Table 4.36 is a summary of issues that need attention.

4.20 TELECOMMUNICATIONS

Telecommunications hardware consists of switchboards, electronics, computers, data transmitters, and receivers. Delicate electronic components must be protected from human actions and weather for smooth operation over long periods of time. Most failures of this type of equipment are caused by environmental factors. If left unprotected from moisture, corrosion of delicate small parts results in malfunction. Most of the telecommunications hardware is placed in buildings and hence are not exposed to corrosive environments. Telecommunications equipment with a longer design life such as cables, connectors, and antennas may be placed outside and buried and are exposed to soils, water, air, and moist weather conditions.

A corrosion issue is possible at telephone facilities that have backup facilities that use diesel fuel generators supplied by USTs. Leaking UST systems can contaminate groundwater supplies and can cause fires, explosions, and vapor hazards. The USTs

TABLE 4.36 Summary of Issues in Corrosion Control and Prevention

Issue	Action
Awareness of corrosion costs and possible savings	Maintaining and updating corrosion cost records and following best engineering practices to reduce costs will be helpful
Change the notion that nothing can be done about inevitable corrosion	The technical personnel can be educated on corrosion prevention and control
Advance design practices for better corrosion management	Proper selection of corrosion-resistant alloys and proper welding procedures to avoid corrosion and cracking-related failures will assist. Coatings and cathodic protection help corrosion control
Change technical practices to achieve corrosion cost savings	Corrosion research on technological needs of electrical utilities industry will improve technical practices
Change policies and management practices to obtain corrosion cost savings	Corrosion prevention strategies and methodologies must be adapted by utilities management and be implemented. EPRI may take part in developing industry standards
Advance life prediction and performance methods	Implementation of life-prediction models for fitness-for-service and risk-based assessment to ensure equipment integrity and remaining life should be carried out.
Advance technology/research development and implementation	Studying the cause of unknown types of corrosion-related failures resulting from new environmental restrictions and aging plant equipment is a must. The results must be used to develop cost effective measures. Use of online corrosion monitoring and inspection to be done
Improve education and training for corrosion control	National Association of Corrosion Engineers (NACE) provided educational courses and certification for corrosion technicians, engineers, and technologists. EPRI gives workshops for engineers on some topics. General and targeted training and courses for managers and engineering personnel will increase awareness of corrosion problems and the best possible ways to tackle them effectively

may be upgraded involving addition of spill, overfill, and corrosion protection of the UST system.

Telecommunication shelters are built with steel, aluminum, fiberglass, and concrete. Wood or concrete blocks are used for the foundations. Carbon steel shelters need to be painted for corrosion protection. Stainless steel shelters are costly but do not require painting. Aluminum may be chosen because of the favorable weight-to-strength ratio. On roof tops and other mounted structures, the dead weight of the shelter is important for structural purposes. Aluminum is generally corrosion resistant in nonmarine environments. Aluminum is costlier than carbon steel but requires no painting. Fiberglass shelters are corrosion resistant and require no painting. The largest and strongest shelters are those constructed using concrete with steel doors. Because of controlled temperature and humidity, corrosion is not a problem for the equipment in these shelters.

The cabinets for cellular telephones are made of steel or aluminum. A double system of galvanizing and painting steel cabinets is applied for corrosion protection. Surface preparation by grinding and the application of zinc chromate primer is done prior to galvanizing. There are nearly 4000 cabinets for cellular telephones in use, and the estimated cost for corrosion is approximately $4 million.

Transmission of the signal is done from antennas mounted at high places achieved by towers such as tapered steel monopoles. Hot-dipped galvanized steel has been used in making the towers. Painting of the towers is done for aesthetic reasons and not for mitigating corrosion. Guided (wire) towers form the second largest group of towers; they are made from carbon steel and are sandblasted and repainted. The continued operation of the aging guided towers is a major corrosion concern because corrosion of the steel members may affect the structural integrity of the towers.

The single major corrosion problem in the telecommunication industry is the degradation of buried grounding beds and grounding rings around towers and shelters. These copper grounding systems are prone to attack by corrosive soil. Problems occur when the electrical connection between the grounding bed and the structure are interrupted or when the corrosion advances to such an extent that the electrical resistance of the bed becomes too large. To prevent electrical disconnection between the grounding bed and the structure, the traditional mechanical connection must be replaced with CADWELD connections (American Welding Society Designation Termi Welding (TW) Process). Galvanic corrosion because of connection between two dissimilar metals is another factor related to copper ground beds.

The copper cables used for the telecommunications industry's electrical supply are encapsulated in plastic to prevent electrical shorts as well as corrosion protection of the wires. The following anticorrosion protection measures should be taken.

1. Check galvanization condition
2. Check paint condition.
3. Check the oxidizing condition of the structure, bolting parts, and accessories
4. For masts with guide wires, check oxidization of the wires.

4.21 MOTOR VEHICLES

The most important change in materials over the years has been the transition from uncoated mild steel to zinc pre-coated steel and other corrosion-resistant metals. Hot-dip galvanized steels were used in the early stages in the motor vehicle industry. Hot-dipped galvanized steels were corrosion resistant but had a spangled surface that resulted in poor appearance after painting. Hot-dipped galvanized steel is still used on most body parts and interior surface of body outer panels.

From the mid-1980s, most exterior body panels were made of electro-galvanized steel, which was coated on both sides and served both cosmetic and perforation corrosion protection purposes. The zinc coating is smooth enough for painting purposes.

Some coatings referred to as a composite or "piggy bank" coating consisting of a thin layer of zinc or zinc alloy are applied on steel, and an organic coating is applied over the zinc or inside surface. These steels have the corrosion protection capability of electro-galvanized steel and increased perforation corrosion resistance because of barrier coating.

Use of aluminum in the place of steel has two benefits, such as the lighter nature of aluminum compared to steel and better corrosion resistance than steel. New designs and aluminum alloys have resulted in some automobiles made entirely with aluminum including the frame. Polymers have been used to a greater extent as body panels in place of steel. Polymer panels are both corrosion and dent resistant. Automobiles must be designed to use the polymer panels, as the panels do not aid in the structural rigidity of the automobile.

Stainless steel use has increased over the years. Most of the exhaust system uses stainless steel or aluminized stainless steel for corrosion resistance. Fuel systems are made of galvanized or stainless steel. Comparison of a car made in 1978 to one made in 1996 shows the reduction in the use of regular steel and iron from 68% to 55.5%, respectively. The use of high-strength steel, stainless steel, plastics, aluminum, and copper in cars has dramatically increased in the period 1978–1996. These materials have replaced mild steel for greater strength, weight reduction, and corrosion resistance.

Increased temperature and more aggressive environments present in today's higher performance automobiles have led to the use of high-performance alloys for some critical components in automobiles such as flexible couplings used in exhaust systems.

To achieve higher engine efficiency and lower emissions, the exhaust system is operated at higher temperatures leading to increased corrosion rates. At these operating high temperatures, the stainless steel couplings fail before the stipulated warranty period of 10 years. High temperatures and high salt concentration along with movement of the flexible couplings have led to failure because of fatigue, corrosion fatigue, hot salt attack, chloride SCC, pitting, and general corrosion of 316 and 321 stainless steels.

To overcome this corrosion problem, nickel-based super alloys such as Inconel alloy 625 and Incoloy alloy 864 have been used. Although these alloys are more costly than stainless steel, they have shown excellent corrosion resistance in the modern automotive exhaust environment.

The automotive industry has found several other applications where the additional cost of specialty may be beneficial. These areas are manifolds and tailpipes, catalytic converters, high-temperature fasteners, exhaust valves, airbag inflators, and other critical electrical components. The future use of these materials will depend on the benefits found in service and the changes in automotive technology that affect the corrosion conditions encountered by automobiles.

Many improvements have been made in the way that vehicles are finished. The first step in finishing a vehicle is the clear phosphate process. This process involves the treatment of vehicle parts with a mixture of zinc and phosphoric acid and some proprietary additives to clean the surface for painting, which leave a thin layer of zinc phosphate coating. A good understanding and control of parameters of the phosphating process has led to improved coatings for better corrosion resistance and paint adhesion. The systems have been optimized for vehicles with mixed-material bodies such as aluminum, coated steel, and plastic.

The second step in the finishing process is the application of primer paint. Before 1975, air spray atomizers were used for painting, and this method led to good coverage on the outside, but poor inside coverage, which resulted in corrosion. Later in 1976, PPG industries introduced a cathodic electrodeposition (ELPO) primer process, which led to every location on the primed part covered with paint. Other advances in primer technology are using thicker "high-build" primers for increased corrosion protection and flaw-hiding capabilities.

The third step in the finishing process is body sealing and augmentation coatings. The body joints and flanges undergo a finishing process involving body sealing and augmentation coatings. Vehicles have their body joints and exposed flanges sealed to reduce cosmetic and perforation corrosion.

Sealing is a robotic operation that ensures the quality of the operation. Many new augmentation coatings have been developed over the past 35 years, and the sealing process has become a robotic operation to ensure the quality of the sealing job. Several augmentation coatings have been developed over the past 40 years to increase corrosion protection in particular areas of the motor vehicle. Some augmentation coatings are antichip plastisols and urethane that are applied in the rear of the wheel house before painting. A second augmentation coating is the use of waxes applied to the interior body cavities. In early times, the waxes were applied with a handheld airless probe spray. Now the waxes are applied using automated equipment, and this resulted in increased rust-through corrosion resistance.

Application of a top coat is the final step in the finishing process. The topcoat is applied for cosmetic reasons and has little effect on the corrosion performance of the automobile. Advances over the last 40 years have led to better overall paint system performance. The robotic processing and control equipment has led to more uniform paint coverage and superior performance. Simplified vehicle design and optimization of the painting process resulted in increased finish quality, which in turn increases corrosion resistance.

Over the past 40 years, automobile engineers have improved the design to reduce the extent of corrosion. The design improvements consisted of removing crevices and locations where salt and soil can accumulate. Dissimilar metal contacts were removed. The number of "nose over" hoods, hood louvers, tuck-under areas, and other design features that promote chipping and corrosion have been reduced.

These changes as well as material and process changes have resulted in increased corrosion resistance of North American cars. At the present time, automobiles in high corrosion areas are driven for 6 or more years with no sign of corrosion in comparison to duration of 2–3 years in the mid-1970s.

Some field data on corrosion defects such as: (i) perforation; (ii) surface rust; (iii) blistering, and (iv) other defects are given in Table 4.37.

TABLE 4.37   Field Data on Corrosion Defects

Year	Perforations (%)	Surface Rust (%)	Blistering (%)	Any Defect (%)
1985	20	78	60	85
1987	7	67	56	80
1989	3	47	34	59
1991	5	62	55	77
1993	3	50	38	60

TABLE 4.38   Automobile Corrosion

Auto Part	Average Rust Area (in.)
Hoods	
Plain steel	0.5
Zn/Zn alloy	0.46
Fenders	
Plain steel	1.12
Zn/Zn alloy	1.00
Doors	
Plain steel	2.75
Zn/Zn alloy	0.50
Quarter panels	
Plain steel	3.7
Zn/Zn alloy	0.5
Deck/Hatch	
Carbon steel	0.7
Zn/Zn alloy	0.25

The data given in the table show a decrease in perforation from 20% to 3%, surface rust from 78% to 50%, blistering from 60% to 38%, and other defects from 85% to 60% in the years from 1985 to 1993.

The effect of changes in materials in automobiles that contained body panels made from carbon steel, steel pre-painted with zinc-rich primer, or steel coated with zinc/zinc alloy. The data in Table 4.38 were obtained from 5- to 6-year-old vehicles in 1985 and 1993.

The data show that Zn/Zn alloy is more corrosion resistant than prepainted steel and plain carbon steel.

Improvements were made in phosphating baths by the addition of Mn, Ni, and Zn resulting in improved coating performance and corrosion resistance. The phosphating by immersion path was found to give better corrosion resistance than a spray system as total immersion could reach tight spaces or interior locations.

The performance of plain steel, prepainted steel, and Zn/Zn alloy coated steels of automobile parts with respect to the number of perforations is given in Table 4.39.

The immersion phosphating bath system and spray phosphating system were used on car models of 1990 and 1998 and 1991 and 1989 to determine the efficacy of

TABLE 4.39   Performance of Plain Steel, Prepainted Steel, and Zn/Zn Coated Steels

	Number of Perforations		
Auto Part	Plain Steel	Prepainted Steel	Zn/Zn Alloy Coated
Hoods	1.75	0.5	0
Fenders	0.30	0.8	0
Doors	–	0.7	0
Quarter panels	2.4	0.3	0.5
Deck/hatch	1.0	0.1	0.0

TABLE 4.40   Two Methods of Phosphating on Automobiles

	Average Defect Size (inches)	
Year	Spray	Immersion
1980	19.8	–
1981	6.2	–
1988	9.0	3.5
1989	3.5	0.2

two methods of phosphating. The resulting data from the two methods are shown in Table 4.40.

It is very obvious from the above data that the total immersion method of phosphating is far superior to the spray method of phosphating.

The use of electrocoat or e-coat paints and primers is another technology that found extensive use in the 1980s. Electrocoated paint is applied on the part to be coated in a paint bath, and electrical current is applied. This method enables painting of the most intricate parts. The electrocoated automobiles were more corrosion resistant. The surface rust was approximately three times higher, and the number of perforations were also higher in vehicles that were treated in the standard conventional manner. Some of these differences may be because of zinc coating on the steel and phosphatizing. However, electroplated paint systems had a noticeable effect on corrosion resistance.

4.22   SHIPS

Corrosion control of ships can be accomplished in (i) the design phase; (ii) manufacturing phase; and (iii) operation phase.

4.22.1   Design

There are several elements that can reduce the amount of corrosion in the ship's lifetime. The design of a ship to have minimal discontinuities, such as sharp corners, will reduce the surfaces where coatings are most likely to fail.

Design can also be made to minimize locations with stress concentration, which can act as crack initiation sites and locations where coatings can crack. The design of the ship should be such that all surfaces of the tank interior can be accessible and that coating and surface inspections can be performed. Crevices that can collect dirt and form corrosion cells should be avoided during the design phase.

Another aspect of design that can influence corrosion prevention is the design of welds. Proper sizing of the welds and planning the sequence of the welds can reduce stress concentrations and distortions of the hull. Past experience has shown that lap joints have been prone to failure on older ships; therefore, butt-welded joints should be used whenever possible. Designs should also avoid intermittent spot welding as this form of welding is more prone to corrosion.

The most important element of corrosion protection is the proper coating selection. The coating should be selected during the design phase on the basis of: (i) the function of the ship; (ii) type of tanks used, and (iii) the expected life of the ship. Because of the high cost of coating application, care should be taken in choosing the proper coating. Possible coating choices are:

Epoxies
- Coal tar epoxy; silicone-modified epoxy; electrodeposition epoxy
- High solids epoxy over a waterborne epoxy zinc primer
- Pure amine epoxy; epoxy amides; epoxy amino/amides
- Hydrocarbon (wax)-modified epoxy amides and epoxy amines
- High solids (low molecular weight epoxy resins) epoxy

Thermoplastics
- Thermal-spray thermoplastics
- 100% solids rust-preventive wax

Others
- Coal tar polyurethane
- Polyurethane (aliphatic polyol) top coats
- Zinc silicates
- Alkyd paints
- Calcium sulfate alkyd

Solvent-free epoxies are much more expensive than coal-tar epoxies or the solvent-borne epoxies previously used in ship construction. Most of the cost in coating a ship is in the cost of grit blasting the steel followed by applying coating. Coal-tar epoxies of solvent-borne epoxies cost $1.80 and $2.80 per square meter while solvent-free epoxy cost $6.60 per square meter. For the amount of coating needed to coat a ship, it will cost $150,000 more than coal-tar epoxy and $120,000 over solvent-borne epoxy. The cost of ship is around $70–80 million, and the use of solvent-free epoxy is not significantly expensive.

The application of solvent-free epoxy has a longer life than coal-tar epoxy, and coal-tar epoxy has to be applied two to three times compared to only one time in the case of solvent-free epoxy coating. To perform recoating, the tanks have to be cleaned and grit-blasted before the application of coating, and such a job on a larger tanker can cost as much as $3 million.

CP or other protection systems should be incorporated, if necessary, in the design phase of the ship. CP system is a secondary defense against corrosion when holidays or cracks form in the coating. CP systems use either sacrificial zinc anodes or impressed-current systems to mitigate corrosion. Other corrosion prevention equipment and materials are inert gases to drive out corrosive gases. Corrosion inhibitors are also used.

There are several elements of ship fabrication and manufacturing that influence the corrosion performance of a ship. The ship classification societies such as Lloyd's Register of Shipping, the American Bureau of Shipping, and the Nippon Kaiji Kyokai have published tolerance standards with which the ships must comply. The tolerances permit a gap of up to 3 mm width and misalignments up to one-half of the plate thickness (53). Keeping the gaps and the misalignments under this level will reduce the possibility of stress concentrations and other possible causes of structural failure. Good painting practices in terms of application and curing, not adherence to the least proper temperatures to ensure a good solvent release in the wet stage, will ensure coatings with low internal stress and hence a longer service life.

Surface preparation before coating is very important. Almost all the coatings used in the marine industry adhere to the metal by mechanical adhesion; thus it is important to have a clean metal surface that readily bonds with the coating to protect the metal. There are two elements in the surface preparation. It is important to clean the metal surface thoroughly by removing any salt, dirt, and chemicals present on the surface. The second step involves the creation of a textured or anchor pattern surface so that the coating can mechanically adhere to the surface. The preferred method of surface preparation is grit blasting. The coating manufacturer will provide information regarding the degree of surface profile that is needed by surface blasting.

The coating is applied after surface preparation of the sample metal or alloy. The quality of coating has a profound effect on the corrosion performance of a ship and hence the directions of the coating manufacturer must be closely followed. One of the most important steps in coating application is hand-finishing, where the painter coats the corners, angles, and edges with a brush by hand. This must be done because surface tension causes drying coatings to draw away from sharp edges. Because of this, coatings tend to be thinner in the corners, angles, and edges; therefore, extra coating must be applied by stripping to ensure proper coating thickness. The coating should not be too thick that can result in solvent and thinner retention, film cracks, and gas pockets.

Human actions such as the actions of owner and crew during the operation of the ship are the last element of corrosion control. Coating represents the most important part of corrosion control on a ship and hence maintaining the integrity of the coating during operation is vital to corrosion control.

Damage to coatings can be caused in many ways such as the following:

1. Wear caused by crew members and equipment moving through the tank.
2. Wear caused by water sloshing in partially filled ballast tanks.

3. Wear caused by mud silt and other debris that accumulate in the tanks.
4. Aggressive corrosion caused by high-temperature cargos.
5. Abrasion of the ballast tanks caused by sloshing sand.

To ensure that a ship operates through its design life, it is necessary that the operator does everything possible to keep the coatings intact and inspect the coatings periodically to enable making timely repairs while the damage is minimal.

Several major changes in corrosion control technologies, environmental legislation, and ship design have led to significant changes in corrosion control approaches in the marine industry during 1975–2000. The most important change is in coatings such as limiting the use of lead, chromates, and certain VOC and the formulation of better performing multipart epoxies and other coatings. This led to high solid epoxies of different types as the primary choice of coatings. These coatings are more effective and more expensive than the coatings used in the past. These coatings require more extensive surface preparation than earlier coatings and increased costs for application and repair.

Another major change is the switch to high-strength steel and other materials with higher strength-to-weight and thickness ratios than the standard carbon steel. The transition to use of high-strength steel resulted in thinner structural elements of ships and allowing the ship hull to have more internal room to hold cargo. The negative aspect is that a small degree of corrosion will affect the structural integrity of the high-strength steel compared to carbon steel. Unprotected high-strength steels are prone to failure even when the corrosion rates are small. This is borne out by several failures of structural components in the late 1970s and early 1980s.

Double-hulled tankers are in use over the single-hulled variety in coastal waters. The space between the inner and outer hulls of the tankers is often used for ballast water, and the coating damage occurs here and hence corrosion. Greater corrosion damage in oil tanks because of the thermos effect has been observed. In double-hulled tankers the temperature is 46–55 °C, which leads to a higher corrosion rate.

One of the elements of corrosion is the actions of the owner and the crew during the operation of the ship. As the coating represents the most important part of corrosion control on a ship, it is vital to maintain the integrity of the coating during operation of the ship. Damage to coating can occur in many different ways including: wear caused by crew members and equipment moving through the tank; wear caused by sloshing in partially filled ballast tanks; wear caused by mud silt and other debris that accumulate in the tanks; aggressive corrosion caused by high-temperature cargos; abrasion of the ballast tanks caused by sloshing sand. To ensure the operation of the ship through its design life, the coating has to be intact along with periodic inspection so that the necessary repairs can be made when the damage is minimal.

From 1975 to 2000, environmental regulations led to avoiding the use of lead, chromates, and VOC. Thus coatings comprised multipart epoxies and high-solid epoxies. The use of epoxies requires extensive surface preparation, which results in higher costs of application and repair.

Another change is the use of high strength steel in the place of carbon steel that allows the structural elements to be thinner. The hull size remains the same, but has more room for cargo. High-strength steel is more sensitive to corrosion than carbon steel. In fact, many failures of structural components have been reported in late 1970s and early 1980s.

Double-hulled tankers are designed to have a ship inside a ship to reduce the risk of sinking of the ship and loss of cargo. The space between the inner and outer hulls is often used for ballast water to balance the tanks. These areas often have coating damage followed by corrosion because of the conditions in the tanks. These are difficult to inspect, and corrosion occurs because of the thermos effect.

In a single-hull tanker, the surrounding seawater keeps it at a lower temperature, which is not the case with double-hulled tankers. On double-hulled tankers, the outer hull acts as an insulator and the ballast (outer) tanks stay at a higher temperature resulting in a higher corrosion rate. The resulting higher temperature will cause the degradation of coating and in particular tar epoxy coatings.

The bottom of crude cargo tanks show pitting corrosion. A small amount of water at the bottom of a crude oil bearing tank and the hydrogen sulfide in the crude oil will result in corrosion. Pitting corrosion occurs at the bottom of the tanks.

The water at the bottom of the tank can cause microbiological corrosion because of the presence of acid-producing and sulfate-reducing bacteria (SRB). Under suitable conditions, the SRB damage steel by reducing sulfates in the crude oil to sulfides, leading to pitting corrosion in the areas under the SRB colonies.

The top of the tank or ullage showed general corrosion. These areas are difficult to inspect and repair. The top portion of the tank experiences general corrosion rather than pitting. Inspection and repairing in these areas is a major concern. The top portion of the crude tanks is made of HT steel, and the flexing/descaling problem is of concern.

Two different mechanisms affect the top of tanks depending on cargo or emptiness. When the tank is filled with crude, the ullage space is filled with inert gas that should limit corrosion. The crude oil will emit corrosive gases such as hydrogen sulfide that can combine with moisture and oxygen to form sulfuric and sulfurous acids that can attack the steel and cause general corrosion.

Although the tank in ballast voyage is filled with inert gas saturated with water vapor, the water vapor on condensation absorbs oxides of sulfur, carbon, and nitrogen to form various acids that attack the steel.

The first double-hulled tanks were protected in the same manner as the single-hulled tankers. This involved no coating to line the crude oil cargo tanks but using only a single-layer tar epoxy coating on the water ballast tanks. The corrosion rate in the single-hulled tank was constant, and repairs and steel replacement were done when the ship was 12 years old. However, a significant amount of corrosion in double-hulled tankers was observed in 5 years.

To prevent corrosion of double-hulled tanks, coatings with corrosion resistance at temperatures of 70–90 °C and resistance to MIC as well as resistance to acid attack in the ullage were found desirable. Because of the flexing of HT steel, the coating needs to be flexible and does not become brittle and break off over time as the ship flexes. The coating should last for 20 years. Solvent-free epoxies satisfied these requirements.

Coatings such as modified epoxies, coal-tar epoxy, or solvent-borne epoxy have been used by many shipbuilders since 1998. The problem with these coatings is that the lifetime of these coatings is such that they can give protection for only 8–10 years, which is less than the design life of the vessels. These coatings are not resistant to the temperatures in double-hulled tankers and the bacteria that causes pitting corrosion on the bottom of tanks. The use of solvent-free epoxy on the top and bottom of a crude tanker should prevent corrosion of the tanker.

4.23   CORROSION CONTROL IN AIRCRAFT

Typical causes and sources of corrosion are as given below. Corrosion can occur during manufacture as well as in operation.

Manufacture of Aircraft	
Basic Design	*Manufacture and Processing*
Poor design	Materials finishing processes
Crevices	Bonding process
Stress	Training
Dissimilar metals	Assembly
Finish system	
Materials selection	

Operator		
Maintenance Problem Areas	*Finish Deterioration*	*Operational Environment*
Neglect	Chipping	Sea coast
Improper repairs	Scratches	Tropical
Poor corrosion control	Breaks around fasteners	Humidity
Program or lack of	Abrasion	Industrial
implementation		
Poor training	Deposits	
	Age	

Accidental Contamination	*Environment in Airplanes*
Lavatory spillage	Condensation
Galley spillage	Animals
Chemical spills	Fish
Mercury	Mercury
Fire residues	Microbial growth

Corrosion control can be accomplished in the design and manufacturing phase as well as in the operation and maintenance phase of the aircraft. Proper design for corrosion control must include the selection of materials, coatings, sealants, and corrosion inhibitors. It is also necessary to avoid dissimilar metal contacts, access for maintenance, and proper drainage.

4.23.1 Material Selection

High-strength aluminum alloys are most widely used in airplanes because of their high strength-to-weight ratio. However, these alloys and the low-alloy, high-strength carbon steels are most susceptible to corrosion. Clad aluminum alloy sheets and plates are used where weight and function permit, while corrosion-resistant alloys and tempers are used to increase the resistance of the alloys to exfoliation corrosion and SCC. For example, aluminum alloy 7055-T7751 plate, which is not susceptible to exfoliation corrosion, has replaced the alloy 7150-T651 plate on upper wing skins. Major structural forgings of aluminum alloys and steel may be shot-peened to improve their fatigue and stress corrosion life. Titanium alloys such as Ti6Al–4V are used in environments such as floor structures under entryways, galleys, and lavatories. Where possible, stainless steels are used. However, a number of highly loaded structural components such as landing gears and flap tracks have to be made of low-alloy, high-strength steel. FRP are corrosion resistant and are widely used. Carbon fiber-reinforced plastics (CFRP) can cause galvanic corrosion in attached aluminum structures.

Application of CFRP is in the Boeing 777 CFRP floor beam design where an aluminum splice channel is used to avoid attaching the floor beam directly to the primary structural frame.

4.23.2 Coating Selection

The most practical way of combating corrosion is the use of appropriate coating. The coating for aluminum alloys consists of an appropriate surface such as an anodized surface with a corrosion-inhibiting primer. Anodizing is done using phosphoric acid. The other corrosion inhibiting primers are Skydrol-resistant epoxies formulated for general use, resistance to fuel and hydraulic fluids, or for use on exterior aerodynamic surface. Exterior surfaces of the fuselage and vertical stabilizer are painted with a Skydrol-resistant, decorative polyurethane topcoat over a urethane-compatible epoxy primer that resists filiform corrosion. Titanium and stainless steel are cadmium plated and primed if they are attached to aluminum or steel parts. This is done to prevent galvanic corrosion of aluminum or the steel.

4.23.3 Drainage

Effective drainage of the entire plane structure is important in preventing fluids from becoming trapped in crevices. The entire lower pressurized fuselage is drained by a system of valved drain holes. The fluids are directed to the drain holes by a system of longitudinal and cross-drained paths through the stringers and frame shear clips.

4.23.4 Sealants

The potential for lap joints or joint crevice corrosion is eliminated by the application of a sealant to the faying surfaces of the joints. A poysulfide sealant is applied to areas such as skin-to-stringer and skin-to-shear tie joints in the lower lobe of the fuselage, longitudinal and circumferential skin splices, skin doublers, the spar web-to-chord and chord-to-skin joints of the wing and empennage, wheel well structure, and pressure bulkheads. High-strength steel and titanium fasteners on the exterior of the air plane and fasteners that penetrate the pressurized portion of the fuselage are installed with a sealant. Fillet seals are used in severe corrosion environments.

Although the design aspects provide most of the corrosion protection for airplanes, corrosion inhibitors are widely used to provide additional protection when used periodically in service. Corrosion inhibitors are applied in areas such as the lobe of the fuselage and most of the aluminum parts. The corrosion inhibitors are petroleum-based compounds that either displace the water or serve as a coating. The water-displacing inhibitors are sprayed onto a structure to penetrate faying surfaces and keep water away from crevices. The application of these inhibitors must be repeated at intervals of every few years. The more viscous heavy-duty inhibitors are also sprayed, which form a much thicker film and have a lesser penetrating ability. These thicker inhibitors are applied on airplane parts that are most prone to corrosion.

The proper corrosion maintenance program should prevent or eliminate the conditions favoring corrosion: (i) trapped moisture; (ii) wet insulation blankets; (iii) plugged drain holes and passages; (iv) chipped or missing paint; (v) loss of protective finish; (vi) corrosive cargo.

Most of the corrosion in aircraft can be avoided by proper and timely application of sealants and corrosion inhibitors. When lavatories and galleys are removed for maintenance or repair, utmost care should be taken when the sealants are applied. Maintenance programs should be thorough such that necessary action can be taken before the corrosion problem becomes uncontrollable. Nondestructive inspection (NDI) techniques such as ultrasonic testing, eddy current testing, optical testing, and radiographic testing may be used to detect flaws in the aircraft parts before they become major defects. When corrosion is detected, it is removed by blending out. Further developments in NDI techniques should lead to detection of smaller flaws and corrosion that is hidden in the structure, which are beyond the capabilities of the existing NDI techniques. Until recently, corrosion control of airplanes was based on "find and fix." In an effort to control corrosion in an economical manner, corrosion is now being managed by a combination of selective blend-out and application of corrosion-inhibitive or water-displacing compounds.

4.24 HAZARDOUS MATERIALS TRANSPORT (HAZMAT)

The corrosive materials that were most often involved in HAZMAT incidents in 1998 consisted of sodium hydroxide solutions, basic inorganic liquids, hydrochloric acid solutions, acidic inorganic liquids, phosphoric acid, caustic alkali liquids, acidic organic liquids, potassium hydroxide solutions, sulfuric acid, cleaning liquids, hypochlorite solutions, basic organic liquids, liquid amines, and ammonia solutions.

Internal corrosion of tankers usually requires mitigation when an oxidizer or a corrosive material is transported. Internal corrosion from settled contamination is limited because of high throughput and movement of the contents during transportation. Internal corrosion of tankers is more of a problem during storage over a long time if the tankers are not properly cleaned in the beginning.

Shipping containers such as drums and pails are subject to corrosion damage and failure when corrosive materials are shipped. Usually, internal corrosion is not a problem when the goods are shipped by the manufacturer as proper container material is used when the containers are transported in a short time. Corrosion can be a problem and result in failure when the contents are stored beyond the material's shelf life. The corrosion failure of containers with hazardous waste is a serious problem. This problem occurs when wastes are mixed or when they are contaminated and stored in noncompatible materials.

External corrosion of tanker trucks and railcar-mounted tanks is a problem. Both general and pitting corrosion from the atmosphere and splash water from the roadway or rail bed can affect the tank's structural integrity. This can be a serious problem in locations where chloride sources are present such as road salt or airborne marine atmosphere and airborne industrial pollution. The common mitigation method involves painting of the tanks.

In the case of tanker trucks and railcar-mounted tanks, linings and corrosion allowances for internal corrosion are used. In the case of transportation of corrosive materials corrosion-resistant alloys are used. In extreme cases, rubber bladder tanks have been used on flat-bed trailers or railcars. External corrosion is controlled with coatings and designs that minimize crevices. One possible solution to minimize crevice corrosion is by placing a horizontal tank with a circular cross-section on legs, thus avoiding direct contact with other surfaces.

4.25 OIL AND GAS EXPLORATION AND PRODUCTION

Corrosion in oil field production environments can be in the range of zero corrosion to severely high corrosion. Crude oil at normal production temperature (less than $120\,^{\circ}C$) without dissolved gases is not corrosive. The economics of controlling corrosion in many oilfields are dependent on efficient separation of crude oil from other contaminants. While the rates of corrosion may vary, the species causing corrosion are nearly universal. Carbon dioxide and hydrogen sulfide gases dissolved in water define most of the corrosion problems in oil and gas production. Other problems include microbiological activity and the accumulation of solids.

The mechanisms of CO_2 corrosion are well defined, but the reality inside a pipeline occurs when CO_2 acts in combination with H_2S, deposited solids, and other environments. Hydrogen sulfide is highly corrosive but in some cases forms a protective sulfide scale that prevents corrosion. Microorganisms can attach to pipe walls and cause corrosion damage. Solids such as formation sand can both erode the pipe internally and can cause underdeposit corrosion, if stagnant.

Oxygen is not found in oil reservoirs, and precautions are taken to prevent oxygen entry into the production environment; however, in many cases a few parts per million of oxygen enter the pipeline and enhance corrosion problems.

External corrosion problems in oil and gas production are similar to those in pipelines, but the economic impact on the total cost of production is limited as the lines are shorter and smaller in diameter. Atmospheric corrosion of structures and vessels is a problem for offshore fields and those operating in marine environments. The improved quality of the protective coatings for offshore environments resulted in reduced frequency of repainting platforms and tanks.

A consequence of the use of advanced technology in oil production from a reservoir results in increase in the corrosivity of the oil production environment. The extent of corrosion increases because: (i) oil, water, and gas are present in the field. Seawater or fresh water is injected downhole to drive oil out of formation. As time passes, the amount of water to the amount of oil increases and the degree of internal corrosion increases. Water injection from seawater or fresh water sources causes "souring" of oilfields with H_2S and increases in corrosion rate. These water sources require biocide injection and deaeration to avoid the introduction of new corrosion pathways into the existing system. Tertiary recovery techniques involve miscible and immiscible gas floods that may contain as much as 100% CO_2. This leads to high corrosivity of the fluids.

Because of the high cost of failure and inability to rehabilitate facilities in deep water, offshore production in deep water requires the use of high alloy steels and more exotic corrosion control measures. A similar need for advanced corrosion control measures is encountered when dealing with high-pressure and high-temperature offshore oil and gas fields where conventional corrosion mitigation is not applicable. Typical costs for various expenses for one large oilfield area are listed in Table 4.41.

Corrosion in oil and gas production varies from one location to another location. Corrosion can be internal corrosion caused by the produced fluids and gases, external corrosion because of exposure to groundwater or seawater, and atmospheric corrosion caused by salt spray and weathering offshore. Of these, internal corrosion mitigation is the most difficult and expensive to mitigate and inspect periodically.

The choice of corrosion control activity would vary greatly with the production environment, area, and company philosophy; therefore, some oil fields will use very little treatment chemicals, although the cost of alternatives such as alloys and plastic liners will fill this void.

TABLE 4.41   Typical Costs for One Large Oilfield

Corrosion Expenses	Cost ($× Thousand)
Inspection, monitoring, and staff costs	9625
Repairs	1350
Corrosion inhibitor	7200
Total	$18.175 million

The extent of internal corrosion in a particular oil field environment is largely a function of the amount of water produced. As a field ages and the water increases, corrosion control will become more costly. Increased water means increased levels of bacteria and hydrogen sulfide, and in cases where miscible gas is reinjected, increased levels of CO_2.

It is useful to consider the case of an installation of a subsea gathering system for a natural gas production field. The pipeline design for a new gas production facility consisted of 20 cm diameter subsea gathering lines (flow lines) emptying into a 19 km, 50 cm diameter subsea transmission gas pipeline. The pipeline was to bring wet gas from an offshore producing area to a dehydration facility on shore. The internal corrosion was estimated to be 300–400 mpy. The corrosion mitigation options considered were: (i) carbon steel treated with a corrosion inhibitor; (ii) internally coated carbon steel with a supplemental corrosion inhibitor; (iii) 22% Cr duplex stainless steel; (iv) 625 corrosion-resistant alloy (CRA). The chance for success was estimated from known field histories of each technique, as well as the analysis of the corrosivity of the system and the level of sophistication required for successful implementation (Table 4.42).

On the basis of the risk factors and economics coated carbon steel with a supplemental corrosion inhibitor was preferred over the duplex stainless steel in spite of higher risk of the coated steel. The material selected for flow lines and trunk lines is shown in Table 4.43.

TABLE 4.42 Field Histories

Option	Chance of Success (%)
Bare carbon steel with inhibitor	65
Coated carbon steel + supplemental inhibitor	90
Duplex stainless steel	95
625 CRA	98

TABLE 4.43 Material Selected for Flow Lines and Trunk Lines

Flow Lines	Pipe and Internal Corrosion Protection	Duplex Stainless Steel Alloy (22% Cr)
	Cathodic protection and external coating	Duplex stainless steel alloy (22% Cr)
Trunk Lines	Pipe and internal corrosion protection	Coated carbon steel with supplemental corrosion inhibitor and corrosion allowance
	Cathodic protection and external coating	Coated carbon steel with supplemental corrosion inhibitor and corrosion allowance

4.26 CORROSION AND ITS PREVENTION IN THE MINING INDUSTRY

The corrosive environment of the mining industry limits the life span of the processing equipment and as a result decreases production and endangers lives of personnel. Some examples of mining equipment prone to corrosion are (i) wire rope; (ii) roof bolts; (iii) pump and piping systems; (iv) mining electronics; and (v) acid mine drainage.

Wire ropes are used extensively in the mining industry to help hoist equipment. Mine workers also depend on the rope for safety. Wire rope undergoes both corrosion and abrasion, which degrade the mechanical properties of the wire and reduce its load-carrying strength and cause its failure. About 66% of the ropes lost strength in the portion of the rope in contact with the shaft environment. Wire ropes need to be periodically examined for structural damage, corrosion, and improper lubrication or dressing. Nondestructive testing of the ropes must be done every 6 months. These ropes are replaced every 18–36 months. These are made of carbon steel but are being replaced with stainless steel and synthetic fiber ropes as carbon steel ropes are prone to corrosion.

Roof bolts provide support in underground mines by tying the lower layer to a stronger layer located above the main roof. These are low-carbon steel bolts and about 120 million/year, and corrosion failure of the bolts is hazardous and can result in loss of lives. In sulfide mines, roof bolts fail within a year because of sulfide SCC.

Corrosion within pump and piping systems is another problem, and general uniform attack is common. Pitting, crevice corrosion, intergranular corrosion, dealloying, galvanic corrosion, and cavitation corrosion are also possible depending on the environment.

Erosion–corrosion in milling is another problem. Particulates in a corrosive medium go through pipes, tanks, and pumps. The particulates erode and remove the protective film on the metal and expose the metal surface to high velocity impingement, thus accelerating corrosion.

Harsh environments in the mines can cause electrical equipment to fail after a short time and it can be avoided by a suitable enclosure.

When pyrite (FeS_2) and other sulfide minerals are oxidized by exposure to oxygen and water, ferrous ions and sulfuric acid are produced. The ferrous ions in turn are converted to hydrated iron oxide and more acid. The pH of the water drops and leads to corrosion of the metal. Thus the pipes, well screens, dams, bridges, water intakes are attacked by acid mine water.

Material selection is the most important general procedure for corrosion prevention. Material selection, protective coatings, corrosion inhibitors, and electrochemical techniques such as CP are some of the corrosion control methods that can be used to alleviate corrosion problems.

Cost–benefit analysis associated with four organic coatings used in the mining industry with surface preparation such as surface blasting or no surface preparation led to the finding of the data included in Table 4.44.

TABLE 4.44   Coating Systems and Initial Cost

Coating System	Initial Cost ("P" Is the Basic Cost of Carbon Steel Requiring Protection)
2-coat alkyd	0.2 P
3-coat vinyl	0.5 P
3-coat epoxy (with blasting)	0.9 P

4.27   PETROLEUM REFINING

High-temperature crude corrosion is a complex problem. There are at least three mechanisms: (i) furnace tubes and transfer lines where corrosion is dependent on velocity and vaporization and is accelerated by naphthenic acid; (ii) vacuum column where corrosion occurs at the condensing temperature, is independent of velocity, and increases with naphthenic acid concentration; (iii) side-cut piping where corrosion is dependent on naphthenic acid content and inhibited somewhat by sulfur compounds.

Blending may be used to reduce the naphthenic acid content of the feed, thereby reducing the corrosion to an acceptable level. Blending of heavy and light crudes can change shear stress parameters and might also reduce corrosion. Blending is also used to increase the sulfur content of the feed and inhibit, to some extent, naphthenic acid corrosion.

Injection of corrosion inhibitors may provide protection for specific fractions that are particularly severe with respect to corrosion. Monitoring needs to be effective to check on the adequacy of the treatment. Process control changes may provide sufficient corrosion control provided there is reduced charge and temperature. For long-term reliability, material selection is the best solution. Above 288 °C with very low naphthenic acid content, cladding with 5–12% chromium steels is suitable for crudes of greater than 1% sulfur. When hydrogen sulfide is present, an alloy with 9% chromium is preferred. Carbon steel is found to be adequate in naphthenic acid corrosion, and low-alloy steels containing up to 12% chromium do not provide better performance over carbon steels. Type 316 stainless steel (>2.5% molybdenum) or Type 317 stainless steel (>3.5% Mo) is found suitable for cladding vacuum and atmospheric columns.

The selection of materials for refinery construction depends on the type of refinery, the type of crude oil to be refined, and the expected service life for each vessel. As with all materials selection, the life-cycle cost is of importance in addition to the purchase price. Table 4.45 lists some of the common alloys and their material costs relative to carbon steel. The costs listed are relative to carbon steel, which is assigned a value of 1.0.

Carbon steel is the most commonly used structural material in refineries primarily because of a combination of strength availability, low cost, and its resistance to fire. The low alloy steels are meant for applications that require higher properties than can be obtained with carbon steels. The workhorse refinery alloys for elevated temperatures greater than 260 °C contain 0.5–0.9% chromium plus molybdenum.

TABLE 4.45 Comparison of the Relative Costs of Various Alloys

Alloy Class	Example	Constituents	Cost Factor
Carbon steel	C10	>94 Fe	1.00
Low-alloy steel	1.25 Cr ½ Mo	1.25 Cr, 0.5 Mo balance Fe	1.25
Fe–Ni–Cr + Mo	Type 316L		5.00
	Alloy 800H		–
	20 Cb-3	3.5 Cu	19.0
Ivi–Cr–Mo	Alloy C-2	–	–
	Alloy 276		30.00
	Alloy C4	–	–
	Alloy 625		31.5
Ni–Cr–Fe	Alloy G		32.00
	Alloy 600		–
Ni–Mo	Alloy 132		58.00
Ni–Cu	Alloy 400	–	–
Nickel	Alloy 200	–	–
Co–base	ULTIMET(R)		136.00
To–base	Ti–6Al–4V		–

Normally, at least 5% of chromium is required to resist oxidation at temperatures in excess of 430 °C. At present, most refineries use 9 Cr–1 Mo tubes in coker heaters. For carbon steel and low alloy steel creep becomes an important consideration at about 430 and 480 °C, respectively. These alloys are used for pressure vessels, piping, exchangers, and heater tubes.

Austenitic steels provide excellent corrosion, oxidation, and sulfidation resistance with high creep resistance, toughness, and strength at temperatures greater than 565 °C. Thus they are used in refineries for heater tubes, heater tube supports, and in amine, fluid catalytic cracking (FCC), catalytic hydro-desulfurization (CHD) sulfur, and hydrogen plants.

Austenitic steels are susceptible to grain boundary chromium carbide precipitation "sensitization" when heated in 540–820 °C range. Whenever sensitization is to be avoided, refineries use the stabilized grades of Type 347 (with Nb) or Type 321 (with Ti).

The susceptibility of austenitic stainless steels to SCC limits their use and requires special precautions during operation and at downtime. At downtime, prevention of stress corrosion may be achieved by either alkali with a dilute soda ash and low-chloride water solutions and/or nitrogen blanketing. Austenitic stainless steels are used for corrosion resistance or resistance to high-temperature hydrogen damage or sulfide damage. Strip-lined, stainless-clad, or lined vessels are used in hydrocracking and hydrotreating services. Austenitic stainless steels are also used in tubing in heat exchangers exposed to corrosive conditions. The ferritic and martensitic stainless steels form the AISI 400 series. The most common alloys from this series found in refineries are types 410, 4015, 405 stainless steels. A common stainless steel for trays and lining in crude service is type 410 stainless steel.

The original saltwater condenser tube made of admiralty brass was found to be susceptible to erosion–corrosion at tube ends. Aluminum brass containing 2% aluminum was more resistant to erosion in saltwater. Inhibition with arsenic is necessary to prevent dezincification as in the case of admiralty brass. The stronger naval brass is selected as the tube material when admiralty brass tubes are used in condensers. Cast brass or bronze alloys for valves and fittings are usually Cu–Sn–Zn compositions, plus lead for machinability. Aluminum bronzes are often used as tube sheet and channel material for exchangers with admiralty brass or titanium tubes exposed to cooling water.

The 70/30 cooper–nickel alloy is used for exchanger tubes when better saltwater corrosion resistance than in aluminum brass is needed, or when high metal temperatures in water-cooled exchangers may cause dezincification in brass. Monel is 67:30 of nickel: copper has very good resistance to saltwater and, under nonoxidizing conditions, to HCl and HF acids. Monel has a better high-temperature resistance to cooling water than does 70/30 Cu–Ni alloy. Monel cladding and Monel trays are commonly specified at the top of crude towers to resist HCl vapor at a temperature below 205 °C. Above 205° nickel-based alloys are attacked by H_2S. For high-temperature strength and/or corrosion resistance, several nickel-based alloys are used for expansion bellows in FCC process units (Alloy 625), stems in flue gas butterfly valves (Alloy x750), and in springs exposed to high-temperature corrosives (Alloy X).

Titanium has excellent resistance to seawater and it is also used for tubing in crude tower overhead condensers. Use of titanium is limited because of the high cost.

4.28 CORROSION CONTROL IN THE CHEMICAL, PETROCHEMICAL, AND PHARMACEUTICAL INDUSTRIES

The types of corrosion experienced in the chemical manufacturing industry, the petrochemical manufacturing industry, and the pharmaceutical manufacturing industry are similar in many respects. The most common types of corrosion encountered are caustic and chloride cracking, oxidation, sulfidation, corrosion under thermal insulation, ammonia cracking, and hydrogen-induced cracking.

The corrosion failure mode with an average frequency of occurrence is given in Table 4.46.

The data in Table 4.47 show the SCC failures of different alloys.

Plant and process design involves the materials of construction, equipment design, process conditions, and recommended operating practices to minimize the risk of corrosion.

In large companies, an internal project team may design the plant, otherwise contractors provide the design. In any case, the corrosion engineer must be involved from the beginning of the project. The materials of construction must be chosen to satisfy the process condition with respect to corrosion and its control. Figure 4.11 shows the phases of corrosion control.

TABLE 4.46   Corrosion Failure Mode with Average Frequency of Occurrence

Failure Mode	Average Frequency (%)
Cracking	36
General corrosion	26
Localized corrosion	20
Temperature effects	7
Velocity effects	5
Voltage effects	3
Hydrogen effects	2
Biological	0
Total	99

TABLE 4.47   Stress Corrosion Cracking Failures of Different Alloys

Material	Average Frequency (%)
Stainless steels	61.4
Steel	30.4
Copper alloys	4.3
Nickel alloys	2.8
Titanium	0.7
Tantalum	0.3
Total	99.9

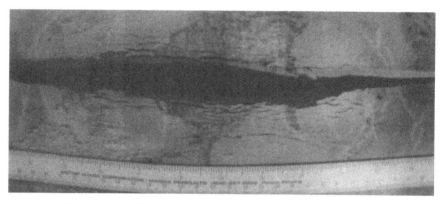

Figure 4.11   SCC colony found on a high-pressure gas pipeline (6).

4.28.1   Corrosion-Resistant Alloys

These alloys are widely used in the chemical process industry. It is always better to use a more corrosion-resistant alloy in spite of its cost as it solves the corrosion problems and saves money in the long run. The relative costs of some alloys used in corrosion control relative to type 316L stainless steel as the reference with a ratio 1.0 are given in Table 4.48.

TABLE 4.48   Relative Costs of Some Alloys Used in Corrosion Control

Alloy	Example	Cost Ratio
Carbon steel	C10	0.2
Low alloy steel	1.25 Cr 0.5 Mo	0.25
Fe–Ni–Cr + Mo	Type 316L	1.00
	Alloy 800 H	–
	ZO cB-3	3.8
Ni–Cr–Mo	Alloy C2	
	Alloy C276	6.0
	Alloy C4	
	Alloy 625	
Ni–Cr–Fe	Alloy G	6.4
	Alloy 600	–
Ni–Mo	Alloy B2	11.6
Ni–Cu	Alloy 400	–
Nickel	Alloy 200	–
Co–base	ULTIMET®	27.2
Ti–base	Ti-6Al-4V	–

Nickel–chromium–molybdenum alloys are used in reactor vessels in the production of acetic acid. These alloys are cost-effective compared to Ni–Cr stainless steels and have good resistance to oxidizing corrosive media; Ni–Mo alloys have good resistance to reducing media. Molybdenum together with the chromium stabilizes the passive film in the presence of chlorides and is particularly effective in increasing resistance to pitting and crevice corrosion.

4.28.2   Piping Design Factors

In piping design, the three conditions that lead to corrosion problems are: (i) water traps; (ii) dead legs; (iii) high velocities.

Water traps are low sections of the piping system where water stagnates and causes corrosion. Pitting corrosion occurs in water traps. This form of corrosion can be countered by minimizing the low sections through slanting the pipe or by installing drain valves at low points that are periodically drained.

Dead legs are the regions of the pipe system where fluid is stagnant. Pitting corrosion can occur in environments where stagnant particles are deposited on a metal surface. Pits can grow and penetrate the metal wall and cause a leak. Pitting is dangerous because a leak can release aggressive or flammable chemicals under high pressure. Dead legs can be minimized in the piping by: (i) eliminating dead ends in piping manifolds; (ii) providing drains; (iii) designing pipes with elbows instead of tees; (iv) placing valves with the shortest dead legs; (v) placing branch lines off from the top rather than from the side.

Velocity effects are cavitation and erosion–corrosion. In general, fluid piping should be designed with large diameters to transport the quantity of material required. It should be borne in mind that sometimes process changes over time may cause a problem leading to erosion–corrosion.

4.28.3 Construction Stage Checks

Inspection is of importance to ensure that fabricators are working according to design codes and quality control. Assembled items are of the right material and specifications for common items such as valves, piping, and welding electrodes must correspond to the standard prescribed specifications. It may be necessary to perform trial runs with the equipment supplied by manufacturers before full-scale operations.

The factors influencing plant and process design are as follows. This has to comply with equipment and process data specifications with the following input:

1. Design philosophy, economic, and political considerations.
2. New materials and equipment design.
3. Data from similar or identical plant.
4. Materials data from manufacturers.
5. Established corrosion design data.
6. Data developed from corrosion testing.
7. Established corrosion design data.
8. Materials data from manufacturers.
9. Data from similar or identical plant.
10. Equipment and process data specifications.

Planned maintenance or regular replacement of plant equipment to avoid failure by corrosion is an essential adjunct to design and constitutes the third phase of control. The factors contributing to a policy of planned maintenance philosophy are: (i) predictable and reasonable rate of corrosion for material; (ii) discounted cash flow advantage over life of plant in using a cheaper less-resistant material; (iii) factors other than corrosion dictating regular maintenance; (iv) no feasible alternative to corrodible material; (v) installed spare preferred policy for reliability of plant.

The design philosophy determines the emphasis placed on controlling corrosion by this means, as opposed to spending resources at the construction stage to prevent corrosion. Planned maintenance should be avoided when maintenance labor costs are high or spares may be difficult to procure.

Planned maintenance consists of periodic shutdowns to inspect the equipment and refurbish or replace equipment that failed because of corrosion or other failure mechanisms.

Regular maintenance of lower grade equipment is preferable over minimum maintenance of expensive or specialty equipment.

Periodic reviews are necessary to determine if the current corrosion control methods are adequate. Coatings on equipment may deteriorate and need to be replaced. Inhibitor dosage has to be adjusted depending on environmental or process changes.

In the pharmaceutical industry, process tanks, pipes, and valves are often electro-polished to reduce adhesion of products and decrease bacterial growth in crevices. It is also necessary to prevent contamination of packages by spilled products, and hence desiccant bags filled with drying agents are placed in packages. The atmosphere in the packages is controlled humidity to avoid corrosion.

The techniques used for monitoring corrosion in a process plant are: (i) visual inspection; (ii) weight loss coupons; (iii) electrical resistance probes; (iv) measurement of corrosion potential; (v) linear polarization measurement; (vi) hydrogen probes; (vii) thickness measurement and crack detection; (viii) visual inspection; (ix) sentinel holes.

In many of these techniques, probes may be affected by oil or paraffin deposits leading to erroneous readings. Corrosion monitors should be located in areas of high corrosion risk. Monitoring should be done according to a schedule. Corrosion monitoring can be continuous or periodic. Continuous corrosion monitoring gives corrosion rates immediately while periodic monitoring gives average corrosion data and may miss an event of short-term upset in the system.

Some of the corrosion monitoring techniques along with the advantages and disadvantages are given in Table 4.49.

TABLE 4.49   Corrosion Monitoring Techniques with Advantages and Disadvantages

Technique	Advantages	Disadvantages
Weight loss	Easy to use; can test multiple specimens	Average corrosion rate; risk of inserting/extracting samples
Radiography	Low initial cost. Permanent record	Health hazards; holes, voids affect X-ray attenuation
Magnetic particle testing	High reliability; no disruption to system; lowers manufacture costs	Limited to certain areas; measures only surface defects
Liquid penetrant testing	Low cost; simple equipment	Measures only near surface
Ultrasonic testing	Automated operation; detects minute defects; electronic recording	No permanent record
Metallography	Assesses equipment condition before repair	High cost
Acoustic emission	Low cost, rapid, online testing; permanent record	Background noise
Remote field eddy current inspection	Automated inspection	Volumetric test; percent wall loss only
Tank floor scanner	Detects corrosion and external/internal defects on aboveground storage tank floors	Follow up with UT. No quantitative data.
Linear polarization resistance	Direct measure of metal loss and corrosion rate; frequent measurement	High cost; corrosion may deteriorate probe; specialized equipment

4.28.4 Remedial Action and Diagnostic Analysis

Corrosion monitoring must be coupled with diagnostic work and remedial action. In some cases of corrosion, remedial action may be obvious or easily deduced. In other cases, diagnostic work precedes a decision or remedial action. The options for remedying corrosion problems in a process plant are: (i) install a CP system; (ii) install an anodic protection system; (iii) change equipment design; (iv) improve feed stock purity; (v) alter process variables; (vi) change the alloy/material; (vii) institute inhibitor additions; (viii) institute planned maintenance.

When the remedial action is not known, diagnostic action should precede remedial action. The root cause of the corrosion failure must be investigated. For example, NDTs such as radiography, magnetic particle testing, liquid penetrant, acoustic emission, and ultrasonic testing may be used to detect cracks in the metal/alloy sample. Options for remedial measures are: (i) complete replacement of the sample/equipment, using a corrosion-resistant alloy, clad material, or application of anodic protection, use of protective coatings or corrosion inhibitors.

4.29 PULP AND PAPER INDUSTRIAL SECTOR

Corrosion control methods used in this industrial sector consist of: (i) equipment design; (ii) process design; (iii) the use of corrosion inhibitors.

4.29.1 Equipment Design

Pulp and paper industrial equipment design consists of proper material selection in conjunction with the process chemicals and prevention of stagnant fluids in the process equipment. In the absence of corrosion, low-alloy carbon steel would be the material of choice if corrosion were not a problem. However, for many processes, stainless steel and even nickel-base and titanium alloys are required for better performance in corrosive environments. At present, current US paper mills are constructed of about one-third carbon steel and two-thirds stainless steels. There are several grades in the group of stainless steels. The relative cost of the stainless steels is dependent on the concentration of the major alloying elements (Cr, Ni, Mo), the volume produced, and the form in which it is supplied such as tube, pipe, plate, or block. It is useful to note that stainless steels are 10–20% more expensive than low-alloy carbon steels.

Stagnant and slow-flowing process fluids in pulping equipment can occur in crevices and water traps. Fluid stagnation can lead to an increase in concentration of chemicals and the localization of a severe environment in which pitting and crevice corrosion may occur. By designing drain holes and easy access to the equipment, stagnant solution can be removed. Slow-flowing liquids containing a solid fraction of pulp may deposit a layer of pulp at the bottom of pipe and reservoirs. Underdeposit corrosion modes such as crevice corrosion and pitting may occur. The equipment if designed with sufficiently fast and/or turbulent flow, deposit formation and hence the resulting corrosion can be controlled. Another useful practice consists of scheduled cleaning and proper equipment maintenance to prevent the buildup of pulp and decrease the amount of stagnant solutions.

4.29.2 Process Design and Corrosion Inhibitors

Corrosion rates in mill equipment also depend on the chemical composition within each section of the process. The chemical composition varies from mill to mill depending on the processes involved, the closure of the mill's systems, the desired paper grade, the speed of the process, and the amount of pulp or paper being produced.

In general, the corrosivity of an environment will increase with an increase in temperature, low pH, dissolved solids content, and chloride and sulfur compound concentrations. Thus by monitoring the process and maintaining the proper temperature, pH, dissolved solids content, chloride, and sulfur compound concentration the extent of corrosion can be controlled.

In some processes, addition of corrosion inhibitors can control the corrosion. Measurements of the process chemistry should be made to ensure that the correct dosage of inhibitor is maintained.

Corrosion control in the pulp and paper industry consists of verification of average corrosion rates, using weight loss coupons, regular inspection, and preventive maintenance. Preventive strategies are considered during design and construction phases when new equipment is considered or existing equipment is refurbished or repaired. Corrosion prevention techniques minimize corrosion initiation altogether, while corrosion control techniques minimize the propagation of ongoing corrosion.

4.29.3 Weight Loss Coupons

Weight loss in coupons exposed to the corrosive medium gives the average corrosion rate in the atmosphere of a pulp or paper plant. This technique is cheap and easy to use as it involves the determination of weight of metal coupons before and after exposure for a known length of time. From the loss in weight of the coupon, the average corrosion rate can be calculated. An advantage of this technique is that several coupons of different materials can be exposed to the environment simultaneously.

By performing weight loss tests, technical personnel can estimate the corrosiveness of an environment over a period of time. If the corrosion rate is determined to be high, preventive measures can be taken to prolong the remaining life of the equipment. The preventive measures can include the addition of a corrosion inhibitor, a change in the process, an application of anticorrosion coatings, or the addition of a CP system.

4.29.4 Inspection and Preventive Maintenance

An additional and simplest and obvious method to monitor the corrosion rate is through visual inspection of all parts of the equipment that are exposed to corrosive environments. These visual inspections may use nondestructive examination (NDE) to identify corrosion pits, crevice corrosion, or wall thinning before they affect or stop the production process or degrade the paper to below its product quality specifications.

Regular periodic inspection is performed as part of maintenance program in plants. The results of the inspections are used to plan equipment repairs during scheduled shutdowns and to take action to maintain optimum production quality and quantity.

4.30 AGRICULTURAL PRODUCTION

Corrosion control and prevention can be done by keeping the equipment clean and dry after each use by applying corrosion-resistant materials or materials with a corrosion allowance, by applying external coatings or paints as well as internal linings or by CP.

Agricultural production occurs by working or using farm land; thus the equipment used to work the fields is exposed to the climate and weather conditions present at that time and place. Rain water or wet products may collect in the corners or ridges of the equipment leading to corrosion. Corrosion may also occur in locations where mud buildup occurs or where waste from vegetables, cattle, or feed can be present.

Corrosion can also occur in fuel and fertilizer storage tanks. Corrosion of tank bottoms, walls, roofs, and roof structures can pose a danger to their structural integrity. Corrosion can result in leaks resulting in loss of valuable product or soil pollution and pollution of water around the storage tank. Fuel leaks and soil pollution must be prevented in a farm environment where farm products depend on good soil quality.

4.30.1 Keeping Equipment Clean/Dry

The obvious method of corrosion control of farm equipment and machinery is to keep it clean and dry after use. Prevention of corrosion under deposits such as mud or waste products prolongs the life of the equipment. Exposure of the equipment to bacteria, fertilizers, cleaning agents, and sanitizing solutions should be minimized. It is also advisable to remove the mud or adhering dirt such as sand particles to decrease wear and possible erosion–corrosion on engines and moving components.

4.30.2 Material Selection

Corrosion-resistant materials may be selected for farming equipment and machinery, but the high cost of high-alloy components make it prohibitive. In general, where possible, painted carbon steel is the primary material of choice for most farm machinery and equipment because of its low cost and ease of maintenance. Nickel alloys are used for augers, which are resistant to corrosion, abrasion, and wear. Stainless steel fittings are used in equipment exposed to corrosive chemical fertilizers or in milking equipment. Fiber-reinforced polymer storage tanks can be used for water storage or to store small amounts of chemical products used in farming.

4.30.3   External Coatings/Paint

Painting the exposed external surfaces of equipment gives good appearance as well as corrosion protection. The selection of the type of coating depends on the use of the equipment. External coatings must withstand the effects of abrasion, weather, and ultraviolet radiation. Surface preparation and painting is an easy method to prevent or delay the onset of corrosion.

For underground structures such as USTs, corrosion control of the external surfaces can be achieved with a combination of CP and a dielectric coating. The external coating must be applied when the tank is new. A buried tank cannot be retrofitted with an external coating unless it is removed from the ground.

4.30.4   Internal Linings

Internal corrosion protection, where required because of contamination or corrosive products in storage tanks, is usually maintained with an internal liner sometimes in combination with galvanic CP. Internal coatings prevent internal corrosion and prolong operational life of the tanks. An example of this is the internal lining in mild steel fertilizer tanks. The linings offer corrosion protection from fumes and condensation in the vapor space and immersion exposure to the stored liquid chemicals.

4.30.5   Cathodic Protection

Aboveground tanks containing stored liquid and is contaminated with water layer should be internally coated and cathodically protected on the bottom and partially along the wall. The external bottom corrosion of the site-fabricated tanks can be controlled with select sand/concrete foundation pads, impervious liners, and CP.

The design of CP for new ASTs and USTs includes consideration of the proximity to other metallic structures and existing CP systems, the type of grounding, the estimated remaining life of the tank, the type and temperature of the stored product, the amount of product stored, the cycling rates, the method of tank bottom plate construction, the type of tank foundation, back-fill soil characteristics. The two types of sacrificial anodes are (i) zinc or magnesium ribbon or ingot anodes, and (ii) impressed-current CP using perimeter deep-buried, angle-drilled anodes or vertical, loop, or string undertank anodes.

4.31   FOOD PROCESSING

A variety of materials are used in food processing such as stainless steels and aluminum alloys. Plastics and other materials such as brass and bronze may be used. Plastics and other metals may be used. Lead and cadmium plated materials impart toxicity to foods. Food contact surfaces must be smooth, no-adsorbent, nonleaching, and insoluble in the food.

Stainless steels are resistant to corrosion in food-processing environments. However, stainless steel is not immune to a chloride-containing environment. Corrosion is a problem when stainless steels are exposed to the chlorine used in sanitizers and hydrochloric acid used in cleaning agents and processing liquids. Corrosion products must be removed before they can contaminate the food products and impede proper cleaning of surfaces.

Stainless steel consumption in the food processing industry is 370,000 ton or about 15.3% of the overall market for stainless steel of 2.4 million tons/year. The estimated cost of stainless steel for the food-processing industry was determined to be nearly $1.8 billion. Aluminum is often used in processing equipment and is less expensive than stainless steel. Aluminum alloys are usually used in processing, handling, and packaging of foods and beverages. Nearly one quarter of all aluminum is used in packaging. High corrosion resistance and the nontoxic nature of aluminum and its salts and freedom from catalytic effects that cause product discoloration are features suited to the food packaging industry.

Aluminum cans contain both internal and external coatings, primarily for decoration and protection of product taste. Oxygen is removed before the can is filled with the product. This prevents oxidation and the potential risk of toxicity.

The shiny appearance, low-weight per volume, favorable mechanical properties such as material strength, ease of forming, and handling are some favorable features of aluminum and hence its use in the food industry. In addition, aluminum has a better corrosion resistance than carbon steel as it readily forms a protective film that prevents further atmospheric corrosion. Aluminum is also lighter than stainless steel and hence its use in beverage cans.

Inhibitors are used in severe and harsh environments encountered in rotary cookers and hydrostatic sterilizers. The medium consists of hot water, steam, and cooling water. A single approach to this problem may not provide the solution. A combination of anodic, cathodic, and filming inhibitors was selected for corrosion prevention depending on the water composition and equipment material.

Coatings used in food-processing plants must withstand high-pressure cleaning and microbial attack. Microbial attack is a major problem in breweries and beverage bottling plants. Antimicrobial additives are used to control bacterial activity and growth. Urethane coatings instead of epoxy coatings are preferred in the food processing industry because they are resistant to cleaning compounds.

4.32 CORROSION FORMS IN THE ELECTRONICS INDUSTRY

Anodic corrosion between components of integrated circuits with voltaic gradients of the order of 10^5–10^6 v/cm occurs in the case of positively biased aluminum metallization. The combination of electric fields, the atmospheric moisture, and the halide contamination leads to corrosion attack on aluminum. Gold and copper metallizations are also prone to corrosion under these conditions.

4.32.1 Cathodic Corrosion

Negatively biased aluminum metallizations, as those with positive bias, can also corrode in the presence of moisture and the high pH produced because of the cathodic water reduction reaction. High pH leads to dissolution of surface passive oxide and aluminum substrate with corresponding increase in conductor resistance.

Electrolytic metal migration occurs in silver-containing compounds. In the presence of moisture and an electric field, silver ions migrate to negatively charged surface and plate out, forming dendrites. The dendrites grow and bridge the gap between the contacts, causing an electric short and an arc. Even very small amounts of dissolved metal can form a large dendrite. Under certain humidity and voltage gradients, a month-long exposure equals 4 years of service in an office environment. Other metals susceptible to metal migration are gold, tin, lead, palladium, and copper.

4.32.2 Pore-Creep in Electrical Contacts and Metallic Joints

To prevent tarnishing of connectors and contacts, a noble metal such as gold is plated on the contact surface. Substrate may corrode at imperfections of plating. When the substrate is copper or silver and is exposed to sulfur or chloride containing environment, corrosion products can creep out of the pores and over gold plating, forming a layer of high-contact resistance.

4.32.3 Fretting Corrosion of Separate Connectors with Tin Finishes

Fretting corrosion in electronic components is manifested as continuous formation and flaking of tin oxide from the mated surface on tin-containing contacts. As the tin is consumed, the problem becomes more severe. The best available solution to this problem is to replace the part.

4.32.4 Galvanic Corrosion

Galvanic corrosion occurs when two dissimilar metals such as aluminum and gold are coupled together as in package-integrated circuits. The polymers used for packaging are porous, and the gaskets around hermetic covers such as ceramic or metal might leak. In humid environments, moisture can permeate to the IC bond pad, creating favorable conditions for galvanic corrosion. Processing-related integrated circuits are exposed to aggressive media used in reactive ion etching (RIE) or wet etching for patterning of aluminum lines, which can lead to corrosive residues. Reactive ion etching of aluminum metallizations utilizes aggressive chlorine-containing gases. If removed untreated from the etcher, patterned structures are covered with aluminum chloride, which on hydrolysis forms hydrochloric acid in the presence of moisture.

4.32.5 Micropitting on Aluminum

Aluminum metallizations, alloyed with copper, form intermetallic compounds such as Al_2Cu along the grain boundaries, which act as cathodic sites relative to aluminum adjacent to the grain boundaries. This leads to dissolution of aluminum matrix in the form of micropitting during the rinsing step after chemical etching.

4.32.6 Corrosion of Aluminum in Chlorinated Media

Both liquid and vapor phase halogenated solvents used for production of ICs and PCs readily corrode the aluminum components. Water contamination of the solvents increases the time-to-corrosion, on the one hand, but increases the corrosion, on the other hand. Dilution of the stabilized solvents with alcohol solvents leads to break-down of halogenated solvent and formation of chloride ions, which corrode aluminum and aluminum–copper alloys.

4.32.7 Solder Corrosion

Lead–tin alloy solder's resistance to corrosion in aqueous and gaseous environments is a function of the alloy composition. The corrosion resistance of the alloy increases with tin content above 2 wt%. Lead forms unstable oxides, which readily react with chlorides, borates, and sulfates.

4.32.8 Corrosion of Magnetic and Magneto-Optic Devices

Corrosion-related failures can occur in advanced magnetic and magneto-optic storage devices, where thin-film metal discs, thin-film inductive heads, and magneto-optic layers are affected. Corrosion occurs in sites where the deposited carbon overcoat is lacking because of the intentional roughening of the disc and where the magnetic cobalt-based layer and nickel–phosphorus substrate become exposed. The potential differences between the noble (positive) carbon and the metal substrate, a galvanic couple may form resulting in rapid galvanic-induced dissolution of the magnetic material.

Magneto-optic devices use very reactive alloys for the recording media. Exposure of magneto-optic films to aqueous solutions or high humidity results in pitting, even during storage in ambient office conditions.

Mitigation of corrosion of electronics equipment by encapsulating the components in plastics has been done. It should be noted that polymers are permeable to moisture. Hermetically sealed ceramic packaging is more successful; care must be taken to prevent moisture and other contaminants from being sealed in. A common useful approach for mitigating corrosion of circuits housed in large chassis consists of using volatile corrosion inhibitors. This will require periodic replacement of the carrier.

4.33 HOME APPLIANCES

The three basic corrosion control methods used in corrosion control of home appli-ances are: (i) corrosion control by sacrificial anodes; (ii) use of corrosion-resistant materials; (iii) corrosion control by coatings and paint.

4.33.1 Corrosion Control by Sacrificial Anodes

The life of a hot water heater can be extended by checking and changing the dete-riorating anode. The direct cost of not having to replace broken water heaters can result in savings of 70% in time and money. In addition, replacing the anodes of the water provides protection from sudden floods, which can result in indirect costs for clean-ups and damages. The indirect cost can potentially be greater than costs related to installing a new tank or retrofitting an old tank. The life expectancy of sacrificial anodes can range from 2 to 3 years to more than 10 years. The benefits of anode maintenance are (i) long-lasting tanks; (ii) less rust buildup; (iii) saving on costly changeovers.

The cost of replacing water heaters was estimated at $460 million/year, the cost of anode replacement was estimated at $780 million per year, and the cost of increased life expectancy of water heaters was estimated at $778 million/year.

The benefits of anode maintenance are longer tank life, less rust buildup, and savings on costly changeovers. However, a cost–benefit analysis may show the cost of replacing anodes could exceed the cost of increased life expectancy or the cost of water heater replacement. Increased life expectancy without increased cost has not been shown to be technically feasible. It is also not known how much life is gained by the replacement of water heater anodes as the water heater can fail because of other reasons such as heating element failure.

4.33.2 Corrosion Control by Corrosion-Resistant Materials

Whenever possible, carbon steel may be used because of its high strength and low cost. However, corrosion-resistant materials such as plastics, galvanized steel, stainless steels, aluminum, and copper–nickel alloys are used in corrosive environments.

Plastics are corrosion resistant and they prolong the product life and durability in hostile environments where major appliances operate. The use of plastics increases the durability and equipment life by 30–40%.

Carbon steel coated with zinc is known as galvanized steel, which protects the cold-rolled steel from corrosion in aqueous and high-temperature environments. Galvanized steel is used in laundry appliances as it provides resistance to laundry detergents. The cost of converting steel into the galvanized condition depends on the facility, but ranges between $50 and $100/metric ton of zinc.

Stainless steels are usually used to design high-efficiency furnace components such as heat exchangers. According to the Specialty Steel Industry of North America the total steel usage for heating and air-conditioning equipment is 81 million kg, and the annual (1998) consumption of stainless steel for the appliance sector was estimated at $315 million/year. This estimate can be attributed to the control of corrosion by using stainless steels.

Aluminum is often used for control panels of appliances. The thin film of aluminum oxide that forms readily on exposure to air gives protection against corrosion. Surface treatments such as anodizing and cladding help to further corrosion resistance.

Copper-nickel alloys are used in tubings and coils of heater and air-conditioning systems because of their high thermal conductivity in heating and cooling applications. Copper–nickel alloys such as 70/30 Cu/Ni and 90/10 Cu/Ni have sufficient erosion–corrosion resistance in water compared to pure copper.

4.33.3   Corrosion Control by Coatings and Paint

Liquid coatings (paints), powder coatings, porcelain enamel coatings are used to coat appliances. Pretreatments that affect the performance level are used for surface cleaning and adhesion of coatings. Pretreatment systems consist of iron phosphate and zinc phosphate. The cost of phosphate pretreatment depends on the following factors: (i) continuous or intermittent manufacturing; (ii) geometry of the part; (iii) control of chemical processing.

Liquid coatings are used for refrigerators as they are corrosion resistant and are of low cost. In laundry washers and dryers, a primer coating is applied on galvanized steel.

Powder coatings are organic coatings and are used primarily on boiler and furnace steel sheets. These coatings are applied by depositing a mist of powder on the product sample in the presence of an electrostatic field, followed by baking. Powder coatings are more popular and widely recognized because of their advantage with respect to environmental acceptability, quality of coating, coating requirement, and cost.

Porcelain enamel is used mainly in high-level performance appliances. The porcelain enamel coatings are more scratch resistant and heat resistant than the thinner liquid and powder coatings; however, porcelain enamel is porous. Holidays are sometimes found in the porcelain (glass). Magnesium, zinc, and aluminum anodes are used in combination with porcelain enamel coatings in water heating systems to act as a sacrificial anode.

4.34   DEFENSE

Corrosion of military equipment and facilities has been a significant and ongoing problem. Large yearly costs are incurred to protect these assets from corrosion, affecting procurement, maintenance and operations. The effect of corrosion is a problem that is becoming more serious as the acquisition of new equipment slows and more reliance is placed on modifications and upgrades to extend the life of the current systems. As the intention to operate the aging fleets of aircraft, ships, land combat vehicles, and submarines continues into the 21st century the potential detrimental effects of corrosion on the cost of ownership, safety and readiness must be taken into account. The effects of corrosion of DOD equipment will continue to deteriorate unless and until new technologies can be used. Within DOD, the annual costs are very difficult to ascertain; however, from available data obtained from the individual services (Army, Air Force, Navy, and Marine Corps) the total estimated cost of corrosion to the DOD is approximately $20 billion for systems and infrastructure.

4.34.1   Army

This major branch of the armed forces owns and operates a range of facilities and equipment. These include buildings, vehicles, trucks, aircraft, helicopters, missiles, and weapons storage facilities. Corrosion costs are a significant burden for the Army, including the cost of maintenance of weapons systems.

A major portion of the corrosion cost of $2 billion is attributed to Army ground vehicles such as:

1. Abraham Tank Systems – MI Abrams
2. Bradley Fighting Vehicle Systems
   M2 Infantry Fighting Vehicles (IFV)
   M3 Cavalry fighting Vehicles (CFV)
   Multiple-Launch Rocket System (MLRS)
   Command and Control Vehicles (C2V)

Bradley Carrier Systems
Bradley Fire Support Vehicles
   Medium Tactical Vehicles (MTV)
   2 ½ Ton Cargo Trucks
   5 Ton Cargo Trucks
High-Mobility Multipurpose Wheeled Vehicles (HMMWV)
Light Armored Vehicles

In general, very little attention is given to corrosion and corrosion control of army vehicles. Little has been done to incorporate corrosion protection and control in the design and manufacture of army vehicles. The significant shortcomings are:

1. Use of 1010 carbon steel without galvanizing or any protective coatings.
2. Presence of many galvanic couples and the use of more than 2800 rivets that may act as possible locations for corrosion.
3. Use of painting procedures that are not state of the art.
4. Use of paint that provides little corrosion protection, such as the chemical agent-resistant coatings that deteriorate rapidly in the presence of a corrosive environment.

Corrosion of HMMWV and other vehicles cost about 2–$2.5 billion/year and corrosion also affected the readiness of the vehicles. Corrosion damage was greater than 65% of the vehicle acquisition cost and the vehicles requiring corrosion repairs were out of service for about 2–12 months. Vehicles as old as 5 years had to be scrapped and replaced by new vehicles.

An analysis of the corrosion control deficiencies of the HMMWV indicated that the corrosion problems in these vehicles are a result of design mistakes. A glaring fault is that the frame is built of 1010 carbon steel without galvanizing or any other form of corrosion protection. Holes were drilled into the sides of the frame members and no holes at the bottom of the frame to allow for drainage. This allowed for stagnation of the water in the interior of the frame leading to corrosion from inside out. The lack of drainage on a vehicle designed to be used in water up to 1.5 m (60 in.) deep shows poor design of the vehicle.

Many of the components of the vehicle such as fasteners, handles, brackets, and frame were made from 1010 carbon steel, which is prone to corrosion with ease. These parts corroded on almost every HMMWV in service leading to extensive repairs and maintenance. Vehicle parts such as the engine compartment, suspension, and steering, body, underbody, and other miscellaneous parts such as welded seams, fuel tank assemblies, nuts, bolts, fasteners, and frame suffered corrosion and the percentage of vehicles affected was in the range of 13–76%.

Most of the corrosion problems in HMMWV vehicles could have been eliminated by using galvanizing and coatings. Other problems could have been avoided by using polymers and other alternate materials.

Much of the body of the HMMWV is made of aircraft grade aluminum while the frame and doors are made of 1010 carbon steel. The whole vehicle is secured with more than 2800 rivets, which gives high strength-to-weight ratio. Each rivet is a preferential site for corrosion.

Electrodeposition of coatings assures complete coverage of the surface including otherwise inaccessible areas. On the HMMWV, e-coating technology was not used for coating application, and coating was applied by spraying.

Chemical agent-resistant coatings (CARC) consist of a surface cleaning, epoxy primer, epoxy interior topcoat, and a polyurethane exterior top coat. The purpose of this coating was to provide resistance to chemical penetration of the coating and to aid in decontamination of the vehicle in case of chemical attack. Other purposes of the coating were to provide corrosion protection as well as camouflage protection.

The CARC Paint System did not perform as well as expected (Table 4.50).

CARC coating is difficult to apply. It is also difficult to achieve the required thickness levels. Field repair of CARC coating was found to be difficult. CARC also contains some VOC, which may be environmentally unacceptable.

According to the TAACOM CPC acquisition document dated September 16, 1993, the HMMWV vehicle is supposed to have a service life of 20 years with a minimum of 15 years. The vehicles need to be corrosion resistant for extended periods in corrosive environments involving one or more of the following conditions: high humidity, salt spray, road deicing salts, gravel impingement, atmospheric contamination, and temperature extremes. There shall be no corrosion past stage one, nor corrosion

TABLE 4.50   CARC Paint System Performance

Category	Number of HMMWV	Number of HMMWV with Deteriorated Coating	
		Number	Percent
Army 1	17	4	24
Army 2	13	9	69
Army 3	9	3	33
Army 4	11	11	100
Marine Corps	2	30	75
Marine Corps	40	30	75
WI National Guard	29	13	45

impairment of fit or function. Corrosion control shall be achieved by a combination of design features, materials selection such as composites, galvanized steel, E-coat, coil coating, production techniques, process controls, inspection, and documentation. The minimum requirement is galvanizing of ferrous components in accordance with the galvanizing policy, appropriate pretreatment, and E-coat primer. Subsequent use of rust-proofing materials, such as MIL-C-46164, is not a substitute for any of these minimum requirements.

The MTVs of the first batch did not meet the corrosion protection requirements. Corrosion on the cabs was found in 3 years although the contract specified that corrosion of the vehicles should not occur for 10 years. Rather than replacing the MTVs, damage was repaired to give a life of 10 years free from corrosion.

The army subjected one of the 4995 trucks to corrosion tests. The corrosion damage occurred in 60 areas, and the procedures were modified, and about 3751 trucks were produced with galvanized steel cabs.

Corrosion also occurs in Howitzer firing platforms. Severe galvanic corrosion because of dissimilar contact in the areas on the platform took place where water could collect. The total cost of the remedy of the corrosion problem is estimated to be nearly $9 million. Another corrosion problem experienced with the towed M198 Howitzer can be solved by drilling a drain hole at the lowest point.

Nearly 8–22% of overhaul and repair costs of helicopters are because of corrosion. In 1998, nearly 44 billion was spent on corrosion control of helicopters.

4.34.2 Navy

The fleet consists of ships, submarines, aircraft weapons, and facilities, and the total cost of corrosion is estimated to be $2 billion/year.

The surfaces of ships are exposed to extremely aggressive environments. An extensive corrosion control program is required to maintain the fleet during dry-dock cycles. The primary defense against corrosion is the diligent use of protective coatings. In addition to coatings, CP is used for protection of the underwater hull. The cost of CP is lower than coatings. The traditional coatings have a design life of 10–15 years after which the old coating is removed and a new coating is applied. Between major maintenance cycles, there is an annual maintenance demand for continuous coating maintenance.

Table 4.51 lists the annual coating maintenance of Navy surface ships.

One of the major tools to prevent corrosion on naval aircraft is painting. During the period from August 1995 through 1996, 341% more paint than necessary was used for the prevention and control of corrosion damage of the F-18 aircraft and F-14 squadrons, by painting large sections of their aircraft every 56 days.

4.34.3 Air Force

The total cost of direct corrosion maintenance to the Air Force for 1997 was estimated at nearly $800 million. The majority of the cost can be attributed to repair ($573 million) and paint ($146 million). Corrosion maintenance consists of inspection for

TABLE 4.51   Annual Coating Maintenance of Navy Surface Ships

Activity	Man-Years Per Slip
Topside and freeboard (enamel, silicone alkyd)	9.0
Flight decks and topside decks (nonskid)	4.0
Bilges/wet space corrosion	4.5
Machinery space/passageways (enamel, silicone alkyd)	2.25
Interior bulkheads and decks (chlorinated alkyd)	3.00
Superstructure, catwalks, mixing/fan room corrosion (epoxy)	3.25
Total	26.00

corrosion, repair maintenance because of corrosion, washing sealant application and removal, and all coating application and removal.

Comparison of 1997 fleet costs with 1990 fleet costs showed the effect of aging on weapon system costs. The corrosion problems with A-10, C-130, and F-16 have been resolved. The cost of corrosion maintenance of B-1 and E-3 fleets increased.

A severe corrosion problem occurs in fuselage joints, where the voluminous corrosion products at the contact or faying surfaces of the lap joint cause deformation of the skin. Pillowing phenomenon occurs in lap joints. Because of the stress fatigue and stress corrosion, cracks can nucleate near the fastener holes, jeopardizing the structural integrity of the fuselage. Other corrosion problems on KC-135 aircraft are dissimilar metal corrosion, lap joint corrosion on the 7178 upper wing skin, lap joint corrosion on the 7075-T6 fuselage crown section, and SCC of the 7075-T6 forged frame sections.

As a result of the numerous corrosion problems of the KC-135, the Air Force in the United States expended considerable effort to develop methods to control corrosion of the KC-135 ranging from characterizing the type and extent of corrosion to developing new nondestructive techniques, to developing methods to slow down corrosion with corrosion preventative compounds (CPCs), and to developing predictive models.

Severe corrosion occurred in aluminum alloy components of the KC-135 Stratotanker aircraft. Corrosion is because of low utilization, and the majority of the time is spent on the ground being exposed to corrosive atmospheric environments. The KC-13 aircraft was built with aluminum alloys 2024-T3 and T-4, and 7075-T6 and 7178-T6, which are all susceptible to corrosion and SCC. The original construction was without any sealant in the lap joints and fuselage skins that had spot-welded doublers attached to them. The upper wing skins that are made of highly corrosion-susceptible aluminum alloy 7178 were attached with high-strength steel fasteners, causing dissimilar metal corrosion in certain areas.

4.35   PREVENTIVE STRATEGIES

The aim of preventive strategies is to use the opportunities to improve corrosion control in all economic sectors, which will result in increased integrity, durability, and savings. Benefits, approaches, and some specific recommendations are made for the following opportunities for improved corrosion practices.

1. Preventive strategies in nontechnical areas:
   (a) Increase awareness of the considerable corrosion costs and potential savings.
   (b) Change the misconception that nothing can be done about corrosion.
   (c) Change policies, regulations, standards, and management practices to increase corrosion cost-savings through sound corrosion management.
   (d) Improve education and training of personnel in recognition of corrosion control.
2. Preventive strategies in technical areas:
   (a) Advance design practices for better corrosion management.
   (b) Advance life prediction and performance assessment methods.
   (c) Advance corrosion technology through research, development, and implementation.

REFERENCES

1. HH Uhlig, *Corrosion* **6**:29 (1950).
2. G. Okamoto *Report of the Committee on Corrosion and Protection: a Survey of the Cost of Corrosion in Japan*, Japan Society of Corrosion Engineering and Japan Association of Corrosion Control, 1977.
3. T. Shibata,"JSCE Activities for Cost of Corrosion in Japan", *EuroCorr 2000*, London, UK, Sept. 2000.
4. NACE *International Coating Inspector Training and Certification Program (course notes)*, NACE International, 1996.
5. *Corrosion Costs and Preventive Strategies in the United States*, U.S. Dept. of Transportation, Federal Highway Administration, Springfield, VA, 2002.
6. Corrosion Costs *and* Preventive Strategies *in the United States*, U.S. Dept. of Transportation, Federal Highway Administration, Springfield, VA, 2002, p. C7.
7. LM Smith, GR Tinkelberg, *Report no. FWHA-RD-94-100*, U.S. Federal Highway Administration, Turner Fairbank Highway Research Center, McLean, VA, 1995.
8. 1997 Economic Census, www.census.gov//prod/www/abs/97economic/html, Oct. 2000.
9. *A User's Guide for Hot Dip Galvanizing*, NACE International Publications TPC9, Houston, TX, 1983.
10. 1997 Economic Census, www.census.gov, Oct. 2000.
11. *Contributors of Plastics to the U.S. Economy*, Society of Plastics Industry, 1997.
12. Mineral Industry Surveys, http://minerals.ugsg.gov/minerals/titanium, Dec. 1999.
13. Corrosion Inhibitors: A Report by Publications Resource Group, Business Communications Company, July 1999.
14. 1998 *Statistics – U.S. Shipments of Polyethylene Pipe, Tube and Conduit*, Plastics Pipe Institute, 1999.

15. DB McDonald, DW Pfeifer, GT Blake, *Report no. FHWA-RD-96-085*, FHWA, 1996.

16. YP Virmani, GG Clemena, *Report no. FHWA-RD-98-088*, FHWA, Sept. 1998.

17. DB McDonald, MR Sherman, DW Pfeifer, YP Virmani, *Concrete International*, 1995, pp. 65–70.

18. YP Virmani, GG Clemena, Corrosion Protection-Concrete Bridges Report *no. FHWA-RD-98-088*, FWHA, Sept. 1998, p. 72.

19. NG Thompson, DR Lankard, *Report no. FHWA-RD-96-207*, FHWA, May 1997.

20. NG Thompson, DR Lakard, *Report no. FHWA-RD-99-096*, FHWA, Sept. 1999.

21. YP Virmani, "Corrosion Inhibitors for Bridge Members", *FHWA Technical Note on Corrosion Protection Systems*, Structures Division, Office of Engineering Research and Development, Dec. 1997.

22. NG Thompson, DR Lankard, M Yunovich, *Proceedings for Evaluation of Corrosion-Inhibiting Admixtures for Structural Concrete*, NCHRP Final Report, NCHRP Project 10–45, June 2000.

23. NS Berke, DW Pfeifer, TG Weil, *Concrete International*, Dec. 1988.

24. NG Thompson, DR Lankard, MM Sprinkel, *Report no. FHWA-RD-91-092*, FHWA, 1992.

25. K Babaei, NM Hawkins, *Concrete International*, p. 56–66, 1988.

26. PD Carter, *Concrete International*, Nov. 1989.

27. MM Sprinkel, RE Weyers, AR Sellars, *Transportation Research Record*, 1304 Highway Maintenance Operations and Research, 1991.

28. RE Weyers, EP Larson, Concrete Repair, *Rehabilitation and Protection, Proc. of Intl. Conf., Univ. of Dundee, Scotland, June 27–28*, 1996, RK Dhir, MR Jones, eds, E&FN Spon, London, pp. 29-38.

29. *Cathodic Protection of Reinforced-Concrete Bridge Elements, a State-of-the Art Report*, Report no. SHRP-S-337, Strategic Highway Research Program, National Research Council, Washington, DC, 1993.

30. GG Clemena, DR Jackson, Virginia Transportation Research Council, *FHWA/VTRC Report no. 00-R3*, July 1999.

31. "FHWA Showcases Innovative Method of Protecting Bridges from Corrosion", *FHWA Memorandum*, Nov. 1999.

32. "Highway Deicing: Comparing Salt and Calcium Magnesium Acetate", *Special Report 235*, Transportation Research Board, National Research Council, Washington, DC, 1991. p. 11.

33. BR Applemar, *Journal of Protective Coatings & Linings* 60–68 (1993).

34. LM Smith, GR Tinklenberg, "Lead-Containing Paint Removal, Containment and Disposal", *Report no. FHWA-RD-94-100*, FHWA, Washington, DC, Feb. 1995, pp. 109–124.

35. GH Breevort, AH Roebuck, *Material Performance* **23** (4): 31–35 (1993).

36. M Funahashi, *Corrosion 2013*, Paper no. 2117, National Association of Corrosion Engineers, 2013.

37. PG Eller, RW Szempruch, JW McLard, *Summary of Plutonium Oxide and Metal Storage Package Failures, LA-UR-99-2896*, Los Alamos National Laboratory, Los Alamos 1999.

38. C Andrade, M Cruz Alonso *ASTM STP 1194*, G Cragnolino, N Sridhar (eds.), American Society for Testing and Materials, Philadelphia, 1994.

39. *Analysis of the Total System Life Cycle Cost of the Civilian Radioactive Waste Management Program*, DIE/RW-10, U.S. Dept. of Energy, Office of Civilian Radioactive Waste Management, Washington, DC, Dec. 1998.

40. PE Myers, *Aboveground Storage Tanks*, McGraw-Hill, New York, 1997.

41. *API Recommended Practice 651, Cathodic Protection of Aboveground Storage Tanks*, 2nd ed., ANSI/API Std. 651-1991), Dec. 1997.

42. 6th *Annual Report on Petroleum Industry Environmental Performance* (PIEP), American Petroleum Institute, 1998.

43. DA Lytle et al., *JAWWA* **90** (3): 74-88 (1998).

44. I Wagner, *Proceedings of Seminar on Internal Corrosion Control Development and Research Needs*, AWWA, 1989, pp. 1–17.

45. *Internal Corrosion of Water Distribution Systems: Cooperative Research Report*, 2nd ed., American Water Works Association Research Foundation, Denver, CO, 1996.

46. WE Neuman, *Journal of New England Waterworks Association* **109**(1):57–60 (1995).

47. JE Singley et al., "Corrosion Prevention and Control in Water Treatment and Supply Systems" *Pollution Technology Review*, no. 122, Noyes Publications, Park Ridge, NJ, 1985.

48. WC Robinson, *Corrosion/91, Paper 522*, NACE, Cincinnati, OH, 1991.

49. Water Industry Database, Utility Profiles, *Report by the American Water Works Association*, AWWA, Denver, CO, 1992.

50. JS Clift, *Corrosion Protection for Water Transmission Pipe: a Practical Guide for Long-Term Performance*, Price Brothers Company, Dayton, OH, Jan. 2000.

51. BP Boffardi, A Sherondy, *Corrosion/91, Paper 445*, NACE, Cincinnati, OH, 1991.

52. A Cohen, RJ Myers, *Corrosion/84*, NACE Annual Conference, 1984.

53. *Standard Guide for Steel Hull Construction Tolerances*, ASTM F1053-87, American Society for Testing and Materials, Oct. 1997.

54. JA Beavers, GH Koch, WE Berry, *Corrosion of Metals in Marine Environments*, Metals and Ceramics Information Center, Battelle Columbus Div., Columbus, OH, June 1986.

5

CONSEQUENCES OF CORROSION

5.1 INTRODUCTION

The three major consequences of corrosion are the following:

1. Economic costs involved in rectifying the corrosion damage.
2. Life-threatening accidents resulting in loss of lives.
3. Environmental damage threatening the ecosystem.

5.2 CORROSION STUDIES

5.2.1 The Battelle-NBS Study

Corrosion is very costly and has a major impact on the economies of industrial nations. The Battelle-NBS study (1) pointed out the severe impact of corrosion on the US economy. The estimates based on the Battelle-NBS study report that the annual cost of corrosion in the United States alone was approximately $70 billion, which was between 4% and 5% of the gross national product (GNP). A limited study in 1995 (2) updating the 1975 cost estimates, estimated the total cost of corrosion at approximately $300 billion. This staggering total corrosion loss resulted from equipment and structure replacement, loss of product, maintenance and repair, the need for excess capacity and redundant equipment, corrosion control, designated technical support, design, insurance, and parts and equipment inventories. Other national studies such as in the United Kingdom (3), Japan (4), Australia, and (5) Kuwait (6) investigated their respective corrosion costs. All these studies emphasized

Challenges in Corrosion: Costs, Causes, Consequences, and Control, First Edition. V. S. Sastri.
© 2015 John Wiley & Sons, Inc. Published 2015 by John Wiley & Sons, Inc.

the financial losses because of corrosion with no concern toward preventive strategies along with cost–benefit considerations.

The efforts of the aforementioned studies ranged from formal and extensive to informal and modest. It is estimated that the annual cost of corrosion ranges from 1% to 5% of each country's GNP.

Uhlig's study (7) attempted to measure the costs of corroding structures to both the owner/operator (direct cost) and to others (indirect costs). The total cost of corrosion to owner/operators was estimated by summing the cost estimates for corrosion prevention products and services used in the entire US economy such as coatings, inhibitors, corrosion-resistant metals, and cathodic protection and multiplied these totals by their respective prices. Domestic water heater replacement, automobile internal combustion engine repairs, and replacement of automobile mufflers were selected as examples to estimate the cost to private consumers/users. Adding both the direct and indirect costs, the annual cost of corrosion to the United States was estimated to be $5.5 billion or 2.1% of the 1949 GNP. This method was used in Japan and estimated the cost of corrosion at $9.2 billion equivalent to 1–2% of the Japanese GNP.

The Hoar study (United Kingdom, 1970 (3)) identified the sources for the cost of corrosion by sectors of the economy. The study estimated the annual total corrosion cost in the United Kingdom to be approximately 3.5% of their GNP.

The Battelle-NBS study (1) used an economic input/output analysis to estimate the cost of corrosion in the United States. In the model, the US economy was divided into 130 industrial sectors. For each sector, estimates were made on the costs of corrosion prevention, as well as the cost of repair and replacement because of corrosion. The following direct costs were included in the study: replacement of equipment or buildings; loss of product; maintenance and repair; excess capacity; redundant equipment; corrosion control such as inhibitors; organic and metallic coatings; engineering research and development testing; design; insurance; parts and equipment inventory. The final results of the Battelle-NBS study for the base year of 1975 were:

1. The total US cost of metallic corrosion was estimated to be $70 billion, which amounts to 4.2% of GNP in 1975.

2. Fifteen percent or $10 billion was estimated to be avoidable by using corrosion control technology that is available.

The total cost of corrosion in the United States has been estimated by (i) adding the costs of corrosion control methods and services and (ii) estimating the total cost by extrapolating the corrosion costs of representative industrial sectors to the entire US economy. The latter method incorporates all the costs that the first method is likely to miss such as the cost of corrosion management, the cost of direct services related to the owner/operator, and the cost of loss of the capital because of corrosion.

The annual cost of corrosion consists of direct costs and indirect costs. The direct costs related to corrosion consist of: (i) costs of design, manufacturing, and construction; (ii) material selection; (iii) additional material such as increased wall thickness for corrosion allowance; (iv) materials such as coatings, inhibitors, sealants, cathodic protection to prevent corrosion; (v) application including cost of labor and equipment.

TABLE 5.1   Direct and Indirect Costs of Corrosion

Item	Cost ($ Million)
Direct Costs	
Paint	2000
Metallic coatings and electroplate	472
Corrosion-resistant metals	852
Boiler and other water treatment	66
Underground pipe maintenance and replacement	600
Indirect Costs	
Domestic water heater replacement	225
Automobile engine repairs	1030
Automobile muffler replacement	66

TABLE 5.2   Corrosion Costs in the United Kingdom by Major Industry

Industrial Sector	Estimated Cost	
	£ × Million	Percent
Building and construction	250	18
Food	40	3
General engineering	110	8
Government agencies and departments	55	4
Marine	280	21
Metal refining and semifabrication	15	1
Oil and chemical	180	13
Power	60	4
Transport	350	26
Water	25	2
Total	£ 1365	100

Cost of management: (i) corrosion inspection; (ii) corrosion-related maintenance; (iii) repairs because of corrosion; (iv) replacement of corroded parts; (v) inventory of backup components; (vi) rehabilitation; (vii) loss of productive time.

Table 5.1 shows direct and indirect costs.

Corrosion costs in the United Kingdom by major industry are shown in Table 5.2. Factors that could lower the cost of corrosion are the following:

1. Better dissemination of existing corrosion control information.
2. Improved protective treatments.
3. Closer control over the application of existing protective measures.
4. Improved designs with existing materials.
5. Greater awareness of corrosion hazards by the users.

6. Use of new and improved materials.
7. Cost-effectiveness analysis of materials and protective treatments leading to procurement on the basis of life-cycle costs.
8. Previous feedback on service performance.
9. Improved specifications for protective treatments.
10. Basic research on corrosion mechanisms.
11. Improved communications between government departments.
12. Improved storage facilities.
13. Information on corrosion sensitivity of equipment.
14. Better nondestructive testing techniques.
15. Standardization of components.
16. More frequent or longer duration maintenance periods.

The single most important factor to reduce corrosion costs in the United Kingdom was better dissemination of existing corrosion control information and measures.

The preventive strategies to reduce corrosion costs consisted of: (i) information dissemination and corrosion awareness; (ii) education and training; (iii) research and development.

Although a great amount of corrosion control information and strategies were available in the United Kingdom, only certain industries such as oil and chemical industries and aircraft and nuclear power industries paid attention to corrosion in the design stage.

These industries either needed corrosion control to enable a process to work or intended to avoid accidents because of corrosion damage.

The four principal reasons for corrosion problems are the following:

1. Lack of foresight.
2. Lack of information dissemination.
3. Minimum initial capital outlay.
4. Lack of basic knowledge.

The UK report recommended to: (i) establish a national corrosion and protection center; (ii) provide education and training; (iii) provide better research opportunities and avenues; (iv) develop closer links between technical and trade organizations.

A survey on the cost of corrosion was conducted in 1977 and a report was published by the Committee on Corrosion and Protection (4). The annual cost of corrosion to Japan was 2.5 trillion yen (US $9.2 billion) in 1974. The GNP of Japan in 1974 was 136 trillion yen, and the cost of corrosion was 1–2% of Japan's GNP for 1974. This cost is a direct cost. Indirect costs would be much higher. Japan's committee estimated the cost of corrosion by: (i) corrosion protection products and services and (ii) corrosion cost by industry sector. The Uhlig method determined the costs primarily on the basis of the cost of corrosion protection products and services such as coatings, inhibitors, corrosion-resistant materials, and cathodic protection. The

TABLE 5.3 Costs of Corrosion Protection Methods in Japan

Corrosion Protection Method	Cost (Yen × Billion)
Paints and protective coatings	1595
Surface treatment	648
Corrosion-resistant materials	239
Rust-preventive oils	16
Corrosion inhibitors	16
Cathodic protection	16
Research and development	22
Total	2551

total costs amounted to approximately 2.5 trillion yen (US $9.2 billion). Paints and coatings amounted to nearly $3.1 billion. Surface treatments and corrosion-resistant materials accounted for nearly one-quarter and one-tenth of the costs, respectively. All other corrosion control methods accounted for less than 5% of the costs. The costs to prevent or control corrosion by Uhlig's methodology yielded the data provided in Table 5.3 with respect to the Japanese context.

The Hoar method applied to determine the cost of corrosion in the Japanese industrial sector resulted in the data mentioned in Table 5.4.

The disparity between the two estimates is large. The Uhlig method's estimate is 1.5 trillion yen higher than the value obtained by the Hoar method. The difference in the two estimates is in part because of omission of some costs by the Hoar method. In the normal course, the Hoar method is expected to give a higher value than the value estimated by Uhlig's methodology. The differences between the two methods may be attributed to the difficulties and uncertainties in the investigation of the costs of corrosion.

The Japanese report recommended to: (i) establish corrosion prevention service center of technical experts; (ii) increase two-way communications between academic institutions and industry and (iii) enhance training of engineers; (iv) increase awareness from elementary school to university level for saving material resources and conservation of the environment; (v) monitor and inspect, in terms of research and development, equipment and machines for corrosion prevention control.

TABLE 5.4 Cost of Corrosion in the Japanese Industrial Sector (Hoar Method)

Industry Sector	Corrosion Cost (Yen × Billion)	Percent of Total
Energy	60	6
Transportation	195	19
Building	175	17
Chemical industry	154	15
Metal production	27	3
Machinery and manufacturing	433	42
Total	1043	100%

According to the Battelle-NBS report (1, 2, 8), the total US cost of metallic corrosion was estimated to be $70 billion, which comprised 4.2% of GNP in 1975, and 15% or $10 billion was estimated to be avoidable by using the available corrosion prevention technology. The final results based on NBS analysis of uncertainty in the Battelle input/output model estimated the metallic corrosion cost of $82 billion or 4.9% of $1.677 trillion GNP in 1975. Nearly $33 billion or 2% of GNP was estimated to be avoidable.

The corrosion cost analysis of four sectors, namely: (i) federal government; (ii) personally owned automobiles; (iii) electric power industry; and (iv) energy and material loses was done.

The agencies involved in the federal government sector were the US Department of Defense, the National Aeronautics and Space Administration (NASA), the US Coast Guard, the US Government Services Administration, the legislative branch, and the National Bureau of Standards (NBS).

The total cost of corrosion to the US federal government was estimated to be $8 billion, comprising capital costs of $6 billion and maintenance costs of $2 billion. The total corrosion cost was estimated at $8 billion of which 20% or $1.6 billion was estimated to be avoidable. The total capital and corrosion maintenance costs of government assets are given in Table 5.5.

The total corrosion costs to the electric power industry in the generation and distribution of power were estimated to be about $4 billion out of which the annual corrosion-related maintenance costs were estimated at $1.1 billion.

Energy and materials losses because of corrosion were estimated at 3.4% of the country's energy consumption ($1.4 billion). Within the energy sector, coal consumption had a greater impact than oil. Nearly $0.23 billion was found to be avoidable.

Nearly 17% of metallic ores (1.4 billion) resulted from corrosion, and 2.1% of metallic ore demand or $180 million was found to be avoidable.

The Battelle-NBS study found two sources of potential sources of savings in terms of corrosion. Approximately $10 billion was estimated to be avoidable by using more cost-effective currently available technology.

The Battelle-NBS study provided a reference to the following factors:

1. Measure of the severity of corrosion costs.
2. The places where and how the impacts of corrosion are felt.
3. Useful method of analysis of corrosion costs.

TABLE 5.5   Capital and Corrosion Maintenance Costs of Government Assets

Asset	Total Capital ($ × Billion)	Maintenance ($ × Billion)
Aircraft	195	0.99
Ships	56	0.40
Buildings and real estate	144	0.655
Total	395	2.045

4. Bibliography and data base on corrosion economics.
5. Reference point for the impact of corrosion against which other factors affecting the economy can be measured.
6. Basis for technological assessment of the economic effect of proposed means to reduce corrosion costs.
7. Sectors where high affordability and presently unavoidable corrosion costs are encountered.

It is also useful to note that corrosion problems were envisaged in areas of energy, environment, materials conservation, and food production.

In 1982, the Commonwealth Department of Science and Technology initiated a study to determine the feasibility of the establishment in Australia of a National Centre for Corrosion Prevention and Control. The feasibility study consisted of the determination of the annual cost of corrosion to Australia, the need for a national center for corrosion prevention and control, and a review of national corrosion centers in European countries. The study considered organizational structure, technical functions, and the financial structure of the proposed center. The results were presented in a 1983 report entitled "Corrosion in Australia: the Report of the Australian National Centre for Corrosion Prevention and Control Feasibility Study." (5).

The annual cost of corrosion in Australia was estimated to be $2 billion Australian dollars equivalent to 1.5% of Australia's GNP in 1982. Improved technology transfer and establishment of a national corrosion control center were found to be useful in the potential savings in the nation's economy.

The Australian study was patterned after the Battelle-NBS study. An IO model of the Australian national economy was constructed to first represent the real world and secondly to represent the world of optimum corrosion mitigation technology. Differences between the two scenarios were used as estimates of avoidable costs of corrosion and to indicate areas of potential savings.

Some of the drawbacks of the Australian cost estimates were the cost estimates were large and did not include the cost to users and the cost of disruption. A high-pressure gas pipeline may fail because of corrosion resulting in a disruption cost to a third party. Corrosion processes could cause gas pipeline rupture, failure of industrial plants, buildings to deteriorate, and aircraft to crash. It is useful to note that secondary costs resulting from corrosion failure could be large in many cases.

The savings or avoidable cost was estimated to be equal to 1.2% of the Australian GNP in 1982 of which 35% of the savings was because of personal consumption expenditures, 55% because of fixed-capital formation, and 5% because of federal, state, and local government expenditures. The various industry sectors expressed the need for a national center for corrosion prevention and control to lower corrosion costs. The three roles of the corrosion control center are consulting, research, and education.

In 1992, Kuwait conducted an economic assessment of the total cost of corrosion using a modified Battelle-NBS ISO model, and the results were presented in a report (9) in 1995. The total cost of corrosion was estimated at about $1.0 billion (1987 dollars) amounting to 5.2% of Kuwait's 1987 GDP and avoidable cost at $180 million or 18% of the total cost.

The total cost of corrosion in West Germany during 1968–1969 was estimated to be 19 billion Deutschmarks (US $6 billion) of which 4.3 billion DM (US $1.5 billion) was estimated to be avoidable. The corrosion costs amounted to 3% of West German GNP for 1969 of which 25% was avoidable (10).

The cost of corrosion to Finland (11) in 1965 was 150–200 million markaa (US $47 to 62 million for the year 1965).

The painting expenditure in Sweden to combat corrosion was 300–400 million crowns (US $58–75 million) of which 25–35% were found to be avoidable.

The cost of corrosion in India in 1961 was estimated to be 1.54 billion rupees (US $320 million): 25% for paints, varnishes, and lacquers; 20% for metallic coatings and electroplating; and 55% for corrosion-resistant metals (12).

A survey of 148 chemical industries was conducted in 1986 in China and found that the average corrosion cost to be 4% of the annual income (13). The results of another survey of iron and steel complex indicated the corrosion costs to be 1.6% of the total income.

The Chinese initiative in reducing corrosion losses consisted of education and training of personnel, organizing 15 technical courses for industrial personnel, each of 2–4 weeks duration with particular emphasis on corrosion prevention technology. As of 1986, 11 institutions of higher education have established special courses in corrosion and its prevention through which about 400 college graduates are trained annually. During 1980–1985, the State Science and Technology Commission of China and the Chinese Society of Corrosion and Protection organized 15 training courses on a variety of corrosion topics of 2–4 weeks duration for personnel in various industries. Intensive efforts were directed toward the following topics:

1. More efficient inhibitors and better corrosion-resistant steels for use in the oil and gas industry.
2. Coatings and cathodic protection of steel structures used in exploitation of oil in offshore locations, harbors, parts, and vessels.
3. Development of efficient inhibitors for acid cleaning of chemical equipment.
4. Development of heat-resistant alloys and protective coatings.
5. Development of low-alloy steels to resist atmospheric corrosion.
6. Effects of amorphous state and ion implantation on the corrosion resistance of surface layers of materials.
7. Development of titanium and its alloys for use in chemical industries.
8. Development of super alloys and ceramic materials to increase the efficiency of gas turbines (GTs).
9. Promotion of the application of nonmetallic materials in chemical industries.

In addition to the above initiatives, extensive use of cathodic protection technology, protective coatings, and proper selection of materials have been made in the industrial settings. The measures taken to reduce corrosion by the use of corrosion prevention technology resulted in an annual saving of $700 million.

Corrosion costs in Canada (14, 15) are about $3.475 billion out of which $235 million are avoidable by adopting corrosion control measures. So far, there has been no initiative in implementing corrosion control measures to reduce the corrosion costs. This aspect is in the realm of activity of a federal government or a national research laboratory that deals with materials science. The following recommendations have been made to counter the effects of wear and corrosion in Canada: establish a national secretariat whose functions are to:

1. ascertain the corrosion costs in all sectors of industry in detail and arrive at avoidable costs;
2. assemble a directory of all available wear and corrosion experts in Canada including private consultants, academicians, researchers in industry, and government organizations;
3. publish, in consultation with both industry and academicians, as many guides as necessary pertinent to various sectors of industry where wear and corrosion measures are invaluable for distribution;
4. produce, in addition to the guides, booklets and films on wear and "controlling corrosion" with the titles: (i) methods; (ii) advisory services; (iii) economics; (iv) specifications and standards; (v) case studies; and (vi) monitoring for distribution;
5. take field trips by staff to wear and corrosion sites for consultation and education;
6. identify, in cooperation with industry and academia, the areas of wear and corrosion research of utmost urgency and needs and organize consortia for a 50/50 funded research investigation;
7. set up telephone technical advisory and referrals services;
8. set up computer data base with a complimentary suite of expert system;
9. operate on a cost recovery basis and in course of time become self-supporting;
10. the annual budget of the secretariat will be about $250 K and will comprise one technical expert and one administrative assistant.

5.3 CORROSION DAMAGE, DEFECTS, AND FAILURES

One of the consequences of corrosion is the failure of a machine or a system to function according to the specifications or the prescribed standards. The failure of a system to function according to specifications warrants a failure analysis of the system to identify the root causes of failure and suggestions to correct the situation. Corrosion damage, defects, and failures can have major and serious consequences on the operation of a system. Failure of a system with respect to corrosion is a final step or stage before the system undergoes corrosion damage and defects.

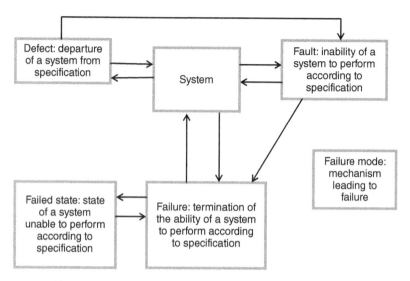

Figure 5.1 Interrelation among defects, failures, and faults (16).

A universal representation describing the interactions between defects, faults, and failures of a system is shown in Figure 5.1 (16):

The arrows in the figure imply that quantifiable relations possibly exist between a defect, a fault, and a consequent failure. A defect in materials science is any microstructural feature representing a disruption in the perfect periodic arrangement of atoms in a crystalline material. The fundamental defects and their distribution in a given material can have a significant impact on the overall properties of the material. While such defects do not constitute flaws in the normal sense of the word they can nonetheless serve as anchors for the initiation of actual faults and subsequent failures. There are four fundamental defect types:

1. Point defects
2. Line defects
3. Planar and surface defects
4. Bulk defects

5.3.1 Point Defects

Point defects or sites are vacancies that are usually occupied by an atom, but are presently unoccupied. If a neighboring atom moves to occupy the vacant site, the vacancy moves in the opposite direction to the site that used to be occupied by the moving atom. The stability of the surrounding crystal structure guarantees that the neighboring atoms will not simply collapse around the vacancy. In some materials, neighboring atoms actually move away from a vacancy, because they can form better bonds with atoms in other directions.

5.3.2   Line Defects

Line defects are dislocations around which some of the atoms of the crystal lattice are misaligned. There are two types of dislocations: the edge dislocation and the screw dislocation. Dislocations are caused by the termination of a plane of atoms in the middle of a crystal. In such a case, the surrounding planes are not straight, but instead bend around the edge of the terminating plane so that the crystal structure is perfectly ordered on either side.

5.3.3   Planar and Surface Defects

An important example of planar defect is the grain boundaries that occur where the crystallographic direction of the lattice abruptly changes. This occurs commonly when two crystals begin growing separately and then meet. Any types of corrosion to be discussed later are directly related to the nature and geometry of grain boundary structures.

5.3.4   Bulk Defects

An example of bulk defects are voids where there are simply no structural atoms. Another example consists of impurities that can cluster together to form small regions of a different phase such as precipitates. Impurities and precipitates also play an important role in the corrosion resistance of metals.

5.3.5   Fault

The growth of a defect into what becomes a fault or a faulty component really depends on many factors, which is predominantly the type of corrosion that is progressing. In the fault-tree analysis context, the fault event of a component is defined as a state transition from the normal state to a faulty state of that component. These state transitions are irreversible, which means that a faulty state does not return to the intended state even if the influences that caused the fault event in the first place disappear.

Corrosion phenomena are irreversible by nature as they change a metal into more stable oxidized states. In fact, the corrosion products can be converted into metals only by complicated and energetically expensive processes that eventually result in molten metals. However, not all corrosion processes lead to undesirable processes and products if, for example, a corrosion allowance is included in the system at the design stage. Some well-known examples of corrosion faults are encountered in electronic components where even very small amounts of surface corrosion can drastically alter the intended behavior of the components.

5.3.6   Connector Corrosion

Connector corrosion is well known and understood as a familiar problem that contributes significantly to electrical wiring failures. Connector corrosion also plays a significant role in several military and commercial aircraft incidents and accidents. For example, fretting corrosion in electronic components is because of the flaking of tin oxide film from a mated surface on tin-containing contacts. The problem becomes more serious and frequent as tin is used in place of gold for purely economic reasons as tin is less costly than gold. The only available solution for this hard-to-diagnose and often intermittent problem is to replace the defective component.

A problem discovered by an air force corrosion engineer was the corrosion of tin-plated electrical connector pins mated with gold-plated sockets. Fretting corrosion between these very small contacts appears to have been implicated in as many as six F-16 fighter aircraft crashes when their main fuel shutoff valves closed uncommanded (17).

Microscopic quantities of corrosion product can cause problems in complex electronic systems. An example of this is the formation of dendrites across circuit channels. In the presence of moisture and an electric field, metal ions can migrate to a negatively charged cathodic surface and plate out forming dendrites. The dendrites can be silver, copper, tin, lead, or a combination of metals. The dendrites can grow and eventually bridge the gaps between the contacts, causing an electric short and possibly arcing and fire. Even a small amount of dissolved metal can result in the formation of a relatively large dendrite.

Dendrite growth can be very rapid. Failures because of dendrite growth have been known to occur in less than 30 min or can occur in several months or longer. The rate of growth of a dendrite depends on the applied voltage, the extent and quantity of contamination, and surface moisture. The amount of contamination required for silver dendrites can be extremely small.

5.3.7 Failure

A failure may be defined as an unsatisfactory condition or a deviation from the original state or condition that is undesirable or unsatisfactory to a particular user or a context. The determination that a condition is unsatisfactory, however, depends on the failure consequences in a given operating context (18).

The exact dividing line of demarcation between satisfactory and unsatisfactory conditions will depend not only on the function of the item in question, but also on the nature of the equipment in which it is installed and the operating context in which the equipment is used. Thus the determination will vary from one operating organization to another. Within a particular organization, however, it is essential that the boundaries between satisfactory and unsatisfactory conditions be defined for each item in clear and unambiguous terms.

The judgment that a condition is unsatisfactory implies that there must be some condition or performance standard on which the judgment is based. However, an unsatisfactory condition can range from the complete inability of an item to perform its intended function to some physical evidence that it will soon be unable to function properly. For maintenance purposes, failures must therefore be further classified as: (i) functional failures and (ii) potential failures.

5.3.7.1 Functional Failure A functional failure is the inability of a system to perform and satisfy a specified performance standard. A total and complete loss of function is clearly a functional failure. However, it also includes the inability of a system to function at the level of performance that has been specified as satisfactory.

To define a functional failure of a component or a system, a clear and total understanding of the functions of the system is necessary. It is very important to determine all the functions of every item in the system that are significant in a given operational context, as a functional failure can be defined only in these terms.

5.3.7.2 Potential Failure When a particular function has been defined, it is possible to identify the indication that the particular failure is imminent. Under these circumstances, it might be possible or advisable to remove the suspected component or system from operational service as it is only in these terms that the functional failure mechanism can be identified followed by its elimination. The failure that occurs under these conditions is termed as potential failure.

A potential failure is a definite identifiable condition that indicates when functional failure is imminent. The fact that potential failures can be identified is an important facet of modern maintenance theory, as it permits maximum use of each system without the consequences associated with functional failures. The units or parts are replaced or repaired at the potential failure stage, so that potential failures preempt functional failures.

In some cases, the identifiable condition that indicates imminent failure is directly related to the performance criterion that defines the functional failure. The following three factors help to identify either a functional or a potential failure.

1. Clear definitions of the functions of a component or a system as they relate to the equipment or operating context in which the item is to be used.
2. A clear definition of the conditions that constitute a functional failure.
3. A clear definition of the conditions that indicate the imminence of the failure.

Thus it is not only important to define the failure but also necessary to probe the exact evidence by which it can be identified.

5.3.7.3 Consequences of Failures There are many consequences of corrosion-based failures ranging from minor failures of equipment, loss of productivity, minor injuries to personnel, and as serious as loss of lives. Failure analysis is the conventional method of relating a failure to its consequences as well as the lessons learned from the failure along with the necessary steps and precautions to be taken to avoid the future occurrence of similar failure. Failures may range from modest cost of replacing a failed component to the possible destruction of a piece of equipment and fatalities. The consequences of a failure determine the priorities of the maintenance or improvements in design to prevent future failures of a similar nature and degree.

The more complex the design of the equipment, the more ways by which the equipment is likely to fail. The four groups of corrosion-based failure consequences are the following:

1. Safety-related issues
2. Operational problems
3. Nonoperational consequences
4. Hidden failure consequences.

The most important consideration in the evaluation of corrosion-related failure is safety, that is, whether the failure causes a loss of life or of function or secondary damage that could have adverse effect on operating safety. A critical failure is any failure that could affect adversely the safety of operation of both the equipment and the operating personnel. The term, direct effect, implies certain limitations. The impact of the failure must be immediate if it is to be considered direct. Further, the consequences must result from a single failure and not from a combination of the failure with another that is yet to occur. If a failure has no resultant effect on the system, it cannot by definition have a direct effect on safety.

It is to be noted that not every critical failure results in an accident. However, the question is not whether such consequences are inevitable, but whether they are possible. Safety considerations are conservative and rigid and they are assessed at the most conservative level. In the absence of proof that a failure cannot affect safety, it is considered by default as a critical issue that requires immediate consideration.

When possible critical failure is envisaged, it is imperative that all possible effort must be made to prevent its occurrence. Quite often, redesigning of one or more vulnerable items is all that is required to avoid potential failure. However, the design and manufacture of new parts and their subsequent use in service equipment can take a long time, of the order of a few months, and sometimes as much as years. Thus temporary measures to rectify the problem are often required.

Once the safety consequences are ruled out, attention turns toward the next set of consequences such as the effect of the failure on the operational capability of the system without any difficulty. A failure has operational consequences whenever a need to correct the failure disrupts the planned operations. The operational consequences consist of the need to abort an operation after a failure occurs, the delay or cancellation of other usual operations to make unanticipated repairs, or the need for operational limitations until the necessary repairs are made. A critical failure may be viewed as a special case of a failure with operational consequences.

In such a case, the consequences are those of economic in nature and consist of imputed cost of lost operational function. Nonoperational consequences consist of many kinds of failures that have no direct adverse effect on the operational capability. This is illustrated by the failure of a navigational unit in a plane equipped with a highly redundant navigation system. As other units ensure availability of the required function, the failed unit can be replaced at some convenient time. Thus the costs, because of such a failure, are limited to the cost of corrective maintenance.

Hidden failures comprise another important class of failures with no immediate consequences as the failures of hidden function items are responsible for the failures. By definition, hidden failures have no direct adverse effects, that is, if they did, these failures would not be hidden. However, the ultimate consequences of hidden failure can be major if the failure is not detected and corrected in time. In other words, the consequence of any hidden function failure is increased exposure to the consequences of a multiple failure.

5.3.7.4 Selected Examples of Failure and Their Prevention
Cathodic protection of water mains: Here two examples of cathodic protection: (i) ductile iron main and (ii) cast-iron-lined main are discussed.

5.3.7.4.1 Ductile Iron Main The valve was buried in 1995 in the soil area of 800–900 Ω cm resistivity thus providing active conditions. The 8-in. valve was epoxy-coated with a zinc anode attached to the top body by an exothermic weld. The valve is a two-part casting fabricated with a sealing ring and 5/8-in. diameter bolts around the flange. The larger bolts create a continuous electrical circuit. This valve failed as a result of a combination of ½-in. diameter bolts holding the top body instead of 5/8-in. bolts and the corrosion affecting the undersized ½-in. bolts that held the valve together. The bolts were uncoated and made of carbon steel. As the valve was epoxy-coated, the corrosion was mainly on the 12-in. coupling bolts and on the undersized ½-in. bolts that held the valve together. These were uncoated carbon steel bolts. The ½-in. bolts did not provide electrical continuity between the top body and the valve as with larger bolts, resulting in the lower part of the valve being unprotected (Fig. 5.2).

In the valve system, bolts are generally the first to corrode. In chambers, the cause of corrosion is easy to recognize and corrective action can be taken in early stages. The chambers are generally filled with water and soil that run into them from the road. Further water from the ditches and along the water main enters the manhole. Leaks can be reduced, but with traffic and shifting of the chamber in the frost cycles

Figure 5.2 Twelve-inch bolt from coupling. (David Raymond, City of Ottawa, Public Works & Services.)

Figure 5.3   Aluminum ladder rungs. (David Raymond, City of Ottawa, Public Works & Services.)

it becomes very difficult to eliminate the inflow of water. Chambers in these areas are found with totally corroded bolt heads. Bolts with stems reduced to pencil thickness are also noted. Rungs of aluminum ladders for access to the chambers have completely disappeared.

Corrective action consists of replacing mild steel bolts with 304 stainless steel bolts coated with a wax-based primer to reduce the corrosion of bolts in flooded chambers.

In the chambers, crevice corrosion, galvanic corrosion, and general corrosion of the aluminum ladder were observed (Fig. 5.3). Crevice corrosion was also observed on couplings, bolts, and valves. The aluminum steps in some valve chambers were attacked by chlorides/phosphates.

5.3.7.4.2   Cast-Iron-Lined Main   A chamber in service for a 10-year period in which rebuilt valves were present was found to be saturated with chlorides and phosphates. The chlorides caused the corrosion of bolts to such an extent that the chambers were leaking. The chambers housed 24- and 16-in. valves along with air drain-out. All the bolts on the valves needed replacement and the aluminum rungs were completely corroded.

The corrosion prevention strategy consisted of wrapping the valve with wax tape, along with magnesium anodes. The chamber was thoroughly cleaned and valves wrapped in wax tape as shown in Figure 5.4. As the chamber was entered for a limited number of times, the corroded ladder was left intact and the chamber entered with the restraining device and/or portable ladder.

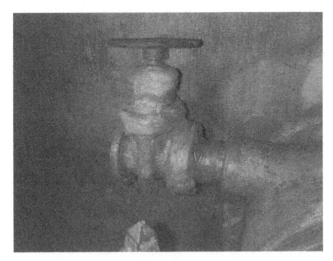

Figure 5.4 Tape-wrapped air valve. (David Raymond, City of Ottawa, Public Works & Services.)

The use of wax taping and magnesium anode resulted in very little corrosion, requiring the normal maintenance and inspection of the chamber for valve integrity.

Corrosion of bolts in areas other than valves was observed as shown in Figures 5.5 and 5.6. The economies of cathodic protection compared with the replacement of the water main are illustrated in Table 5.6.

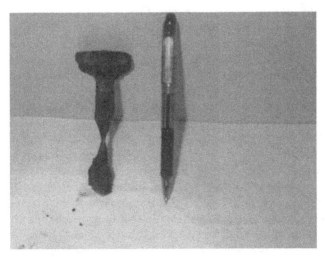

Figure 5.5 Four-inch T-bolt from 8-in valve. (David Raymond, City of Ottawa, Public Works & Services.)

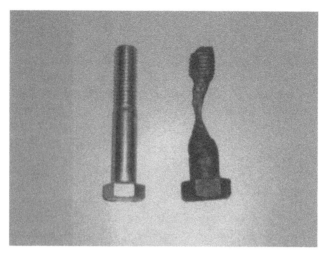

Figure 5.6   Hydrant bolt from base flange. (David Raymond, City of Ottawa, Public Works & Services.)

TABLE 5.6   Cost of Cathodic Protection and Replacement of Water Main

1 km	Replacement ($)	Cathodic Protection	Savings ($)	Percent Cost
6-in. water main	800,000	$20,000 × 3 cycles	740,000	7.5
12-in. water main	1,100,000	$37,500 × 3 cycles	987,500	10.5

5.3.7.4.3   Internal Corrosion of Aluminum Compressed Air Cylinders   The safety of aluminum compressed air cylinders has been an important problem in recent years. Safety procedures regarding their use and maintenance are necessary to ensure public safety. Sustained-load cracking (SLC) is a common failure mode of these cylinders. SLC is a metallurgical anomaly that develops occasionally in high-pressure cylinders made of aluminum alloys.

Visual inspection of the air cylinder made of 6161 aluminum alloy that failed a safety test because of high moisture content (>200 ppm) was done. An optical probe inspection showed internal corrosion. Figure 5.7 shows an out-of-service compressed air cylinder. The cylinder was cut open (Fig. 5.8) and the internal surface was corroded. The corrosion pattern appeared to be caused by condensation. Figure 5.9 shows the bottom portion of the cylinder, and the corrosion pattern appears to be uniformly distributed inside the cylinder. Figure 5.10 is a close-up view of the corrosion surface, which shows white spots. These spots are probably corrosion pits. Pitting corrosion is more like corrosion-induced cracking. Pitting seems to grow intergranularly (Fig. 5.11) and deepens along the boundary of the aluminum grains. Figure 5.12 is an enlarged cross section of a corrosion pit. The oxygen/aluminum ratios were determined at various locations in the cracks shown in Figure 5.12. The oxygen/aluminum

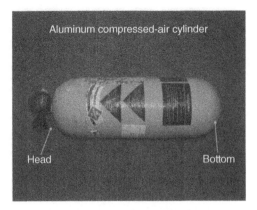

Figure 5.7 Out-of-service compressed air cylinder. (Copyright of her Majesty the Queen in Right of Canada, as represented by the Minister of Natural Resources, 2004, 2006.)

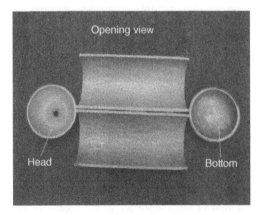

Figure 5.8 Overall view of internal surface. (Copyright of her Majesty the Queen in Right of Canada, as represented by the Minister of Natural Resources, 2004, 2006.)

ratio increases $\sim$1.5–2.0 as the crack location moves toward the crack tip. The ratios 1.5 and 2.0 likely represent Al_2O_3 and AlO_2^-, respectively, thus leading to the conclusion that corrosion-induced cracking occurred.

5.3.7.4.4 Some Common Failure Modes in Aircraft Structures Failure of an aircraft component can have catastrophic consequences such as loss of precious life and aircraft. Failure modes in aircraft are given in Table 5.7.

It is clear from the data in the table that fatigue mode of failure is more common than other modes of failure in aircraft. When the component is no longer able to withstand the imposed stress, failure is likely to occur. Thus failures are associated with stress concentrations, which can occur because of: (i) design defects and the presence

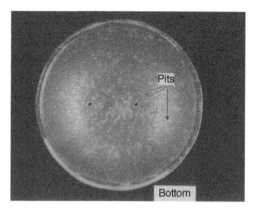

Figure 5.9   Internal surface of the bottom of cylinder. (Copyright of her Majesty the Queen in Right of Canada, as represented by the Minister of Natural Resources, 2004, 2006.)

Figure 5.10   Close-up view of internal surface of the opened cylinder. (Copyright of her Majesty the Queen in Right of Canada, as represented by the Minister of Natural Resources, 2004, 2006.)

of holes, notches, tight fillet radii; (ii) presence of voids, inclusions in microstructure; (iii) corrosion such as pitting, which can cause a local stress concentration.

Example 5.1   A nose undercarriage turning tube failed on landing after only 1300 flight cycles, well below the expected service life. Scanning electron microscopy (SEM) showed the fracture surface to have ductile appearance consistent with a ductile overload. Figure 5.13a shows fatigue striations indicative of a fracture mode, resulting in a fast fracture (overload) on reaching a critical crack length. The striations on the fracture surface consisted of distinct bands of repetitive units of striation spacing as seen in the figure. Measurement of striating spacing and band spacing at various points along the crack from the origin to the end enabled the determination of

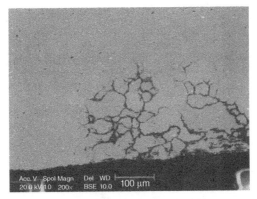

Figure 5.11 SEM photo of intergranular pit. (Copyright of her Majesty the Queen in Right of Canada, as represented by the Minister of Natural Resources, 2004, 2006.)

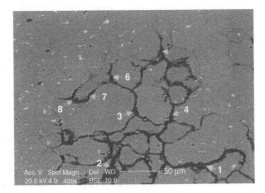

Figure 5.12 SEM photo of enlarged cross section of corrosion pit (19).

crack growth rate graphically. As the striations and bands are related to load cycles, a comparison of load cycles with the anticipated load spectrum enables the determination of the load cycles required to propagate the fatigue crack to the point of failure. The number of cycles that caused the failure was in the range that the component was in service, which indicated that the fatigue crack initiated very close to the beginning of the component's service life.

The origin of the crack was located at a notch on the surface of the tube. It is likely that the notch produced the required stress concentration in the surface of the tube, thereby reducing the initiation time of the crack. Most likely the notch on the surface of the tube occurred during manufacture.

Example 5.2 During routine service inspection corrosion of an upper service wing panel containing an access door was observed. The panel and door were made from an alloy plate to which aluminum catches were attached for securing the door

TABLE 5.7   Failure Modes in 2002

Mode	Failures (%)	
	Engineering Components	Aircraft Components
Corrosion	29	16
Fatigue	25	55
Overload	11	14
High-temperature corrosion	7	2
SCC/Corrosion fatigue/hydrogen embrittlement	6	7
Wear/abrasion/erosion	3	6
Brittle fracture	16	
Creep	3	

in the closed position. Stainless steel shims were fitted between catches and the aluminum plate.

The outer surface of the panel and door after stripping paint and removing the catches is shown in Figure 5.13b. There appears to be no damage to the plate and the door. The examination of the inner surface showed extensive exfoliation corrosion on the panel and the door in the catch positions (Fig. 5.14). Cracking originating from the catch position was observed in the stiffening rib. A cross section of plate containing an extensively corroded area is shown in Figure 5.15.

Exfoliation corrosion occurs when the attack occurs along the grain boundaries, in particular, when they are elongated and form thin platelets. The voluminous corrosion product causes splitting of layers of uncorroded material.

The material conformed to the specification, and the aluminum alloy is known to be subject to exfoliation corrosion. The corrosion was extensive at the catch position, which is attributed to the stainless steel shims fitted below the catches. The paint between aluminum and stainless steel shims deteriorated, resulting in galvanic corrosion with the stainless steel acting as the noble metal.

Example 5.3   A bolt from an aircraft flap control unit fractured in the threaded region of the shank near the shoulder with the head on installation after a major service. The bolt was made from cadmium-plated high-strength steel. The bolt conformed to the specifications and had ultimate tensile strength of ~1400 MPa.

SEM examination showed ductile features (Fig. 5.16) on the center of the bolt and intergranular features on the outer circumference (Fig. 5.17). Both sides of cracking were caused by static overload failure with the ductile features at the center present throughout. The intergranular appearance at the edge is suggestive of embrittlement, leading to premature failure at loads below those expected.

The embrittlement is attributed to cadmium plating (21) on the bolts applied to protect them from corrosion. During cadmium plating, hydrogen is absorbed by steel, and cadmium acts as a barrier for the escape of hydrogen. In high-strength steels

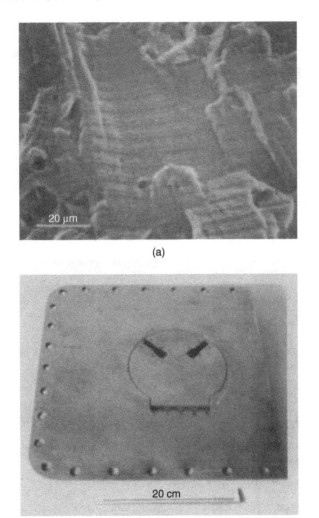

(a)

(b)

Figure 5.13   (a) Fatigue striations on fractured surface (19). (b) Outer surface of wing panel and door after paint stripping and removal of catches (19).

(>1100 MPa) hydrogen embrittlement (HE) occurs, but can be minimized by baking at 175 °C for 24 h or by treatment with 1:1:1 mixture of nitric, acetic, and phosphoric acids.

5.3.7.4.5   Premature Failure of Tie Rods of a Suspension Bridge   Tie rods of a newly built suspension bridge in service for only 6 winter months at temperature −20 °C failed because of cleavage and lack of low-temperature toughness of the steel.

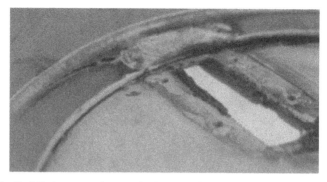

Figure 5.14 Exfoliation corrosion on the inner surface of panel and door around a catch location. (Reproduced by permission of the Society of Petroleum Engineers (20).)

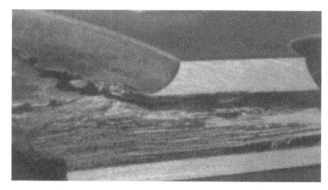

Figure 5.15 Cross section through panel showing exfoliation corrosion (Reproduced by permission of the Society of Petroleum Engineers (20).)

Fatigue crack propagation because of cyclic and uneven loading was also a contributing factor. The use of steels of higher toughness with intrinsic weathering resistance has been advanced as a remedial measure.

The bridge is 150 m in length with three sections. The middle section is 70 m in length sandwiched between four columns that support tie rods. The other two sections are 40 m in length. There are 32 tie bars, which are 90 mm in diameter. The bars are attached to several columns, four overhead columns that rise above and eight columns that drop below to the base of the bridge. Figure 5.18 shows a general view of the bridge. Figure 5.19 shows a close-up view of the exposed tie rods and corrosion of the threaded portions. Figure 5.20 is a close-up view of an upright column with a missing tie rod.

Figure 5.21 is a close-up view of tie rods showing corroded threads, nuts, and washer.

Metallographic examination showed the cracks initiated at the first thread immediately adjacent to the nut. Beach marks were present indicating the presence of fatigue.

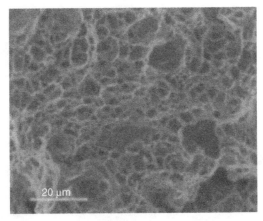

Figure 5.16 Ductile fracture surface of center of bolt. (Reproduced by permission of the Society of Petroleum Engineers (20).)

Figure 5.17 Intergranular region of fracture surface of bolt. (Reproduced by permission of the Society of Petroleum Engineers (20).)

The final area of the fracture showed the presence of a shiny granular appearance, indicating a sudden brittle (cleavage) fracture, which was also confirmed by the SEM photographs showing cleavage steps and river pattern.

Fractured tie rods were sectioned and examined with an optical microscope, and no evidence was found for microcracks at the root of the threaded notch and along the fracture path.

Conclusions

1. The presence of beach marks shows fatigue as a major factor in the failure; relatively small areas of crack propagation zone indicate a short service life.
2. Steel 350w is unable to withstand −20 °C temperature.

Figure 5.18   General view of bridge. (Reproduced by permission of the Society of Petroleum Engineers (20).)

Figure 5.19   Close-up view of exposed tie rods. (Reproduced by permission of the Society of Petroleum Engineers (20).)

3. Fatigue cracks initiated next to a nut on the washer side because of stress concentration when the nut is tightened.
4. Lack of microcracks shows the absence of machining defects.
5. Use of steel (350 W.T., 350 AT, A 370) is recommended. (22)

5.3.7.4.6   Hot Water System Corrosion   A schematic diagram of the domestic hot water system is given in Figure 5.22. Visual inspection led to the following observations:

Figure 5.20 Close-up view of upright column with missing tie rod. (Reproduced by permission of the Society of Petroleum Engineers (20).)

1. Different metals and alloys such as copper galvanized steel, cast iron pipes, and brass joints are used in hot water distribution line.
2. A noticeable temperature difference is observed in the heat exchanger loop where the water in the hot water tank is circulated to a heater. No mechanical pump is used in this loop and the water flow is by convection. Slow water circulation results in a larger temperature difference between the upper and lower pipe sections. A temperature difference is also seen in the heat circulation loop where the circulated hot water is mixed with incoming cold water. Severe corrosion is observed in these areas.
3. In some sections of the pipe, a large area of the copper pipe was in contact with a relatively small area of galvanized iron and steel pipe.

Water samples taken from various locations (23) indicated high iron content in samples taken from locations where the cold water and hot water are mixed, for example, the heat exchange loop and hot water recirculation pump (Fig. 5.22).

Visual examination and the metal ion content of the water samples shows internal corrosion of pipe in various locations in the hot water distribution system that compromises the quality of hot water. The internal pipe corrosion is because of the following factors or a combination of the factors.

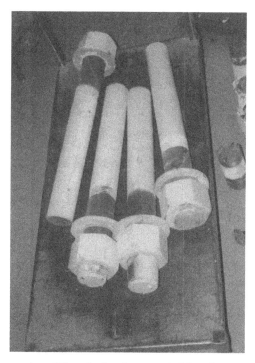

Figure 5.21 Close-up view of tie rods showing corroded threads, nuts, and washer. (Reproduced by permission of the Society of Petroleum Engineers (20).)

GALVANIC CELL FORMATION   When dissimilar metals are in contact in an electrolyte, a galvanic cell is established because of the electrochemical potential difference (24). The consequence of the galvanic cell is the corrosion of the less noble metal. For example, when copper is in contact with iron in the presence of an electrolyte, the iron anode will corrode to generated Fe^{2+} ions, which on further reaction with oxygen forms brown iron oxide while oxygen reduction occurs at the copper cathode. The reactions are

$$Fe \rightarrow Fe^{2+} + 2e^-$$

$$Fe^{2+} + 2H_2O \rightarrow Fe(OH)_2 + 2H^+$$

$$2Fe(OH)_2 + {}^1\!/_2O_2 \rightarrow Fe_2O_3 + 2H_2O$$

$$\text{Cathodic reaction : } {}^1\!/_2O_2 + H_2O + 2e^- \rightarrow 2OH^-$$

The reduced pH in the anode area accelerates the corrosion of iron.

TEMPERATURE DIFFERENCE   The severe corrosion that occurs in heat exchangers and heat circulation pump locations is explained by the temperature effect. A potential

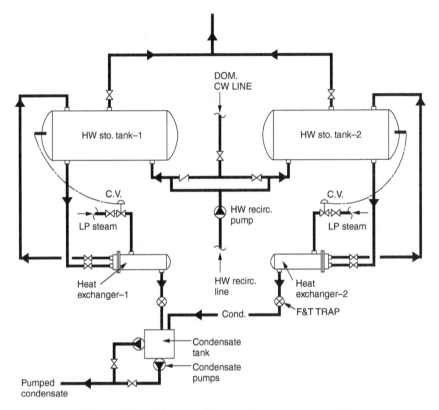

Figure 5.22   Schematic diagram of hot water system (19).

difference can result in a temperature gradient even within the same material. For example, a pipe with hot water at one end and cold water at the other end can result in a potential difference. A temperature difference of 60 °C could result in as much as 720 mV of electrochemical potential difference, which may be enough to cause corrosion in the area of lower temperature. There is no mechanical pumping in the heat exchange system. The circulation of water depends solely on convection, which creates a slow flow and larger temperature difference. All these factors contribute to severe corrosion of iron.

LARGE CATHODE AND SMALL ANODE   This scenario leads to accelerated corrosion of the anode. If a small section of iron or steel pipe is in contact with a large piece of copper pipe, the area effect will accelerate the corrosion of iron pipe that acts as the anode of the galvanic cell.

   In summary, the corrosion of a domestic hot water system in the building is because of the use of dissimilar metal pipes and associated components. The temperature difference in the heat exchanger and heat circulation locations causes severe corrosion. An area effect also contributes to the corrosion. The remedial measures must involve reducing the galvanic cell effect by minimizing the area and temperature effects.

The high amounts of Al and Zn reported may be because of the use of galvanized pipes and a sacrificial anode cathodic protection system in the hot water tank.

5.3.7.4.7 Cathodic Protection of Steel in Concrete   More than 5 billion dollars are spent annually in repairs to concrete structures such as buildings, bridges, parking garages, and other structures. Carbon dioxide enters the concrete and reacts with the lime, forming carbonic acid, and lowers the pH of the medium. Further, the chloride in deicing salts along with oxygen and water creates an aggressive corrosive environment. This results in an electrochemical reaction leading to delamination. The rebar corrodes, and the resulting rust is voluminous, leading to cracking, spalling, and delamination of the concrete. This corrosion process is illustrated in Figure 5.23.

Cathodic protection is one of the methods to mitigate the corrosion of steel in concrete as illustrated in Figure 5.24. Some of the factors to be considered in this regard are: the remaining service life of the structure should be more than 10 years; delamination and spalls should be less than 50% by weight of concrete; half-cell

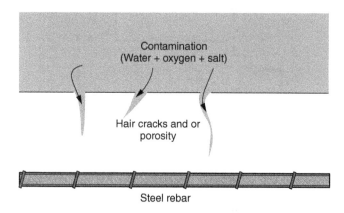

Figure 5.23   Atmospheric contamination (19).

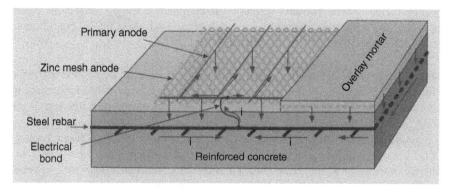

Figure 5.24   Galvanic anode cathodic protection in concrete (19).

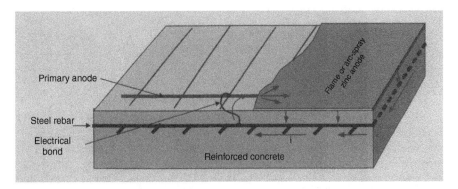

Figure 5.25   Arc-sprayed galvanic layer anode system (19).

potential should be less than −200 mV, indicating the breakdown of passive film; the structure should be sound; the reinforcing bars should be electrically continuous; AC power should be available.

The two types of cathodic protection are: (i) sacrificial and (ii) impressed current systems. The sacrificial anode system typically uses magnesium, zinc, or aluminum and their alloys (Fig. 5.25). These metals or alloys act as anodes when coupled with steel and its alloys. These metals or alloys act as anodes when coupled with steel and preferentially corrode. Magnesium is often used in fresh water media while zinc and aluminum are used in seawater and brackish water media.

Impressed current cathodic protection requires: (i) DC power supply (rectifier); (ii) an inert anode such as catalyzed titanium anode mesh; (iii) wiring conduit; (iv) an embedded silver/silver chloride reference electrode. A schematic diagram of an impressed current cathodic system is shown in Figure 5.26. By an impressed current, the potential of the steel is adjusted to values greater than −850 mV, thus making the steel bar cathodic and prevent the corrosion (25).

5.3.7.4.8   Corrosion of Aluminum Components in the Glass Curtain Wall of a Building   Corrosion was observed on the aluminum pressure plates and dress caps that hold the glazing on a curtain-walled building. The dress caps with L-shaped cross sections were externally clad with thin copper sheet.

Figure 5.27 shows a pressure plate and a dress cap. The aluminum extrusions were hidden with rubber strips and screws (dress caps). The dress caps were clad with copper sheet. The pressure plate with one of the dress caps is shown in Figure 5.27.

In the course of several years after building construction, the copper cladding was separating from the dress caps near the base of the building. In the course of time, the aluminum extrusions were covered with corrosion products (white–gray in color) indicating general corrosion and a pitted aluminum surface (Fig. 5.28). The copper cladding sheet was bulged at a few locations along the dress cap edges, and a typical cross section of the bulge is shown in Figure 5.29. The bulge was caused by the voluminous corrosion product trapped between the aluminum and copper ultimately leading to the deformation of soft copper sheet. The figure shows aluminum

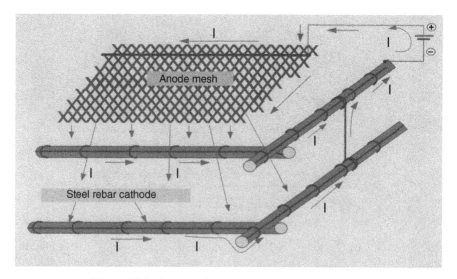

Figure 5.26   Impressed current system line diagram (19).

Figure 5.27   Pressure plate with a dress cap. (Figure originally published in Reference 26. Reproduced with permission of the Canadian Institute of Mining, Metallurgy and Petroleum. www.cim.org.)

is severely corroded at one edge and cracked leading to detached copper sheet at this location. All of the copper cladding of commercially pure architectural grade was in excellent condition with no significant metal loss.

The major damage to dress caps was caused by the galvanic corrosion between copper cladding and aluminum extrusion. In the case of copper-containing solutions

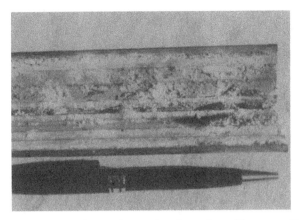

Figure 5.28 Aluminum corrosion products on interior surface of dress cap. (Figure originally published in Reference 26. Reproduced with permission of the Canadian Institute of Mining, Metallurgy and Petroleum. www.cim.org.)

Figure 5.29 Cross section through a bulge in copper cladding on a dress cap. (Figure originally published in Reference 26. Reproduced with permission of the Canadian Institute of Mining, Metallurgy and Petroleum. www.cim.org.)

running over aluminum, electrochemical reactions will cause the corrosion of aluminum and the deposition of copper. These reactions establish local galvanic cells that cause deterioration of aluminum. Copper is less prone to atmospheric corrosion and hence widely used in architectural applications. The slow corrosion rate of copper can be seen by the blue–green runoff seen on the copper roofs of public buildings.

The new dress cap substrate material should be either fiber-reinforced plastic (FRP) or type 316 stainless steel. FRP is a nonmetallic insulator and would eliminate galvanic corrosion. Type 316 stainless steel has good passivity in urban atmospheres and has been successfully used in contact with copper.

Figure 5.30 Hard deposits inside the tubes (20).

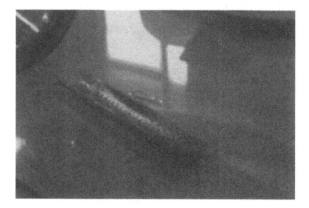

Figure 5.31 Deep pits inside the tube (20).

Contact between aluminum and metallic copper and/or copper-containing solutions must be avoided at all times under any circumstances.

The effects of galvanic (dissimilar metal) corrosion and their consequences must be made known to designers, architects, and engineers in other disciplines (14).

5.3.7.4.9 Corrosion in a Cooling Water System Cooling water tubes are used extensively in industry. Corrosion of cooling water tubes or pipes is a common phenomenon. The heat exchanger was opened to examine the extent of corrosion. Heavy deposits were revealed inside the tubes (Fig. 5.30). Isolated but deep pits were present under the hard deposits (Fig. 5.31).

The measured thickness of the tube, mechanical tests, chemical analysis, and etching showed the tubes to conform to the properties specified for SA179 tubes. Examination of cooling water side tubing showed mild pitting and a hard, sticky uniform deposit inside the tube (Fig. 5.32).

Figure 5.32   Hard deposit inside GE 16 tube (20).

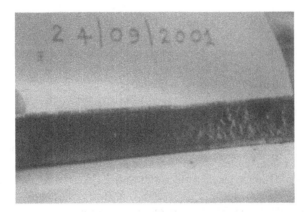

Figure 5.33   Perforated tube (20).

At some points, the deposit was in the form of lumps. Deep pits were also observed below the solid deposits and perforation below the deposit in one case (Fig. 5.33). Measurement of the thickness of the tubes showed no general thinning.

Flow measurements showed flow rates lower than expected, and backwashing was observed. The backwash contained brownish siliceous matter and other solid impurities. All three heat exchangers had silt and dirt deposits and partially plugged tubes. The tube ends were corroded and thinned down. Perforation of the tubes originated from inside (cooling water side). Pitting appeared to be underdeposit corrosion.

A commercial cooling water program during start-up and stabilization of the cooling water tower after major repairs is likely to avoid the adverse results (27).

5.3.7.4.10   Pitting Corrosion of 90/10 Cupronickel Chiller Tubes   The cupronickel tubes in a water chiller suffered from severe corrosion damage during their relatively short service life (28). Figure 5.34 shows the top section of the condenser where a

Figure 5.34 Top section of condenser tube showing corrosion. (Figure originally published in Reference 26. Reproduced with permission of the Canadian Institute of Mining, Metallurgy and Petroleum. www.cim.org.)

large number of tubes are undergoing internal corrosion. Several tube sections were opened and examined by SEM and energy dispersive X-ray to study the corrosion morphology and identify the corrosion products.

Various sections of the opened tubes were examined by optical techniques. Figure 5.35a shows the pitting pattern where several pinhole-sized pits are surrounding a large pit and areas of different colors. Species of different colors consisting of copper salts such as $CuO/Cu(OH)_2$, $CuCl_2$, and Cu/Cu_2O were also present.

Microscopic examination showed pit initiation. Figure 5.35b shows a pit of the size of a pinhole located on the inside surface of the tube. Energy dispersive spectroscopic analysis was done at three locations, namely: (i) surface of the deposit layer; (ii) inside the pit; (iii) outside the pit. The bottom of the pit is shown in Figures 5.35c, d. EDS analysis showed O, Al, Si, Fe, and Cl peaks, and the deposit layer was iron oxide. Peaks because of Cl and S were seen.

EDS analysis at the bottom of the pit showed Cu, Ni, and Fe. The data showed no denickelification inside the pit as both Ni and Cu were found. There was no corrosion product at the bottom of the pit, but Cu was observed at the edge of the pit.

The main mode of attack is pitting of Cu–Ni 90/10 tubes. Impingement because of air bubbles on the tube surface will lead to the destruction of the protective film. Copper oxide ringlets were observed around the pits indicating corrosion occurred in an environment consisting of corrosive ions, moisture, and oxygen as shown in Figure 5.36.

Pitting probably occurs because of the reactions:

Low pH (anodic):

$$Cu - 2e^- \rightarrow Cu^{2+}$$

$$Ni - 2e^- \rightarrow Ni^{2+}$$

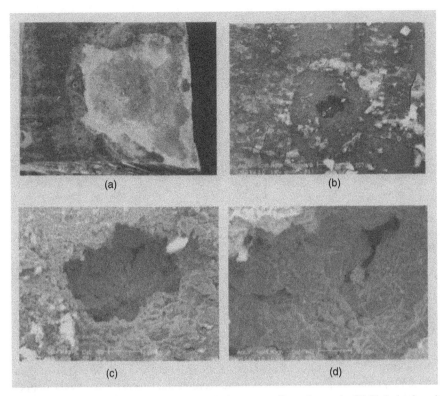

Figure 5.35 (a) Pitting pattern with several pits surrounding a large pit. (b) Pinhole size pit inside the tube. (c) Close-up of pit. (d) Close-up of inside of pit. (Figure originally published in Reference 26. Reproduced with permission of the Canadian Institute of Mining, Metallurgy and Petroleum. www.cim.org.)

Neutral solutions (pH ~7.0) (anodic):

$$Cu-2e^- + H_2O \rightarrow Cu(OH)_2 + 2H^+$$
$$Ni-2e^- + H_2O \rightarrow Ni(OH)_2 + 2H^+$$

Cathodic reactions:

$$\tfrac{1}{2}O_2 + H_2O + 2e^- \rightarrow 2OH^-$$
$$Cu^{2+} + 2e^- \rightarrow Cu \text{ (redeposit)}$$

Low pH reactions lead to dissolution of Cu and Ni. At the beginning of pit initiation copper and nickel hydrolyze leading to low pH.

Cathodic reaction involving oxygen produces a hydroxyl ion.

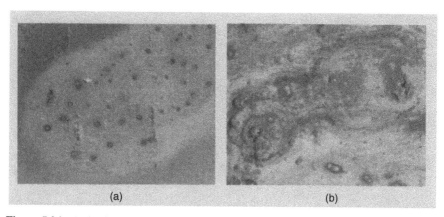

Figure 5.36 (a, b) Copper oxide ringlets round pits. (Figure originally published in Reference 26. Reproduced with permission of the Canadian Institute of Mining, Metallurgy and Petroleum. www.cim.org.)

Pit propagation can take place because of the redeposition of copper. This reaction is driven by low pH and high copper ion concentration inside a pit as observed by the redeposition of copper around a pit.

CONCLUSIONS Pit initiation takes place in areas where oxide film is damaged or broken. Pit propagation could occur without oxygen and is accelerated by the copper redepositing reaction. Thus in such a case, prevention of pit initiation is important.

5.3.7.4.11 High-Temperature Corrosion and Wear Problems High-temperature corrosion and wear occurs in: (i) waste incineration, (ii) fossil energy, (iii) pulp and paper; (iv) petroleum refining; (v) the chemical and petrochemical industries; (vi) mining and smelting operations.

One of the methods to alleviate corrosion and wear and its control is judicial selection of an alloy for the plant design and maintenance followed by a weld metal overlay of the plant equipment to avoid corrosion failures. Some of the industrial applications of uniform overlay technology are given in Table 5.8.

A failed 310 stainless steel tube in a waste-to-energy boiler is shown in Figure 5.37. The appearance of uniform composite tubes with alloy 625 overlay on Cr, Mo steel is shown in Figure 5.38. Figure 5.39 shows the 309 overlay on Cr, Mo boiler tube in service in a coal-fired boiler free from corrosion and cracking for 7 years (30).

5.3.7.4.12 Equipment Cracking Failure Case Studies Equipment fails either alone or in combination with other factors, including substandard materials, improper material selection, poor design, equipment abuse, unexpected stresses or environmental conditions, and poor maintenance practices and/or neglect. Many failures, in one way or another, involve human error to some extent.

TABLE 5.8   Applications of Overlay Technology

Application	Corrosives	Number of Boilers Using the Overlay
Waste-to-energy boilers	Municipal waste containing chloride, sulfur, alkali metals, zinc, and lead	59 (Alloy 625 overlay weld metal used)
Coal-fired boilers	Sulfidation attack, boiler tube wastage (50–60 mpy)	8 (Alloy 625 and 309 SS)
Pulp and paper digesters, Kraft recovery boilers	Thiosulfate and polysulfides, sulfate, chloride	21 (Overload with 309 SS) 11 (309 L on side walls) (625 on floor tubes)

Figure 5.37 Tube that failed in waste-to-energy boiler. (Figure originally published in Reference 29. Reproduced with permission of the Canadian Institute of Mining, Metallurgy and Petroleum. www.cim.org.)

CRANKSHAFT The crankshaft from an eight cylinder 2400 horsepower natural gas-fired engine, operating at 900 rpm, was used to drive a gas compressor. The engine operated well for 5.5 years with regular routine maintenance. The crankshaft was removed for inspection and polishing of the bearing journal surfaces. There was no cracking, but the critical crankshaft dimensions revealed the journal diameters to be 0.03 mm under tolerance. The journals were lightly machined, and a chromium layer applied by electroplating. The engine was reassembled and operated for about 2000 h when the crankshaft failed. On disassembling, the crankshaft was removed and was found to be cracked through the web between the last rod journal and the last main journal next to the fly wheel (Fig. 5.40).

Figure 5.38 Surface appearance of uniform composite tubes with alloy 625 weld overlay. (Figure originally published in Reference 29. Reproduced with permission of the Canadian Institute of Mining, Metallurgy and Petroleum. www.cim.org.)

The crack shown in the figure propagated through the entire cross-sectional area of the crankshaft web. The remaining ligament was broken open to expose the crack faces (Fig. 5.41). The failure was because of a high-cycle, low-stress fatigue crack that initiated in the main journal and propagated through the web to the fillet of the adjacent journal. Ratchet marks, beach marks, and a smooth texture typical of high-cycle fatigue were observed. The metallographic specimen with fatigue fracture, web, and journal zones is shown in Figure 5.42. Metallographic examination (Fig. 5.43) showed a thin layer of chromium on the journal side of the fracture. The fracture initiated at the tip of the chromium layer. By etching, a number of journal fillets were found to have chromium electroplating. In the case of the crankshaft, the fillet regions were not peened, and the chromium electroplating was inadvertently applied to many of the journal fillets, instead of using masking.

FAILURE OF ELECTRIC MOTOR SHAFT   A 350 horsepower electric motor drive shaft broke adjacent to coupling, which was connected to a three-stage compressor. The motor had a speed of 1200 rpm and operated for 5 years. After 3.5 years, the motor shaft was severely damaged. A repair with a weld overlay enabled the motor to run for 1.5 years, and the shaft broke in the region of the weld repair. The failure was consistent with high-cycle fatigue, and the 45° orientation was consistent with torsional fatigue. The fatigue crack propagated through the entire shaft. Figure 5.44 shows that some foreign material was embedded in the shaft below the surface and coincided with the fatigue initiation point. The foreign material was intentionally embedded to save welding and machining time so that the cost of repair could be minimized.

A polished and etched cross section of the fracture profile (F), weld overlay (w), foreign material (K), and original shaft material (S) is shown in Figure 5.45. The shaft material was AISI 1040 carbon steel and did not comply with Ni–Cr–Mo low-alloy steel and was softer than the shaft alloy.

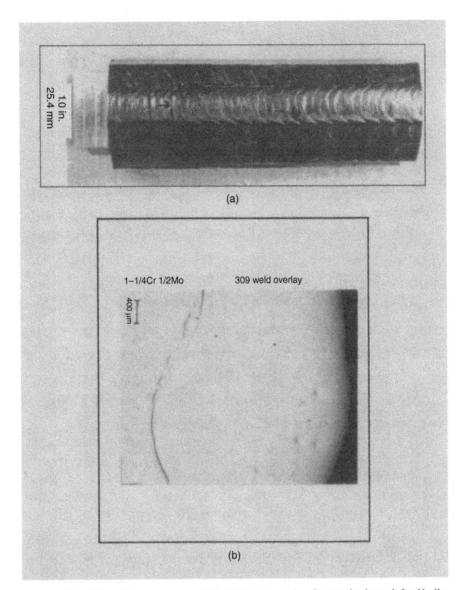

Figure 5.39 309 uniform overlay on 1 ¼ Cr ½ Mo boiler tube after service in coal-fired boiler. (a) Weld overlay with visible weld beads. (b) Weld overlay cross section showing no corrosion and no cracking. (Figure originally published in Reference 29. Reproduced with permission of the Canadian Institute of Mining, Metallurgy and Petroleum. www.cim.org.)

Figure 5.40 Appearance of large crankshaft crack. (Figure originally published in Reference 26. Reproduced with permission of the Canadian Institute of Mining, Metallurgy and Petroleum. www.cim.org.)

Figure 5.41 Fatigue fracture face showing crack growth direction. (Figure originally published in Reference 26. Reproduced with permission of the Canadian Institute of Mining, Metallurgy and Petroleum. www.cim.org.)

The failure appears to be torsional fatigue failure. The foreign substance embedded in the area of the fatigue initiation point is suggestive of the subsurface anomaly acting as a stress raiser responsible for fatigue initiation. The amount of fatigue crack propagation did not result in breaking of the shaft indicative of low applied stresses on the shaft. The problem was intensified by the low strength of the weld overlay material because of its low fatigue resistance. In addition, the shaft material did not comply with the required material properties and standards. The shaft strength was also low and had a detrimental effect on the fatigue strength of the shaft.

FAILURE OF PIPE CLAMP JOINT CONNECTOR The clamp that failed is shown in Figure 5.46. The clamp was used to join two ends of NPS 8 carbon steel steam

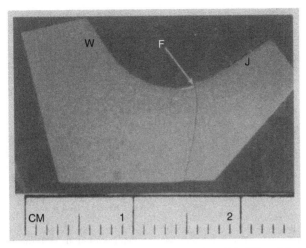

Figure 5.42 Metallographic specimen: fatigue fracture (F), Web (W), and Journal (J). (Figure originally published in Reference 26. Reproduced with permission of the Canadian Institute of Mining, Metallurgy and Petroleum. www.cim.org.)

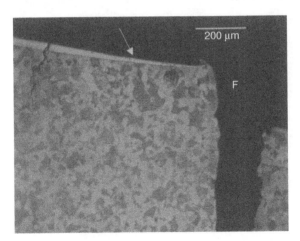

Figure 5.43 Chromium layer at fatigue initiation site. (Figure originally published in Reference 26. Reproduced with permission of the Canadian Institute of Mining, Metallurgy and Petroleum. www.cim.org.)

piping at a pressure of 11 MPa. The failure involved complete fracture of the bottom half-shell. The studs were bent because of the failure.

The failure was preceded by a steam leak, and attempts to stop the leak failed. The clamp fracture and pipe rupture caused extensive damage in the process building and there were no injuries as no one was present in the building.

Figure 5.44 Drive shaft fatigue fracture face, showing initiation point (arrow) and foreign keyway material (K) (19).

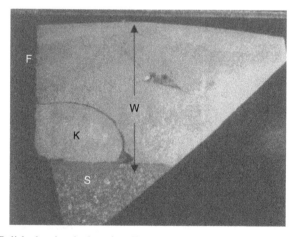

Figure 5.45 Polished and etched section: Fracture profile (F), weld overlay (W), foreign key material (K), and original shaft material (S) (19).

The fracture surfaces were brittle and exhibited a very coarse morphology. Additional cracks in the fractured half of the clamp were observed and the high-strength steel fasteners were free from cracks.

Chemical analysis of the surface deposits showed the presence of iron oxide, sodium carbonate, sodium chloride, and iron sulfide. The high sodium concentration of the scale was attributed to the zeolite ion exchange system used to soften the boiler feed water.

Metallography of the cross section of the failed clamp showed it to be austenitic stainless steel with extensive branched transgranular crack as shown in Figure 5.47.

Figure 5.46 Fracture pipe clamp and fasteners. (Figure originally published in Reference 26. Reproduced with permission of the Canadian Institute of Mining, Metallurgy and Petroleum. www.cim.org.)

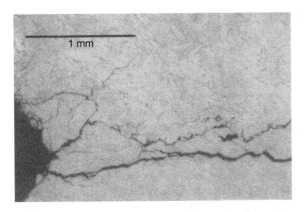

Figure 5.47 Transgranular cracking in failed clamp (19).

The pipe joint was leaking steam and water prior to the failure, and the chemical analysis of the leaking steam and water showed 10% sodium chloride. The high chloride and the highly stressed austenitic stainless steel with atmospheric oxygen created an ideal chloride stress-corrosion cracking (SCC) environment.

The failure could have been prevented by joining the steam piping by welding instead of using high-pressure clamps. It can be stated that the root cause of the failure is human error in choosing stainless steel.

FAILURE OF A STEEL COAL-CONVEYOR DRIVE SHAFT This failure was because of a transverse crack that passed through the right keyway near the center of the keyway (Sastri et al. (19), p. 499, Fig. 7.50). The shaft was cut to open the crack and expose the corroded mating surfaces. Two sets of crack arrest marks were concentric to each

Figure 5.48   Dark red and greenish spots on the outside surface (19).

corner of the keyway (Sastri et al. (19), p. 500, Fig. 7.51). Transverse cross sections had secondary cracks (Sastri et al. (19), p. 500, Fig. 7.52). The conveyor drive shaft failed because of corrosion fatigue in bending. The failure was initiated at both corners of the keyway. Many cracks were observed, indicating the presence of a large number of stress concentrator sites resulting from pitting. Fatigue cracks initiate at sites of maximum stress and minimum local strength.

The failure of the conveyor drive shaft can be prevented by the use of protective coatings or other means of shielding the shaft from a corrosive medium.

FAILURE OF COPPER PIPE IN A SPRINKLER SYSTEM   Straight sections and T-sections were examined along with the analysis of water for corrosion products for microbiological activity. The straight sections of the pipe showed dark red and greenish spots on the surface (Fig. 5.48); through-wall pits were present in the red spots. The T-sections did not have red spots (Fig. 5.49), but one sample had a crack (Fig. 5.50).

Figure 5.49   The exterior surface without red spots (19).

Figure 5.50   Visible crack on the exterior surface (19).

Figure 5.51   Internal surface of pipe with thick black deposit (19).

Internal examination of pipe sections showed thick black adherent scale (Fig. 5.51). Green deposits were also present (Fig. 5.52).

Some areas in the pipe showed the presence of both green deposits and shiny copper (Fig. 5.53). Pits were present in areas where copper is surrounded by green deposits (Fig. 5.54). The green deposits indicate significant corrosion. The black scale was carefully removed to identify the morphology of the pits (Fig. 5.55). The pits were located at the bottom of a hole. The morphology of the pits appeared to be characteristic of microbiologically influenced corrosion (Fig. 5.56).

Figure 5.52   Green deposits of copper compounds (19).

Figure 5.53   Shiny copper and green deposits (19).

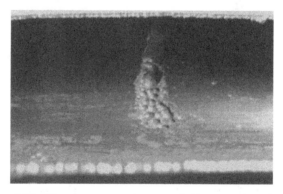

Figure 5.54   Pits where copper is surrounded by green deposits (19).

The T-sections were free from pitting, and the inside surface was clean. The water samples contained a significant amount of iron-reducing bacterial activity. A significant extent of microbiologically induced corrosion because of the presence of bacteria

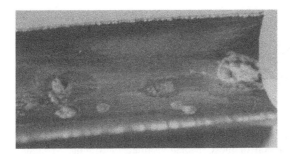

Figure 5.55   Well-developed pits impinging on one another (19).

Figure 5.56   Pits typical of microbiologically induced corrosion (19).

in the water resulted in saucer-sized pits, smooth sided pits, and bright shiny copper to matte-red clean areas. The black deposits – corrosion products from carbon steel – may cause underdeposit corrosion resulting in failure. Treatment of water with biocide may minimize microbiologically induced corrosion.

FAILURE OF ROCK BOLTS   A typical Swellex rock bolt shown in Figure 5.57 has the following modes of support: (i) beam building; (ii) suspension; (iii) pressure arch; (iv) support of discrete bolts.

The corrosion modes of rock bolts are: (i) uniform corrosion; (ii) pitting or crevice corrosion; (iii) galvanic corrosion.

The factors involved in corrosion are: low pH mine water because of oxidation of iron sulfides, or bacterial oxidation such as sulfate-reducing bacteria, carbon dioxide, chloride, and sulfate; mine air also contains SO_2 and NO_2.

A bolt fails when it can no longer provide the support it is designed for. Fracture is the separation of a solid body into two or more parts under the action of stress. It consists of crack initiation and propagation. Figure 5.58 shows the visual examination

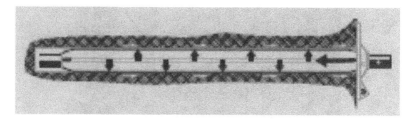

Figure 5.57 Swellex rock bolt. (Figure originally published in Reference 31. Reproduced with permission of the Canadian Institute of Mining, Metallurgy and Petroleum. www.cim.org.)

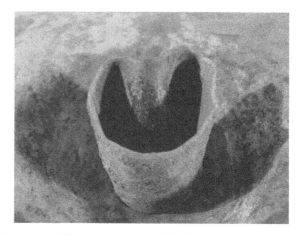

Figure 5.58 Visual examination of the fracture of a rock bolt (19).

of the fracture of a Swellex bolt. The failure plane at 45° and localized reduction of the surface were observed.

Brittle fracture is characterized by a rapid rate of crack propagation with very little deformation and is demonstrated by cleavage. The visual appearance of a fracture is shown in Figure 5.58.

Figure 5.59 shows a failed bolt, and the dimples are shallow with respect to the surface. The dimples are parallel to the direction of the fracture and thus denoted as shear fracture.

Figure 5.60 shows the fracture surface of a bolt from a hard rock mine.

The close-up of dimpled areas is shown in Figure 5.61. The dimples are shallow and inclined characteristic of ductile fracture. No sign of cleavage or quasi-cleavage is observed. The absence of river pattern excludes the possibility of brittle fracture.

Figure 5.62 shows an extremely irregular attack of the external surface of a longitudinal section of the rock bolt. The presence of an oxide layer in Figure 5.62 and the fragmentation of the metallic surface in Figure 5.63 support the analysis

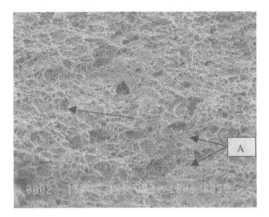

Figure 5.59   Failed bolt with dimples (19).

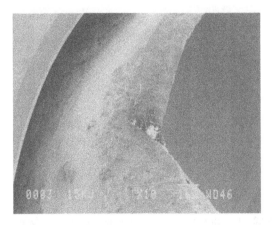

Figure 5.60   Fractured surface of a bolt from hard rock mine (19).

that for this rock bolt, corrosion was severe and probably contributed to the failure of the rock bolt. Figure 5.64 is a micrograph that shows typical equiaxial grains free from plastic deformation. However, metallography through the fracture surface shows the presence of elongated grains developed during shearing at failure (14, 27) (Fig. 5.65).

FAILURE ANALYSIS OF 316 L STAINLESS STEEL TUBING OF A HIGH-PRESSURE STEEL CONDENSER   Corrosion problems occurred during ammonia recovery. Figure 5.66 shows the simplified view of a high-pressure still condenser. The HPS condenser is made of 316 L seamless tubes of 25 mm diameter, 2.2 mm thick, and 6.4 in. long. The tube ends are joined to the tube sheet (100 mm thick) to form a bundle that is welded to the shell to make up the shell-tube condenser. The tubes are weld sealed at the

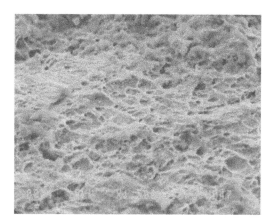

Figure 5.61   Close-up of dimpled surface (19).

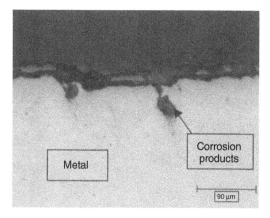

Figure 5.62   Longitudinal section showing extreme localized attack (19).

top of the tube sheet as shown in Figure 5.67. The condenser is installed vertically with severe operating conditions at the inlet of the top tube sheet as the temperature reaches 150 °C and pressure 250 psi.

Corrosion resulted in leak failures of many tubes and led to several shutdowns. The corrosion attack was at the top 100 mm where the tube was expanded (rolled) to join the top tube sheet with the rest of the tube intact. The rolled end of the tube is the corrosion end.

Repairing the condenser by plugging leaking tubes was unsuccessful as seen by the corrosion of plugs, top tube sheet, and seal welds. Plugging the tubes caused turbulent flow, which eroded the passive film on the steel plugs and caused severe corrosion of the plugs, top tube sheet, and steel welds.

A section of the tube with the inner surface covered by a mixed black/brown scale is shown in Figure 5.68. Circumferential cracks growing in a zigzag fashion

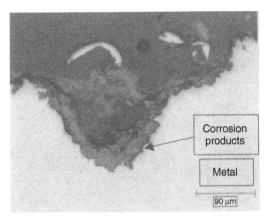

Figure 5.63 Fragmentation of metallic surface (19).

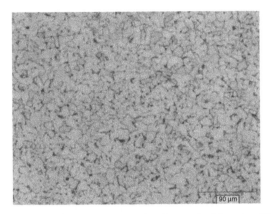

Figure 5.64 Fracture surface showing equiaxial grains (19).

and chloride SCC propagation in the axial direction are shown in Figures 5.69 and 5.70, respectively. Multiple crack initiation on the outer surface of the tube with crack growth from top to bottom is seen in Figure 5.71a. The cracks showed significant branching. In some cases, wide cracks appeared to develop into wide pits. Figure 5.71b shows one of the cracks that penetrated the whole wall thickness and resulted in a leak failure accompanied by significant plastic deformation.

In conclusion, the tubes suffered both internal and external corrosion attack. The corrosion was in the region within the joint between the tube and tube sheet. Inlet erosion–corrosion was observed in the tube because of the fluid that contained CO_2 and $(NH_4)_2CO_3$ in solution. SCC and crevice corrosion were evident on the external surface of the tube at the rolled end. It was concluded that SCC occurred because of chloride in the shielded area in the absence of proper venting. Some cracks grew and led to leakage failure. Poor venting and tube-end overrolling were thought to be the cause for the degradation of the tube in such a short time.

Figure 5.65   Figure showing elongated grains (19).

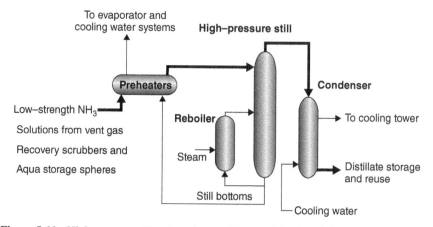

Figure 5.66   High-pressure still and condenser. (Figure originally published in Reference 26. Reproduced with permission of the Canadian Institute of Mining, Metallurgy and Petroleum. www.cim.org.)

Failure of a Landing Gear Steel Pin   A landing gear pin failure occurred in the cylindrical part of the pin whose outer surface was corroded (Sastri et al. (19), p. 515, Fig. 7.78). The fracture surface with convergent chevron patterns at the initiation site and a cover of corrosion product on the surface is shown in Sastri et al. (19, p. 516, Fig. 7.80). Corrosion pits were present at the exterior surface of the cylindrical pin. Pitting corrosion was also present at the fracture initiation site.

The fracture surface profile is shown in Sastri et al. (19, p. 515, Fig. 7.78). The presence of a circumferential ridge and depression in the cylindrical surface is to be noted in both the broken and reference pins. Comparing both the reference and broken pins it was concluded that the fracture of the broken pin initiated at the circumferential

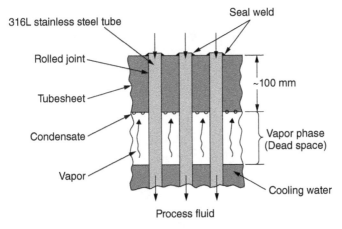

Figure 5.67 Top portion of HPS condenser. (Figure originally published in Reference 26. Reproduced with permission of the Canadian Institute of Mining, Metallurgy and Petroleum. www.cim.org.)

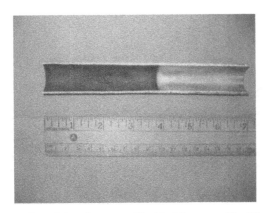

Figure 5.68 Inner surface of the tube covered by black/brown oxide. (Figure originally published in Reference 26. Reproduced with permission of the Canadian Institute of Mining, Metallurgy and Petroleum. www.cim.org.)

depression. Macroetching with 50% HCl for 30 s enabled the identification of the fracture initiation site.

A transverse cross section through the fracture initiation site was examined by metallography. The fracture surface profile was relatively flat and there was no crack branching. The microstructure showed dark-etching-tempered Martensite. There was no plastic deformation at the fracture initiation site.

It is concluded (28) that the pin failed because of fatigue initiated at the outside cylindrical surface where wear and pitting corrosion occurred. The fracture initiated

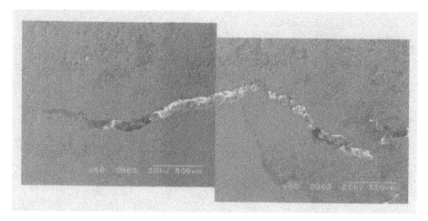

Figure 5.69 Circumferential crack in the tube. (Figure originally published in Reference 26. Reproduced with permission of the Canadian Institute of Mining, Metallurgy and Petroleum. www.cim.org.)

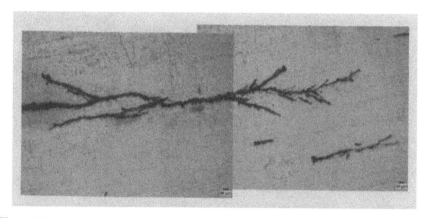

Figure 5.70 Chloride stress corrosion cracking in stainless steel. (Figure originally published in Reference 26. Reproduced with permission of the Canadian Institute of Mining, Metallurgy and Petroleum. www.cim.org.)

at a shallow circumferential groove and corrosion of the fracture occurred after rupture. There was no evidence of SCC.

HYDROGEN-INDUCED CRACKING (HIC) This mode of failure of a component involves hydrogen atoms absorbed in the steel leading to the development of internal cracks in low-strength steels. The cracks generally lie parallel to the rolling plane and the surfaces of the steel component. Residual or applied stress is not needed for HIC development.

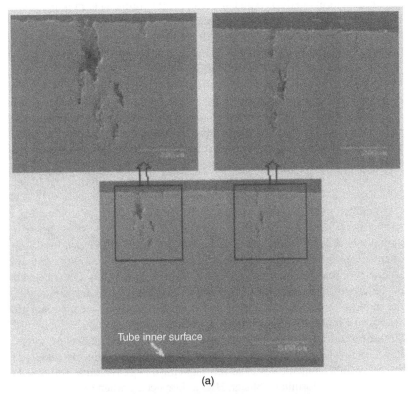

(a)

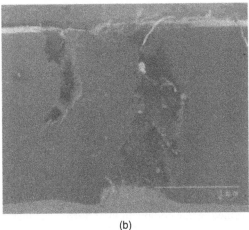

(b)

Figure 5.71 (a) Multiple initiation of stress corrosion cracks. (Figure originally published in Reference 26. Reproduced with permission of the Canadian Institute of Mining, Metallurgy and Petroleum. www.cim.org). (b) Crack that penetrated wall thickness leading to failure. (Figure originally published in Reference 26. Reproduced with permission of the Canadian Institute of Mining, Metallurgy and Petroleum. www.cim.org.)

Blisters are a form of HIC in which the buildup of hydrogen gas pressure at the initiated cracks or preexisting laminations results in localized deformation and bulging of the steel to the closest surface. High hydrogen pressures occur in blisters. Blisters occur when the hydrogen-induced crack is unable to propagate and is unable to link up with HIC on adjacent planes in the steel.

Stepwise cracking (SWC) is a form of HIC in which adjacent cracks in different planes link up in a stepwise manner. This can lead to through-wall cracking and the loss of vessel integrity. SWC does not require applied or residual stress. The hydrogen gas pressure in the HIC cracks can be as high as several hundreds of atmospheres and generate stress at the tips of HIC cracks, leading to plastic deformation of the steel. The linkup of cracks leads to HE cracking.

The presence of tensile stress in the component may cause individual ligaments of HIC to form a stack of through-thickness array. This is a necessary precursor to stress-oriented hydrogen-induced cracking (SOHIC). SOHIC is common in heat affected zones (HAZs) of welds in low-strength steels.

Examples of SOHIC failure in spiral-welded line pipe are common forms of failures. Failures were also encountered in electric-resistant welded pipes because of HIC. Failures of seamless line pipes have also been reported (30). HIC-related damage and failures of 20 line pipes in H_2S and CO_2 atmospheres have been reported (19). Despite the availability of adequately HIC-resistant line pipe since the late 1970s, failures are still occurring in pipelines constructed since then.

HIC occurs when hydrogen atoms diffusing through a line pipe steel get trapped as H_2 molecules in the inhomogeneities in the steel. A planar gas-filled defect is created, and it grows parallel with the vessel surfaces and traps more H atoms. The defect may grow to the extent of forming a blister. HIC failure occurs when the defects result in a blister. HIC failure occurs when the defects or blisters link up inside or outside the vessel surfaces with the possible linkage by SWC or SOHIC.

NITRATE STRESS CORROSION CRACKING (SCC) OF CARBON STEEL IN THE HEAT RECOVERY STEAM GENERATORS OF A COGENERATION PLANT   The carbon steel casings of heat recovery steam generators (HRSGs) in a cogeneration plant developed cracks that originated on the inside surfaces. The HRSGs receive hot exhaust from the primary GTs, which are fueled with pipeline natural gas. The GTs were initially operating without any suppression of nitrogen oxides.

After about 2 years of operation, cracks were found to penetrate through ¼-inch-thick casings. Steel samples cut from the casings showed: (i) fine cracks propagating through the steel from inside surface; (ii) presence of large amounts of nitrate on the inside surfaces. Nitrate ions originated from the nitric acid on the walls and the nitric acid formed by the combination of nitrogen oxides and moisture.

The outer surface of a steel plate cut from HRSG casing had a crack running perpendicularly to the butt weld. The majority of the cracks associated with welds joining the plates was perpendicular to the welds and ran through the weld. Repair-welded areas also cracked. Metallography showed small branched cracks originating at the inner surface and running across the plate. Highly branched intergranular cracks led to the conclusion that the cause of the failure is intergranular SCC because of nitrate.

TABLE 5.9 Corrosion Resistance of Some Steels

Alloy	Corrosion Resistance	PREN	Cost Ratio
Steel	Poor	–	1.0
Steel epoxy coated	Poor	–	1.7–2.0
S30400	Good/fair	18	3.8
S30453	Good/fair	21	4.3
S31600	Good/better	25	4.4
S31653	Good/better	27	4.7
Duplex 31803	Very good	36	4.4

5.3.7.4.13 Performance of Stainless Steel Rebar in Concrete Corrosion protection of steel rebar can be achieved by: (i) selection of corrosion-resistant steel; (ii) use of coatings; (iii) addition of corrosion inhibitors such as calcium nitrite to concrete mix; (iv) addition of concrete sealers; (iv) use of membranes; (v) use of thicker concrete overlay; (vi) cathodic protection.

Some of the reinforcing steel bar alloys are: cold-worked steel; epoxy-coated steel; 30400 stainless steel; 30453 stainless; 31600 stainless; 31653 stainless; Duplex S31803. Table 5.9 shows the corrosion resistance of some steels.

Carbon steel and 304 stainless steels were tested in 0.2% and 0.5% chloride and exposed for 2 years in 80% relative humidity. The carbon steel failed, and the stainless steel was free from cracking and corrosion. The concrete also remained intact without cracking. Life-cycle costs for the Öland Bridge in Sweden were calculated in the case of carbon steel and 304, 316 stainless steels for a time period of 120 years. For a duration of 18 years, the cost of carbon steel is less than type 304 stainless steel, which in turn is less than 316 stainless steel. For a life between 18 and 120 years, the costs of stainless steels are far lesser than carbon steel (19).

5.3.7.4.14 Corrosion of an Oil Storage Tank A 3 mm hole developed at the bottom of an oil storage tank after 25 months of service. The premature failure of the tank raised questions about the construction material of the tank as well as the corrosive species such as water in the contents of the tank.

The top and bottom sections of the tank were subjected to spectrographic analysis. The analysis showed it to be SAE-AISI 1006 or 1010 steel. The bottom section (70 cm long and 33 cm wide) was corroded to an extent of 80%. The surface corrosion along the edge appeared to be smeared and brushed as a result of cutting. The underlying metal had a blue–gray mill scale.

About 80% of the surface area was corroded. Some areas had deep corrosion that penetrated past the mill scale and into the metal substrate. Pitting of a severe nature is present when one moves beyond the mill scale.

Pitting was present at the bottom of the tank. The area with the hole is the most severely pitted surface. The pits were deep ($\sim$150 μm) with an external diameter of 50 μm.

The following equations represent the anodic reactions:

$$Fe^0 - 2e^- \rightarrow Fe^{2+}$$

$$\tfrac{1}{2}O_2 + H_2O + 2e^- \rightarrow 2OH^-$$

$$Fe^0 + \tfrac{1}{2}O_2 + H_2O \rightarrow Fe^{2+} + 2OH^-$$

$$Fe^{2+} + 2OH^- \rightarrow Fe(OH)_2$$

Pit propagation can become autocatalytic when one of the products of the reaction accelerates it. Pitting occurs in three stages: namely, initiation, propagation, and termination.

Pit initiation depends on: (i) imperfections or defects in the metal oxide, passive, or protective layer between the metal and the environment; (ii) exposure of the metal to an aggressive medium.

Pit propagation consists of corrosion driven by the potential differences between the anodic base of the pit and the surrounding cathodic area. Furthermore, the reaction products autocatalyze the propagation and make the environment more aggressive.

Pit propagation may terminate because of an increased resistance to local cell. This can occur because of the filling up of the pit with corrosion product or the filming of cathodic area with corrosion product. Pitting will cease when the metal is dry and may reinitiate when the metal becomes wet. Pitting tends to occur at defects or imperfections in the passive layer. Defects may be naturally occurring or as a result of mechanical damage. In fluid containers, stagnation of liquid favors pitting. The rate of pitting decreases as the number of pits increase following the availability of a reduced cathodic area.

The storage tank failed by pitting corrosion from the inside because of the presence of water. The source of water is not clear, although it is speculated that condensation of water vapor could have been responsible for it (19).

5.3.7.4.15 Underground Corrosion of Water Pipes in Cities A total of 17 pipe failures and soil samples were studied by metallurgical and analytical evaluation and the conclusions are: in almost all the failures, including the failure of the service saddle on an asbestos–cement 10-in. water pipe, corrosion from the soil side was to a great extent responsible for the failures.

The corrosion rates of cast iron pipes were in the range 0.21–0.58 mm/year in a typical environment with the average rate of 0.41 mm/year. The failure appeared to be some form of pitting, leading to perforation of the pipe walls. Graphitic corrosion involving selective leaching of steel matrix resulting in a loss of strength and mechanical properties ultimately leading to pipe rupture was observed. Both localized corrosion and graphitization resulted in wall thinning and cracks in the pipe.

Loss in mechanical properties such as reduced tensile strength of cast iron pipes contributed to the pipe failures.

The corrosion rate of ductile iron pipes in soils was in the range 0.62–2.5 mm/year with an average of 1.11 mm/year, which is twice the corrosion rate of cast iron pipe.

Figure 5.72   Cast iron pipe with large hole (19).

Failure of ductile iron piping was solely because of pitting corrosion resulting in pipe wall perforation.

There was no indication of pipe failure because of defects in the pipe material. In the majority of cases, corrosion of both ductile iron and cast iron pipes from the water side was negligible.

Deterioration of water pipe fittings such as service saddles contributes to water main failures. This type of failure may be because of galvanic corrosion as the pipe is cast or ductile iron, saddle is bronze, steel, or cast iron, and the service line is made of copper.

Chemical analysis of the soils where failures of pipe occurred showed the soils to be corrosive. The soils consisted of wet salty clay containing iron and resistivity in the range 820–4200 Ω cm. A strong relationship between the corrosion rate of buried cast and ductile iron pipes and soil resistivity was found.

Thin bituminous coatings on the ductile and cast iron pipes peeled and flaked. In addition, rust spots were present.

Figures 5.72 and 5.73 show the corrosive attack on samples of cast iron pipe and ductile iron pipe buried under the soil for 20 and 9 years, respectively. The large hole in cast iron pipe (Fig. 5.72) and the corrosion pit and perforation in ductile iron pipe (Fig. 5.73) show the severity of soil corrosion. It is suggested that cathodic protection can reduce the extent of corrosion of iron pipes.

5.3.7.4.16   Corrosion in Drilling and Well Stimulation   Acidizing is the most commonly used procedure for stimulating oil and gas wells. The main purpose of acid treatment is to dissolve rock or other plugging solids. The choice of the acid depends on the following:

1. Rock-dissolving capacity
2. Contact time
3. Solubility of the reaction products

Figure 5.73   Ductile iron pipe with large corrosion pit (19).

4. Extent of metal corrosion
5. Compatibility of acid with reservoir fluids
6. Density and viscosity of spent fluids

As common with any acidification, the temperature, acid concentration, amount of acid, velocity of injection, viscosity of acid, and fluid loss properties of formation have profound effect on the reaction rate.

Acidizing involves treatment with mineral acids such as HCl or HF or organic acids such as acetic or formic acid. The acids are generally used in combination with inhibitors. The general reactions involved are:

$$2HCl + CaCO_3 \rightarrow CaCl_2 + H_2O + CO_2$$
$$4HCl + CaMg(CO_3)_2 \rightarrow CaCl_2 + MgCl_2 + 2H_2O + 2CO_2$$
$$SiO_2 + 4HF \rightarrow SiF_4 + 2H_2O$$
$$2HF + SiF_4 \rightarrow H_2SiF_6$$
$$2CH_3COOH + CaCO_3 \rightarrow Ca(CH_3COO)_2 + H_2O + CO_2$$
$$2HCOOH + CaCO_3 \rightarrow Ca(HCOO)_2 + H_2O + CO_2$$

Iron control agents such as citric acid and organophosphorus compounds form iron complexes and prevent the formation of ferric hydroxide precipitate.

During acid treatment, sludge consisting of asphaltenes, resin, paraffin, and other high molecular weight hydrocarbons is formed. Addition of oil-soluble surfactants can prevent the formation of sludge.

The most commonly used metals/alloys in drilling and well stimulation are low-alloy carbon steel (API grades I-55, L-80, N-80 coiled tubing); quality tubing (grades QT-70, 80, 1000); chrome alloy steels (ASM grade Cr-13, duplex steel).

Low-alloy steels are used for wellbore tubulars. Corrosion limits for jointed tubing are set at <0.05 lb/ft^2 of tubing surface area representing a loss of 1/1000 in. in wall thickness. Pitting deeper than 0.381 mm is not acceptable.

The weight loss of 0.02 lb/ft^2 and zero pitting tendency should be met in the case of low-alloy steel.

The guidelines for high-alloy tubular materials are set at 0.05 lb/ft^2 chrome and duplex steels are used to prevent bottom hole corrosion. Acid corrosion causes significant loss in metal along with pitting

$$Fe^0 \rightarrow Fe^{2+} + 2e^-$$

$$2H^+ + 2e^- \rightarrow H_2$$

The inhibitors used in acid systems are organic amines, which form an inhibitor film because of Fe–N interactions. Other inhibitors such as acetylenic alcohols (propargyl alcohol) may be used.

The mechanism of inhibition by acetylenic alcohols consists of chemisorption of the alcohol on the metal followed by subsequent polymerization. For maximum inhibition, the hydroxyl group in the alcohol should be in the β position to the acetylenic function, and the acetylenic function should be terminal.

The synergistic blends consist of amines, acetylenic alcohols, surfactants, and oils to make a readily dispersed product.

CORROSION IN UNDERBALANCED DRILLING OPERATIONS The underbalanced drilling technique minimizes the harmful effects of drilling fluid invasion by reducing the hydrostatic pressure of the drilling fluid (32). This method requires nitrogen, which can be accomplished by using nitrogen membranes. When nitrogen membranes are used, the impact of injecting 5–8% of oxygen into the drilling fluid must be considered. The drilling fluids are water-based saline solutions that cause corrosion of casing, drill pipe, and bottom hole assembly.

Oxygen along with CO_2 or H_2S causes severe corrosion.

$$Fe \rightarrow Fe^{2+} + 2e \quad \text{Anodic reaction}$$

$$O_2 + 2H_2O + 4e \rightarrow 4OH^- \quad \text{Cathodic reaction}$$

Above pH 4, the precipitation of ferric hydroxide can occur. The corrosion because of oxygen may also result in pitting.

Hydrogen cracking may occur when moist hydrogen sulfide is present and the steel is high-strength steel under tensile stress. Corrosion inhibitors such as Nowcorr 800 consisting of a film-forming amine and a component that reacts with H_2S combat sulfide stress cracking (SSC).

Batch treatment of the inhibitor is based on the formula:

$$\text{Volume of inhibitors} = 12 \text{ l inhibitor} \times \text{inch diameter} \times \text{mile}$$

An additional 50% volume of inhibitor is added to compensate for the loss of inhibitor because of adsorption on solids.

In continuous treatment, Nowcorr 800 is added to the drilling fluid or to the acid blends. The concentration of the inhibitor depends on downhole conditions, temperature, total pressure, and partial pressure of H_2S.

In the case of underbalanced acid wash treatments in sour environments, Nowcorr 800 is used in addition to the acid corrosion inhibitor such as Nowsco's AI-275 and CI-30.

5.3.7.4.17 Environmental Cracking Environmental cracking refers to corrosion cracking caused by a combination of conditions that can result in SCC, corrosion fatigue, and HE. Stresses that cause environmental cracking arise from cold work, welding, grinding, thermal treatment or may be externally applied stresses during service and must be tensile to be damaging.

Stress variables: mean stress, maximum stress, minimum stress, constant load/constant strain, strain rate, plane stress/plane strain, biaxial, cyclic frequency wave shape.

The origin of stress can be: (i) intentional or applied stress consisting of quenching, thermal cycling, thermal expansion, vibration rotation, bolting, dead load, pressure, residual stress consisting of shearing, punching and cutting, bending, crimping and riveting, welding, machining, grinding, and products of corrosion reaction.

Stress cells can be present in a single piece of metal where a portion of the metal's microstructure possesses more stored strain energy than the rest of the metal. Metal atoms are at their lowest strain energy state when situated in a regular crystal array. Applied stresses include cyclic stressing consisting of cyclic frequencies and wave shapes. Although less well defined, the most important contribution to the modes of corrosion, such as SCC, are residual stresses. Applied stresses are usually less than half the yield stress or lower. The residual stresses are usually in the range of yield stress. Quantifying such residual stresses is often omitted in design with the erroneous conclusion that such stresses are irrelevant to design and performance.

Accumulation of corrosion products can result in stresses at restricted geometries where the specific volume of the corrosion product is greater than the corroding metal. These stresses can cause cracks to initiate and grow. Stresses from expanding corrosion products can cause adjacent metals to flow plastically, as occurs in nuclear steam generators in a process called "denting." Denting results from the corrosion of the carbon steel support plates and the buildup of corrosion product in the crevices between the tubes and the tube support plates. This process is known as "denting" as, when seen from the inside of tubes, these deformations seem to produce dents at the tube sheet locations. Similar stresses from the buildup of steel corrosion products cause the degradation of reinforced concrete. An equivalent expansion ratio of 3.0–3.2 has been measured because of the formation of corrosion products on steel bars embedded in concrete (33).

"Pack rust" is an example of the tremendous forces created by expanding steel corrosion products. The effect of pack rust that developed on an important steel bridge under repair has been documented (Roberge (16), Fig. 1.10, p. 14). The force of expansion was sufficient in this case to break three of the bridge rivets. This type of localized corrosion is known to be a serious derating factor when the load bearing capacity of a bridge or of any other infrastructure component is evaluated during inspection.

In some cases, the deformation because of the corrosion of aluminum in lap joints of commercial airlines is accompanied by a bulging (pillowing) between rivets, because of the increased volume of the corrosion products over the original material. This phenomenon was thought to be the primary cause of the Aloha accident in which a 19-year-old Boeing 737 lost a major portion of the upper fuselage (Fig. 5.74) in full flight at 24,000 ft (34). The "pillowing" phenomenon in which the faying surfaces are forced apart is schematically shown in Figure 5.75.

The major corrosion product identified in corroded fuselage joints is hydrated alumina, $Al(OH)_3$, with a particularly high volume expansion (6.0) relative to aluminum (0.9). Such a buildup of voluminous corrosion products can result in an undesirable increase in stress levels near the critical fastener holes and subsequent fracture because of the high tensile stresses resulting from the "pillowing."

Corrosion failures are very environmentally context specific, as well as consequentially context specific. In most of the industrial equipment context, a corrosion rate of 25 μm/year for steel is acceptable whereas such an amount of rust in the food industry would not be acceptable.

Another example consists of the development of a burial site for storing radioactive waste. A corrosion failure would occur if a minimum amount of radioactivity would leach in the groundwater after about 10 to 1 million years. Failure in this case has

Figure 5.74   Boeing 737 that lost major portion of fuselage (16).

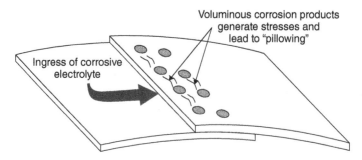

Figure 5.75 Pillowing of faying surfaces (16).

nothing to do with the integrity of the storage container but everything to do with the transport of radioactive species in the surrounding environment. This particular problem is challenging as it is not easy to monitor the performance of containers of radioactive waste because of the long times involved and their relative inaccessibility associated with radioactivity (35).

The F-16 fighter aircraft crashes and the Aloha incident described earlier are good examples of documented corrosion-related failures. Now we turn our attention to some corrosion-related accidents involving great loss of life.

5.3.7.4.18   The Bhopal Accident   This is probably the site of one of the greatest industrial disasters in history. Between 1977 and 1984, United Carbide India Limited, located within a crowded working class neighborhood in Bhopal, was licensed by the Madhya Pradesh State government to manufacture phosgene, monomethylamine, methylisocyanate, and the pesticide carbaryl, also known as Sevin.

In the early morning of December 3, 1984, water inadvertently entered the methylisocyanate tank, where >40 metric tons of methyl isocyanate were being stored. The addition of water to the tank caused a runaway chemical reaction, resulting in a rapid rise in pressure and temperature. The heat generated by the reaction in the presence of higher than normal concentrations of chloroform and an iron catalyst, which, combined with the corrosion of the stainless steel tank wall, resulted in a reaction of such momentum that poisonous gases formed and could not be contained by safety systems (36).

Methylisocyanate and other reaction products in liquid and vapor form escaped from the plant into the plant surroundings. There was no warning for the people in the vicinity of the plant as the emergency sirens had been switched off. The effect on the people living near the plant was immediate and lethal. Many died in their beds, others staggered from their homes, blinded and choking to death in the street. It is estimated that at least 3000 people died as a result of this accident, while figures for the number of people injured currently range from 200,000 to 600,000 with an estimated 500,000 typically quoted. The plant was closed down after the accident.

The Bhopal accident was the result of a combination of legal, technological, organizational, and human errors. The immediate cause of the chemical reaction was the seepage of 500 l of water into the methylisocyanate storage tank. The consequences of this reaction were exacerbated by the failure of containment and safety measures and by a complete absence of community information and emergency measures.

The long-term effects were aggravated by the absence of systems to care for and compensate the victims. Further, the safety standards and maintenance procedures at the plant were deteriorating and ignored for a long time.

5.3.7.4.19 Carlsbad Pipeline Explosion On August 19, 2000, a 75-cm diameter natural gas transmission pipeline operated by El Paso Natural Gas Company ruptured adjacent to the Pecos River near Carlsbad, New Mexico. The released gas ignited and burned for nearly an hour. Twelve people camping under a concrete-decked steel bridge that supported the pipeline across the river were killed and their three vehicles destroyed. Two nearby steel suspension bridges for gas pipelines crossing the river were damaged extensively amounting to a million dollars in property and other damages or losses (37).

The force of the rupture and the violent ignition of the escaping gas resulted in a 16 m wide and 34 m rupture along the pipe. A 15-m section of the pipe was ejected from the crater in three pieces measuring nearly 1, 6, and 8 m in length. The 8-m-long piece was found nearly 90 m northwest of the crater in the direction of the suspension bridge. Visual examination of the pipeline that remained in the crater as well as the three ejected pieces was done. All three pieces showed evidence of internal corrosion damage with one of the pieces showing significantly more corrosion damage than the other two pieces. Pits were observed inside the piece with the greatest extent of corrosion, and the pipe wall showed significant thinning. At one location, through-wall perforation was visible. No significant corrosion damage was visible on the outside portion of all three pieces or on the two ends of the pipeline lying in the crater. Samples were cut from the ruptured pipeline segments for further analysis by the Safety Board's Materials Laboratory in Washington, DC.

The drip between the closest block valve and the rupture site was removed from the pipeline and visual examination showed the drip to contain a blackish oil powdery grainy material. At the area of the large amount of the deposit, ~4 m from the drip opening, the deposit filled ~70% of the cross-sectional area of the drip. There was no significant amount of the material underneath and several centimeters away from the siphon drain at the closed end of the drip. There was no significant corrosion in the drip.

Interconnecting pits were observed inside the pipe in the ruptured area. The pits showed striations and undercutting features that are associated with microbial corrosion. A pit profile showed the chloride in the pits increased steadily from top to bottom. Sulfate-reducing, acid producing, general aerobic, and anaerobic bacteria were present in the deposits taken from two pit areas in the piece of line where internal corrosion was noted after the accident ~630 m downstream of the rupture site.

5.3.7.4.20 Guadalajara Sewer Explosion This corrosion failure is because of a combination of human error and shared responsibilities. The explosion killed 215 people and caused a series of blasts that damaged 1600 nearby buildings and injured about 1500 people. About nine separate explosions were heard ripping a jagged trench that ran about 2 km. The trench was contiguous with the city sewer system and the open holes at least 6 m deep and 3 m across. In many locations, much larger craters of 50 m in diameter were evident with numerous vehicles buried or toppled into the craters. An eyewitness said that a bus was "swallowed up by the hole."

The damage costs were estimated at $75 million. The sewer explosion was traced to the installation of a water pipe by a contractor several years before the explosion that leaked water on a gasoline line lying underneath. The cathodically protected gasoline pipeline had a hole within a cavity and an eroded area, which were all in a longitudinal direction. A second hole did not perforate the internal wall. The galvanized water pipe obviously had suffered stray current corrosion effects that were visible in pits of different sizes (38). The subsequent corrosion of the gasoline pipeline, in turn, caused leakage of gasoline into a main sewer line. The Mexican attorney general sought negligent homicide charges against four officials of Pemex, which is a government-owned company. Three representatives of the regional sewer system and the city's mayor were also charged.

5.3.7.4.21   Crashes of El Al Boeing 747 and Chinese Airlines Boeing 747-200F   On October 4, 1992, an El Al 747 freighter crashed in Amsterdam, killing all four people on board and more than 50 people on the ground. The cause of the crash was attributed to the loss of the number three and number four engines from the wing, which in turn caused a complete loss of control of the airplane. The reason for the separation of number three engine was found to be the breakage of the fuse pin. The fuse pin was designed to break when an engine seizes in flight, producing a large amount of torque. Both the engines were stripped of the right wing causing the Boeing 747-200 freighter to crash as it approached the airport (39).

There are other instances in which a Boeing 747 crashed. In December 1991, a Chinese Airlines Boeing 747-200F freighter crashed shortly after takeoff. A possible reason for the shearing away of the two right engines was that corrosion pits and fatigue weakened the fuse pins that held the strut to the wings. Constant variation of the pressure coupled with corrosion is a favorable force that can cause corrosion pits to grow into cracks such as the 4.3-cm crack found in one of the fuse pins of the El Al 747.

In both the El Al crash and the China airlines crash, the no. 3 and no.4 engines on the right side of the plane ripped away from the fuselage. It is believed that in the El Al crash, the inboard fuse pin failed because of corrosion cracking and fatigue, which caused the outboard fuse pin, already weakened by a crack, to fail. With these two pins malfunctioning, the no. 3 engine tore off the plane in such a manner that it may have taken no. 4 engine with it. Boeing began distributing a safety bulletin pertaining to the inspection of all fuse pins on their 74-100/200/300 versions that used Pratt & Whitney and Rolls-Royce engines. Both the El Al and China airline planes were Boeing 747-200 types with Pratt & Whitney engines.

The design of the fuse pin has been used since 1982 and in a 7-year period there have been 15 incidents of cracked pins. It was discovered that the pin failures resulted from the lack of primer, cadmium plating, and a corrosion inhibition compound. Since the El Al 747 crash, Boeing has also been trying to upgrade the 747 fleet. Specific targets on these models included fabricating new parts for the pylon-to-wing attachment for the Pratt & Whitney engines and the cost and time efficiency of inspection protocol.

5.3.7.4.22 Nuclear Reactor with a Hole in the Head On March 6, 2002, personnel repairing one of the five cracked control rod drive mechanism (CRDM) nozzles at Davis-Besse Nuclear Plant, Oak Harbor, Ohio, discovered extensive damage to the reactor vessel head. The reactor vessel head is a dome-shaped structure made from carbon steel housing the reactor core. The reactor vessel head is placed such that it can be removed when the reactor is shut down to allow spent nuclear fuel to be replaced with fresh fuel. The CRDM nozzles connect motors mounted on a platform above the reactor vessel head to control rods inside the reactor vessel. Reactor operators withdraw control rods from the reactor core to start the operation of the plant and insert the control rods to shut down the operation of the reactor.

The reactor core at the Davis-Besse Nuclear Plant sits within a metal pot designed to withstand pressures up to 17 MPa. The reactor vessel has 15-cm thick carbon steel walls and has adequate strength. The water used for cooling the reactor contains boric acid, which is corrosive to carbon steel. Hence, the inner surface of the reactor vessel is covered with a 0.6-mm thick layer of stainless steel. But water routinely leaked on to the reactor vessel's outer surface.

Because the outer surface is made of carbon steel without stainless steel protection, boric acid attacked the carbon steel until it reached the back side of the inner liner. High pressure inside the reactor vessel pushed the stainless steel outward into the cavity formed by the boric acid. Bending of stainless steel without breaking was observed. The cooling water remained inside the reactor vessel because of the thin layer of stainless steel. The plant's owner ignored many warning signs over the years, which led to this situation.

This corrosion problem exposed the problems within the staff of the regulatory commission, which wanted prompt inspections of all the 68 plants wherein this problem may occur. The regulatory commission relented and gave the owners permission to delay, leaving enough time for the hole in the lid to grow. Plants are generally designed with emergency equipment to cope with leaks, but the designs do not envisage failure of thick steel in that location.

A subsequent investigation by the Commission's inspector general found poor communications within the staff of the agency itself. The Commission had a photograph taken during a refueling shutdown break in 2000 that showed evidence of corrosion damage, but the responsible personnel failed to correct the situation.

5.3.7.4.23 Piping Rupture Caused by Flow Accelerated Corrosion A piping rupture caused by flow accelerated corrosion occurred at Mihama-3 at 3:28 pm on August 9, 2004, killing four and injuring seven people. One of the injured men died later making a total of five fatalities. The rupture was in the condensate system, upstream of the feed-water pumps. According to Japan's Nuclear and Industrial Safety Agency, the rupture was 60 cm in size. The pipe wall at the rupture site had thinned from 10 to 1.5 mm in thickness.

Mihama-3 is an 826 MW Mitsubishi-built pressurized water reactor (PWR) plant situated in Mihama, Japan, 320 km west of Tokyo. The carbon steel pipe carried the high-temperature steam at high pressure and the pipe was not inspected since the inception of the plant in 1976. In April 2003, Nihon Arm, a maintenance subcontractor informed Kansai Electric Power Company, the plant owner, that there could be a problem. Then the power company scheduled an ultrasonic inspection for August 2004. Four days before the scheduled inspection, superheated steam blew the 60-cm wide hole in the pipe. The steam that escaped was not in contact with the nuclear reactor and hence no nuclear contamination has been reported.

In response to the accident, Japan's Nuclear and Industry Safety Agency ordered four other power companies that owned nuclear plants with the same type of PWRs to conduct ultrasonic inspections of their pipes. The ultrasonic inspections were to involve nearly half of the country's 52 nuclear power plants.

Japan has the world's third largest nuclear power industry, after the United States and France. The government was planning to build 11 more reactors in the decade, thus increasing Japan's reliance on home-based nuclear power to 40% of electricity needs. However, there was local opposition to nuclear power generating projects in view of the most deadly history of nuclear power in Japan.

5.3.7.4.24 Sinking Ships On December 12, 1999, the Maltese-registered tanker *Erika* broke in two, some 70 km from the French coast off Brittany while carrying ~30,000 tons of heavy fuel oil. About 19,000 tons were spilled, which is equal to the amount of oil lost worldwide in 1998. The sunken bow section still contained 6400 tons of cargo and the stern had a further 4700 tons. The bow section sank within 24 h. The stern section sank on December 13 while under tow.

The economic consequences of the incident have been felt across the region; a drop in the income from tourism, loss of income from fishing, and a ban on the trade of sea products, including oysters and crabs, have added to the discomfort of local populations.

Corrosion problems had been apparent on *Erika* since 1994 with details readily available to port state control authorities and potential charterers. In addition, there were numerous deficiencies in her firefighting and inert gas systems, pointing to a potential explosion risk on the tanker. Lloyd's list reported that severe corrosion had been discovered just weeks before the incident. However, no immediate remedial action had been taken.

According to the US Coast Guard records publicly available and obtained by Lloyd's List, the *Erika* had been inspected in a variety of US ports on several occasions since 1994. The certificate of financial responsibility, a document legally required by tankers wishing to trade in US waters, had expired in March 1999 and not renewed as of November 30, 1999. During an inspection in Portland in July 1994, holes were discovered in the main deck coatings indicating that signs of corrosion were already in place less than 5 years prior to the catastrophe. The observation showed the presence of holes in the port side and starboard inert gas system risers, which are critical items of safety equipment. Malfunctions would tend to increase vulnerability to explosions.

Many of *Erika*'s problems were simply patched up, rather than properly repaired. In an inspection in 1997 in New Orleans, the US Coast guard ordered that no cargo operations requiring inert gas systems should be conducted until permanent repairs are done. Pinhole leaks remained in the fire main, contrary to *Safety of Life at Sea* convention regulations. There was yet more evidence of corrosion, with the ship's watertight doors not sealing properly and wasting on the door coamings. *Erika* switched from Bureau Veritas to Registro Italiano Navale in 1998, which authorized her to continue operations despite the French society's order for a full inspection.

Another example of major losses because of corrosion that could have been prevented and that was brought to public attention on numerous occasions since the 1960s is related to design, construction, and operating practices of bulk carriers. In 1991, 44 bulk carriers were either lost or critically damaged and more than 120 seamen lost their lives (40).

A highly visible case is the *MV Kirki* built in Spain in 1969 according to Danish designs. In 1990, while operating off the coast of Australia, the entire bow section became detached from the vessel. Luckily, no lives were lost; there was little pollution; and the vessel was salvaged. All through this period, neither coatings nor cathodic protection inside ballast tanks were used. The evidence for the failure is that serious corrosion had greatly reduced the thickness of the plate and that this combined with poor design to fatigue loading had led the failure. This case led to an Australian government report entitled, *Ships of Shame*. The *MV Kirki* was not an isolated case. There have been many others involving large catastrophic failures, although in many of these cases there was little or no hard evidence on what actually caused the ships to go to the bottom.

5.3.7.4.25   The Flixborough Explosion: SCC of a Chemical Reactor   The explosion in June 1974 at Flixborough was the largest ever peacetime explosion in the United Kingdom. There were 25 fatalities, as well as the near complete destruction of the NYPRO plant in North Lincolnshire by blast and then fire. This catastrophic explosion has been traced to the failure of a bypass assembly introduced into a train of six cyclohexane oxidation reactors after one of the reactors was removed owing to the development of a leak. The leaking reactor, like the others, was constructed of 12.3-mm mild steel plate with 3 mm stainless steel bonded to it, and it developed a vertical crack in the mild steel outer layer of the reactor from which cyclohexane leaked, leading to the removal of the reactor. One of the factors contributing to the crack was SCC, resulting from the presence of nitrates contained in the contaminated river water being used to cool a leaking flange.

5.3.7.4.26   Swimming Pool Roof Collapse   In 1985, 12 people were killed when the concrete roof of a swimming pool collapsed after only 13 years of use. The roof was supported by stainless steel rods in tension, which failed because of SCC. There have been other incidents associated with the use of stainless steel in safety critical load bearing applications in the environment created by modern indoor swimming pools and leisure centers. The collapse of the ceiling above a swimming pool showed how a simple structural concept could be sensitive to the loss, through corrosion, of support from one of many hangers.

The Federal Materials Testing Institute, based in Duebendorf, Switzerland, and the Federal Materials Research and Testing Institute of Berlin concluded that the collapse of the swimming pool roof was the result of chloride-induced SCC. The steel rods had been pitted, causing the roof to cave in. The roof collapsed in a zipper-like fashion, starting with the corroded rods. The collapse continued as the remaining rods were unable to bear the increased load. The chloride was either already present in the concrete or came from the pool through water vapor. Chloride can overcome the passivity of the natural oxide film on the steel surface. The inspection of safety-critical stainless steel components for SCC and loss of the section by pitting should be viewed as a priority. The following inspection procedure has been recommended to avoid future accidents: compile an inventory of all stainless steel components in the pool building identifying their grade, location, and function. SCC may be difficult to detect in the early stages. Pitting and brown staining, varying from a pale, dry discoloration to wet pustules may indicate SCC.

Yearly inspection or twice in a year inspection of stainless steel components especially load bearing components is desirable. Where staining or corrosion is found, corrosion products may be removed and the loss of cross section and integrity assessed. Load bearing or other safely components should be tested for SCC. When necessary, components should be more corrosion-resistant stainless steel parts.

5.3.7.4.27 Pipeline Failures   The history of pipeline safety has been reviewed (41). Compared to other forms of transportation, pipelines are inherently safer; however, pipeline failures can have serious consequences. For example, in June 1999, a pipeline rupture in Bellingham, WA, USA, spilled 946,000l (250,000 gal.) of gasoline into a creek. When the gasoline ignited, three people were killed, eight more were injured, several buildings were damaged, and the banks of the creek were destroyed along a 2.4-km section. In July 2000, a natural gas pipeline ruptured in Carlsbad, New Mexico, and when the gas ignited, 12 people were killed.

Table 5.10 summarizes the pipeline accidents and injuries between 1989 and 1998 (41).

Table 5.11 summarizes the major accidents reported to the US Department of Transportation by the operators for the 6-year period between 1994 and 1999. The data show that for transmission pipeline systems, inclusive of hazardous liquid and natural gas, approximately 25% of all reported accidents were because of corrosion (see Table 5.10). Of the hazardous liquid pipeline accidents caused by corrosion, 65% were because of external corrosion and 34% were because of internal corrosion.

In the case of natural gas transmission pipelines, 36% of the accidents were caused by external corrosion and 63% were caused by internal corrosion. In the case of natural gas distribution pipeline accidents, only 4% of the total accidents were caused by corrosion, and the majority of these were caused by external corrosion. The corrosion-related accidents in pipelines carrying natural gas and hazardous liquids are summarized in Table 5.11.

TABLE 5.10   Pipeline Accidents and Injuries Between 1989 and 1998

Period	Number of Accidents		
1989–1998	Natural gas distribution	Natural gas transmission	Hazardous liquid
	900	500	680
1989–1998	Average number of major accidents per 16,000 km of Pipeline		
	5	15	43
1989–1998	Number of injuries because of major accidents		
	700	100	105
1989–1998	Number of fatalities		
	165	22	18

TABLE 5.11   Summary of Corrosion-Related Accidents in Pipelines

Accidents	Hazardous Liquid Transmission	Natural Gas Transmission	Natural Gas Distribution
Total because of corrosion	271	114	26
Total accidents	1116	448	708
% of total because of corrosion	24.3	25.4	3.7
% because of external corrosion	64.9	36.0	84.6
% because of internal corrosion	33.6	63.2	3.8
% corrosion accident, cause not specified	1.5	0.9	11.5

In a summary report for incidents between 1985 and 1994, corrosion accounted for 28.5% of pipeline incidents on natural gas gathering and transmission pipelines. For incidents between 1986 and 1996 corrosion accounted for 25.1% of pipeline incidents on hazardous liquid pipelines.

In addition to the reported accidents, an average of 8000 leaks per year are repaired on natural gas transmission pipelines, and 1600 spills per year are repaired and cleaned up for liquid product lines.

5.4   AGE-RELIABILITY CHARACTERISTICS

Extensive studies on failure patterns to improve maintenance strategies and operational procedures identified six basic patterns.

Pattern A. Referred to as a bathtub curve with three identifiable regions, namely: (i) the initial period of high probability of failure; (ii) region of constant and low probability of failure; (iii) a wear-out region of high probability of failure.

Pattern B. Consists of constant or gradually increasing failure rate with a pronounced wear-out region.

Pattern C. Shows gradually increasing failure probability with no wear-out region.

Pattern D. Starts with low failure probability followed by an increase to constant level.

Pattern E. Shows a constant probability of failure.

Pattern F. Shows initial high mortality followed by constant or very slowly increasing failure probability. This pattern is particularly applicable to electronic equipment. The six patterns of failure (18) are depicted in Roberge 16, Fig. 2.5, p. 43.

The probability of failure depends on two factors, namely: (i) the specific form of corrosion and its rate; (ii) the possible effectiveness of corrosion inspection or monitoring.

The input of corrosion experts is useful in identifying the forms of corrosion and the key factors affecting the propagation of the corrosion rate. The ranking of process equipment may be done by internal probability of failure analysis. This method consists of analysis of equipment process and inspection parameters, and then ranking the equipment on a scale of one to three with "one" as the highest priority. This procedure requires a fair degree of engineering judgment and experience.

The probability of failure approach is based on a set of rules that are dependent on inspection reports/histories, knowledge of corrosion processes, and knowledge of normal and upset conditions. The equipment rankings may change with time and require periodic updating as a result of gaining additional knowledge, changes in process conditions, and aging of equipment. Maximum benefits of the procedure depend on a fixed equipment inspection schedule that allows the capture, documentation, retrieval of inspection, maintenance, and corrosion failure mechanism information.

The probability of detection (POD) concept has been known as an important concept since 1973 and has been used in the design of engineering instrumentation by NASA. The POD concept and methodology are widely used in nondestructive evaluation (NDE) methods, in particular, "fracture control" of engineering hardware and systems.

5.5 HISTORICAL IMPLICATIONS OF CORROSION

In 1761, the 32-gun frigate – HMS Alarm – had its hull covered with a thin copper sheathing to reduce the damage caused by the wood-boring shipworm and the speed-killing barnacle growth on ship hulls. After 2 years of operation, the sheathing became detached from the hull in many places because the iron nails used to fasten the copper to the timbers were corroded to such an extent that the corroded iron nails were hardly visible. Closer inspection showed that the iron nails wrapped in brown paper remained intact. This observation shows that iron should not be in contact with copper in a seawater environment if severe corrosion is to be avoided. This type of corrosion by two dissimilar metals in contact is known as galvanic corrosion or bimetallic or dissimilar metal corrosion.

Dissimilar corrosion was encountered in 1769 when Commodore the Hon. John Byron began a circumnavigation of the globe in his coppered ship, *Dolphin*. In addition to the prospect of the uncharted Coral Sea reef scything through the hull, the ship's captain heard repeated thumps below the stern windows of the cabin. John Byron recorded in his log that he feared the ship's very loose rudder would drop off at any time, for its iron printles, which were in contact with the copper sheathing, were corroded away to needle thickness. With no prospect of repair, this was another worry in addition to the multitude of others with which he had to contend.

Bimetallic corrosion and other forms of corrosion continued to cause service failures. In 1962, a report was sent to the British Ministry of Defense stating that a copper alloy end plate had fallen off a seawater evaporator in a submarine because the steel bolts with which it was secured had effectively dissolved through galvanic action. In 1982, the nose wheels failed on two Royal Navy Sea Harriers that had returned from the Falklands War. Studies showed that the galvanic action was responsible for the corrosion that occurred between the magnesium wheel alloy and its stainless steel bearing.

In general, *corrosion* means *rust*, an almost universal object of hatred. Rust is specifically referred to as the corrosion of iron. But corrosion is a destructive phenomenon that affects almost all metals. The Roman philosopher Pliny the Elder (AD 23-AD 79) wrote about *ferrum corrumpitur* or "spoiled iron." (42) At this time, the Roman Empire was established as the world's foremost civilization because of its use of iron in weaponry and other artifacts.

Corrosion has its beneficial consequences. Techniques such as bluing and gilding were used to protect steel objects as heat treatment resulted in protective oxide films (43).

Corrosion costs society in three major ways: (i) it is very expensive monetarily; (ii) it is a major waste of natural resources at a time of increased concern over damage to the environment; (iii) it can result in fatal accidents.

The cost of corrosion has already been dealt with in the first chapter. The annual cost of corrosion in the United States was estimated in 1949 to be $5 billion, 2.1% of GNP. The cost of corrosion in the United Kingdom was estimated to be £1.365 million (1971 prices). By 1975, the estimated cost of corrosion in the United States had risen to $70 billion, 4.2% of GNP. The cost of corrosion consists of the cost of replacement as well as indirect costs and may occur for any of the following reasons: (i) lost production shutdown or failure; (ii) maintenance; (iii) compliance with environmental and consumer regulations; (iv) loss of product quality in a plant owing to contamination from corrosion of the materials used to make the production line.; (v) high fuel and energy costs as a result of steam, fuel, water, or compressed air leakage from corroded pipes.

Serious problems arose in the United States in 1981 when the report of the Nuclear Regulatory Commission to Congress in Washington stated that the "vast majority" of steam-generating PWRs were suffering from failures of stainless steel cooling tubes. Though this corrosion did not predict any danger to the public, it was estimated that the maintenance bill would top $6 billion. Extra working capital because of increased labor requirements and larger stocks of spare parts is one of the consequences of corrosion.

The above factors illustrate quite clearly that to maximize profit, no manager can afford to ignore corrosion.

5.6 SOCIAL IMPLICATIONS OF CORROSION

In the history of the use of materials, the past 170 years have been closely associated with alloys of metals such as iron, aluminum, and copper. Our highly developed civilization could not exist without them. Yet corrosion is their Achilles' heel. The degradation of metallic materials is very largely because of corrosion, and in a society that focuses more and more on dollar cost, we have seen how expensive corrosion is. It is difficult to understand why corrosion has not been pursued as earnestly as the cure for AIDS. The answer to this question depends partly on the way society is structured. The scientific and engineering communities are strongly structured and rely for their financing on longstanding practices. Corrosion falls between traditional disciplines, so it is frequently considered to be out of the mainstream and somehow less important. Virtually all metals suffer corrosion, so its effects permeate nearly every aspect of human endeavor, and this fact alone makes the study of corrosion and its control more important, not less. Thus we find that corrosion is mostly ignored by chemists and electrochemists, who, not being familiar with metallurgy, do not wish to use systems involving real alloys. Although studied as a minor part of mechanical engineering courses, corrosion is the main reason why engineered systems will ultimately fail. The following examples provide the evidence for corrosion-based failures.

5.7 THE NUCLEAR INDUSTRY

This industry has most of the corrosion problems of other industries and some that are all of its own. Right from the start, the potential for disaster was recognized and tackled by using high-grade materials in many parts of the systems. Zirconium alloys were needed, which had their own corrosion problems and solutions. Growing worldwide demands for acceptable environmental performance have alienated others to the cause of nuclear power, in particular, after events at Three Mile Island and Chernobyl.

This industry too had its share of corrosion costs. For boiler reactors' capacity factor losses because of corrosion problems averaged over 6% between 1980 and 1991, reaching a peak value of 18% in 1982. It is estimated that corrosion problems have cost the nuclear utility industry more than $5 billion since 1980. In addition, repairs and mitigation cost the average US light water reactor >$0.5 billion in the industry with radiation exposures of about 100 rem per year.

Long-term storage of high-level nuclear waste has been a difficult problem. Corrosion behavior has long been known to be aggravated under high-level irradiation. The task of designing suitable containers that will maintain high-level waste in a safe condition for ten thousand years has stretched designers to the limit. Projection of behavior even 10 years into the future has been difficult enough, but significant progress has now been made in the resolution of the problem (44).

5.8   FOSSIL FUEL ENERGY SYSTEMS

Acid rain was recognized as a serious environmental pollutant long before the greenhouse effect and shown to involve gaseous pollutants such as carbon dioxide, nitrogen dioxide, and sulfur dioxide. Acid rain intensified by acid-generating sulfur compounds from fossil fuel power stations had a devastating effect on the ecology of large areas. In the United States, the Environmental Protection Agency (EPA) oversees the regulation of air pollution and sets up a rigorous enforcement timetable. The polluting industries identified are pulp and paper, municipal incinerators, chemical plants, and coal-fired electric power plants. The electric power industry was targeted by the EPA to reduce the amounts of air pollutants. The first flue gas desulfurization or scrubber system at a coal-fired plant in the United States was installed and operated in 1968. By 1995, 110 sites with 261 units were required to comply with EPA regulations.

5.9   THE AEROSPACE INDUSTRY

The safety record of this sector has been enviable because the aerospace industry is based on a manufacturing environment with a very high and proper regard to optimizing corrosion performance. In a span of 50 years, passenger travel has progressed from the 50 seater comet airliner to 800 and even 1000 passenger aircraft. In the early stages of the world's first jet-engine metal fuselage, there were many fatal crashes because of fatigue failure caused by cracks emanating from the window openings. At present, there are very few fatal accidents because of corrosion failure, despite total reliance on high-performance light alloys of reactive aluminum. This does not mean that corrosion problems have been banished from the aerospace industry. On the contrary, there are many cases of corrosion failures. Correct materials management programs have been adopted from the beginning to the end of the project, such as careful materials performance evaluation prior to use; adoption of good design principles; application of effective barrier coatings; regular and efficient inspection and maintenance schedules.

5.10   THE ELECTRICAL AND ELECTRONICS INDUSTRY

This facet of industry might appear to be free from corrosion. On the contrary, there are problems such as the corrosion of aluminum used for tracks on most devices. There is the high probability of strong galvanic interaction with gold used for connectors. Many critical systems must function in severe environments such as a temperature range below zero to 40 °C, 100% humidity, airborne particulate matter and insects. This is certainly a challenging problem.

Corrosion of power cables such as "sheath damage" or "water in the cable" is also a challenging problem. Some of the causes of damage to the protective sheath range from gunshot damage and pinholes caused by lightning strikes on overhead lines to rodent damage, attack by gophers and squirrels, and damage by termites or other insects. Problems with buried cables were solved by replacement.

About 75 years ago, the main material for cable sheathing was lead, protected from corrosion by asphalt-impregnated jute coating. Even then corrosion problems prevailed. Later on, extruded or drawn aluminum sheaths were in use, soon to compete with polythene. A large number of failures in the 1960s caused by water in the cable led to the development of sophisticated methods of sheathing involving coated aluminum and clad metals of copper adjacent to a number of steels (45).

5.11   THE MARINE AND OFFSHORE INDUSTRY

The marine environment is probably the most aggressive environment in which metals are chosen to operate. The corrosion performance is mixed. Great achievements have been made offshore since the development of a new generation of deep-water platforms in the North Sea oilfields during the 1970s. This technology has been used in deeper waters. Over 200 of the largest steel structures have been created and immersed in violent seas, and continued operation without failure for 25 years is surprising and true. The record is successful but for the Piper Alpha disaster in 1988 when 167 men died because of a fire following a mechanical failure.

The battle against corrosion was first waged with hulls made of puddled iron starting with the construction in the Scottish Clyde shipyard of the fast passage barge, *Vulcan*, in 1819. In 1822, *Manby* was the first iron steamship to make an international voyage under power. Both marine fouling and corrosion problems were faced by iron hulls of the ships (46). In spite of the difficulties, many of the iron hulls gave long years of service and survived well, even when lost or abandoned, as in the example of the submarine *Holland I*.

In spite of many advances made in materials (47) and coatings technology, civilian merchant fleets have suffered badly from poor maintenance by owners leading to major disasters, most of which is attributable to corrosion and poor maintenance at every level. The loss of the 170,000 ton *Derbyshire* in 1980 with all 44 crew members is speculated to be because of poor design, accentuated by corrosion.

In the 1990–1991 time span, more than 30 bulk carriers were lost or damaged and more than 300 crew members died. A surveyor found corrosion from original 12 mm thickness down to 3–5 mm of steel during 10–15 years of life. For a long time, the owners cut their ship maintenance costs, which led to extensive corrosion and loss of lives.

In 1993, the oil tanker *Braer* went aground, discharging about a 100,000 ton cargo onto the ecosystem of Shetland Islands to the north of Scotland. The accident resulted from the failure of a marine propulsion system after seawater was ingested in the storm. The blame for this rests on the cost-cutting practices endemic in the industry.

Navy-related problems may be attributed more to the complexity of modern military systems than to cost cutting. This problem may become significant when the cuts to defense expenses are enforced (48).

5.12 THE AUTOMOBILE INDUSTRY

Corrosion of automobiles is well documented (49–51). Environmental factors such as the deicing salts on the roads and temperature fluctuations have profound effects on automobiles. The cars are expected to survive the grinding effect of gravel, offer comfortable protection to the passengers, should be easily and economically repairable on damage with a reasonable service life.

The total cost of corrosion of personally owned automobiles was determined to range from $6 to $14 billion and avoidable costs were estimated to range between $2 and 8 billion. The automobile sector had a significantly higher cost than any other sector, and the cost of this sector was the single most significant driving factor in estimating the total corrosion cost for the entire United States.

A code for Canada in 1981 specified: (i) a car body should last 1.5 years or 60,000 km before suffering cosmetic corrosion, 5 years (200,000 km) for perforation corrosion, and 6 years (240,000 km) for structural corrosion. The code projected for North America in 1990 was to have no cosmetic in 5 years and no perforation corrosion in 10 years. These code requirements resulted in the use of precoated steel, especially galvanized steel, which in 1993 satisfied the requirements (52).

In 1981, about 3 million passenger cars and commercial vehicles were being scrapped in Europe because of three causes: accidents, obsolescence, and corrosion. McArthur showed a correlation between the number of serious injuries suffered in road accidents and the age as well as the amount of corrosion of the vehicle (53). A corroded vehicle was less able to absorb the energy of impact than that of a new vehicle. The average life of a motor car has increased over the years since the auto industry had at last reacted to the public demand for more corrosion protection.

Rust proofing treatments of cars have not been very successful, and the Office of the Attorney General of New York State claimed that consumers were defrauded of $11 million annually because of poor quality rust treatments.

5.13 BRIDGES

In spite of the tendency to rust, iron either alone or as reinforcement in concrete has been the basis of civil engineering projects since the world's first iron bridge was constructed over River Severn at Coalbrookdale in England by Abraham Darby in 1780. Many wrought iron bridges performed well with modest but regular maintenance.

The Royal Albert Bridge at Saltash, Cornwall, built in 1859 by Brunel to carry a main railway line across the River Tamar between Devon and Cornwall is free from corrosion and has been given a clean bill of health until 2035. Brunel's other bridges are Clifton Suspension Bridge and Telford's Menai Straits Bridge, a suspension bridge, have stood the test of time. The Telford Bridge originally opened in 1826, was reconstructed in 1938–1940 to meet the increased traffic demands, and this was the first large bridge to be metal sprayed in the United Kingdom. This serves as a good example of the benefits of protective coating in hostile environments.

In the past, designers applied age-old principles born out of sound engineering practice. The results were beneficial and the structures lasted a longer time without any problems. In today's society, restrictions such as financial and highly competitive tenders have paved the way to use the most advanced and often untried engineering techniques to save money. Operating margins have been trimmed to such a fine extent that whenever a problem arises, as in a novel situation, severe penalties ensue. Thus, the advantages of modern engineering design seem to have created a plethora of problems that underpin the conclusions of Hoar (54).

Some examples that highlight the conclusions of the Hoar Report are as follows.

Example 5.4   In the United Kingdom, the innovative suspension bridge over the River Severn was heralded in 1966. In 1978, broken wires within the raked steel hangers were revealed by cracking in the paintwork. The steel deck assumed an innovative design in which aerodynamic closed box sections intended to eliminate marine atmosphere from inside the boxes was implemented. In a short time, cracking of the steel box-welds had occurred, and by 1983, concern was expressed over the future of the link between England and South Wales. The problem was attributed to unexpectedly high traffic loading, aggravated by corrosion because of the salt-laden atmosphere of the Severn Estuary. The cost of repair in 1983 was £30 million. A new bridge was eventually constructed.

Corrosion problems were experienced with the 350-m long Pelham Bridge in Lincoln. The bridge was constructed in 1957 when the problem because of deicing salts was not envisaged. Penetration of the deicing salt caused severe corrosion of the steel reinforcement in the concrete road bridge decks. Many such cases were reported in England. The cost of repairing was considered to be high.

A report from the New York Department of Transport revealed that 95% of all bridges in New York would be deficient if maintenance remained at the same level as in 1981. This degree of deterioration is attributed to the use of de-icer salts. In the snow belt of the United States, the use of de-icing salts rose from 0.6 million ton in 1950 to 10.5 million ton in 1988. The motor car damage was estimated at $5 billion.

On December 15, 1967, Point Pleasant Bridge in Ohio collapsed, killing 46 people. The cause of the disaster was a stress corrosion crack 2.5 mm deep in the head of an eyebar. The metal had low resistance to fracture once a notch had been initiated. Failure led to the collapse of the bridge.

The bridge at the Charing Cross Railway in central London was subject to corrosion from dog's urine. Repeated urination by dogs caused crevice corrosion in a part of the structure that was not easily accessible. In 1979, the City of Westminster, London, reported a problem of falling lamp posts. The culprits were dogs that urinated at the base of the lamp posts.

5.14 BIOMEDICAL ENGINEERING

Corrosion of prosthesis or other medical implants is an important subject matter as fatigue and corrosion fatigue of artificial heart valves or hip joints can be fatal. A heart valve suffers stress cycling at 60–100 cycles per minute, 24 h a day for many years. Instances of fatigue failure of heart valves have been reported. A ball-and-socket hip joint made from a combination of titanium and stainless steels experiences severe mechanical loads during jogging or even when descending stairs. It is to be noted that the working environment of the hip joint insert is a warm saline solution that would stretch the insert to its limit. The implants that are supposed to last 20 years had to be replaced after only two or three years. These two examples indicate the seriousness with which corrosion impacts present medicine and how the effects are far more serious to people than simply the monetary value.

5.15 THE DEFENSE INDUSTRY

Failure of military weapons in action because of corrosion failure is serious and may be dealt with by reliability, a wider issue that arose from the needs of the defense industry. There is an old saying that the kingdom may be lost for the sake of a nail in a horse's shoe. Cascading consequences from apparently trivial situations can sometimes lead to a serious event or an incident like a war.

Many defense systems are idle for years with the expectation that they will function effectively when the need arises. It is not possible to check out the weapons for proper function as in the case of a torpedo or missile without actually firing. The weapons for war are complex systems manufactured with lightweight corrosion-prone alloys operating in corrosive and widely variable atmospheres. Sometimes the unlikely combination of circumstances results in a failure at the critical time as it happened to the British ammunition in India.

5.16 CORROSION AND ENVIRONMENTAL IMPLICATIONS

Acid rain, the greenhouse effect, and the depletion of the ozone layer are some of the consequences of environmental pollution. The change in social perspective has led to international agreement and legislation with respect to environmental damage. Control of corrosion is an integral part of the pressure because when properly implemented, engineering systems can function efficiently for a longer duration with less wastage of material and energy sources and minimum pollution.

It is interesting to note that in the United Kingdom, 1 ton of steel is converted into oxide rust every 90 s. Apart from the wastage of metal, the energy required to produce 1 ton of steel from iron ore is equal to the energy requirement of a family for 3 months. The international opinion of the Exxon Valdez spillage in Alaska and that of an aircraft accident did not prevent Braer oil tanker spillage. Pollution is still allowed to happen because of the improper attitudes and lack of international action.

In almost all aspects of industry, technology is available to use alloys in ways that are efficient and environmentally sound. The corrosion environments in which materials must operate are tough, but suitable materials selection with efficient barrier coatings has provided excellent solutions. Solid waste disposal is a good alternative to landfill, which can generate energy as a by-product provided gaseous emissions are controlled. The available modern materials can be used in incinerators and thereby combat high-temperature corrosion (55–58).

Rail transport has not been popular in the Western world but the need to reduce automobile emissions and the resulting environmental damage has led to the development of electric-powered mass rail transit systems. The corrosion problems associated with high-current land-based systems in which current leaks into nearby metallic structures leading to corrosion are well understood and within the scope of existing technology (59).

The pollution-has-to-pay policy has been firmly established in the United States, but not in other parts of the world. This leads to a short-sighted view of saving initial costs without caring for the consequences. It is advisable to invest more in the cost of initial design so that the product has a long life and low maintenance.

Sometimes the need to be environmentally acceptable may lead to new problems. For instance, ozone was suggested to replace biocides with no data available on the performance in the chlorination of water (60). Corrosion control techniques can have both favorable as well as ill effects and hence one has to exert balanced judgment before embarking on a corrosion prevention method. Organotin antifouling coatings on ships were effective, but they polluted the seawater and hence were banned from further use. The use of cadmium as a sacrificial anode is restricted because of its toxicity. Large amounts of zinc are used to protect steel platforms in the sheltered and shallow waters of the sea, and the effects of zinc on the contamination of waters are not known.

Corrosion and its control in society parallels engineering at large. Both theoretical and laboratory work by scientists have helped to lay the groundwork. Large engineering systems are too complicated for accurate performance predictions. So far, corrosion and its control have been successful: bridges and building structures do stay up, aircraft safety is high, and cars can survive the rigors of harsh winters and salt-laden roads.

The well-established corrosion science of materials forms the foundation, but the practical aspect is the pivotal part that needs to be integrated to achieve maximum effective corrosion control. It is crucial to stress that cost cutting in times of financial stringency will certainly increase the probability of serious corrosion failures.

REFERENCES

1. Economic Effects of Metallic Corrosion in the United States, NBS Special Publication 511-1, SD stock no. SN-003-003-01926-7, 1978 and *Economic Effects of Metallic Corrosion in the United States*, Appendix B, NBS Special Publication 511-2, SD stock no. SN-003-003-01927-5, 1978.

2. *Economic Effects of Metallic Corrosion in the United States: Update*, Battelle, 1995.

3. TP Hoar, Report of the Committee on Corrosion Protection: a Survey of Corrosion Protection in the United Kingdom, 1971.

4. G Okamoto, Report of the Committee on Corrosion and Corrosion Protection: a Survey of the Cost of Corrosion in Japan, Japan Society of Corrosion Engineering and Japan Association of Corrosion Control, 1977.

5. BW Cherry, BS Skerry, *Corrosion in Australia: the Report of the Australian National Centre for Corrosion Prevention and Control Feasibility Study*, 1983.

6. F Al-Kharafi, A Al-Hashem, F Martrouk, Economic Effects of Metallic Corrosion in the State of Kuwait, KISR Publications, Report no. 4761, Dec 1995.

7. HH Uhlig, *Corrosion* **6**:29 (1952).

8. JH Payer, WK Boyd, DG Lippold, WH Fisher, Materials Performance, May–Nov 1980.

9. F Al-Kharafi, A Al-Hashem, F Martrouk, Economic Effects of Metallic Corrosion in the State of Kuwait, KISR Publications, Final Report no. 4761, Dec 1995.

10. D Behrens, *British Corrosion Journal* **10**:122 (1967).

11. S Lindberg, *Kemiam Teollusius (Finland)* **24**:234 (1967).

12. KS Rajagopalan, *Report on Metallic Corrosion in India*, CSIR, 1962.

13. R. Zhu, *Approaches to Reducing Corrosion Costs*, NACE, 1986.

14. VS Sastri, GR Hoey, RW Revie, *CIM Bulletin* **87**:87 (1994).

15. VS Sastri, Proposal to Establish a National Secretariat for Wear Reduction and Corrosion Control in Canada, *MTL Report*, Ottawa, 1992.

16. PR Roberge, *Corrosion Inspection and Monitoring*, Wiley-Interscience, Hoboken NJ, 2007.

17. DH Horne, Corrosion 2000, NACE International, Houston, TX, Paper no. 719, 2000.

18. FS Nowlan, HF Heap, Reliability Centered Maintenance, AD-A066-579, National Technical Information Service, Washington, DC, 1978.

19. VS Sastri, E Ghali, M Elboujdaini, *Corrosion Prevention and Protection: Practical Solutions*, John Wiley & Sons, 2007.

20. CF Smith, FE Dollarhide, JN Byth, *Journal of Petroleum Technology*, pp. 737–746, 1978.

21. SJ Findlay, ND Harrison, *Materials Today*, pp. 18–25 (2002).

22. E Ghali, M Krishnadev, *Engineering Failure Analysis*, **13** (1): 117–126 (2006).

23. National Printing Bureau, Domestic Water Systems Preliminary Investigations Report, Mansour Keenan and Associates Limited, Jan 2003.

24. MG Fontana, Corrosion Engineering, pp. 43, 1986.

25. MY El-Shazly, Cathodic Protection of Steel in Concrete, 24th Annual Conference on Corrosion Problems in Industry, Egyptian Corrosion Society, Dec 5–8 2005.

26. COM: Conference of Metallurgists, 1999, Canadian Institute of Mining, Metallurgy and Petroleum, Montreal, QC, 2005.

27. J Hadjigeorgiou, F Charette, *Rock Bolting for Underground Excavations*, Chap. 63, Hustrilid , Bullock (eds), Society of Mining Engineers, pp. 547–554, 2001.

28. M Zamanzadeh, E Larkin, D Gibbon, *A Re-Examination of Failure Analysis and Root Cause Determination*, Mateo Associates, Pittsburgh, PA, 2004.

29. COM: Conference of Metallurgists, Canadian Institute of Mining, Metallurgy and Petroleum, Montreal, QC, 1997.

30. MG Hay, MD Stead, Canadian Region Western Conference, National Association of Corrosion Engineers, Calgary, Canada, 1994-02-07/10, 1994.

31. CW Crow, SS Minor, Paper no. 82-33-15, 33rd Annual Technical Meeting of the Petroleum Society of the Canadian Institute of Mining, Calgary, AB, 1982.

32. RH Hausler, *Materials Performance* **13**:16–22 (1974).

33. K Suda, S Misra, K Motohashi, *Corrosion Science* **35**:1543–1549 (1993).

34. D Miller, *Materials Performance* **29**:10–11 (1990).

35. RW Staehle, *Uhlig's Handbook of Corrosion*, Wiley, New York, pp. 27–84, 2000.

36. D Weir, *The Bhopal Syndrome, Pesticides, Environment and Health*, Sierra Club Books, San Francisco, CA, 1987.

37. Natural Gas Pipeline Rupture and Fire Near Carlsbad, New Mexico, August 19, 2000, NTSB/PAR-03/01, National Transportation Safety Board, Washington, DC, 2003.

38. JM Malo, V Salinas, J Uruchurtu, *Materials Performance* **33**:63 (1994).

39. 4X-AXG Boeing 747-258F El Al 04.10.92, Bijlmermeer. Accident report 92–11, Netherlands Aviation Safety Board, Amsterdam, 1992.

40. M Hamer, *New Scientist* **146**:5 (1991).

41. Pipeline Safety, Report no. GAO/RCED-OU-128, Report to Ranking Minority Member, Committee on Commerce, House of Representatives, May 2000.

42. Pliny, *Natural History of the World*, Heineman, London, 1938.

43. AR Williams, *Metals and Materials* **2**:485–489 (1986).

44. D Shoesmith, B Ideka, F King, *Modelling Aqueous Corrosion*, KR Trethewey, PR Roberge (eds.), Kluwer Academic, Dordrecht, The Netherlands, pp. 201–238, XXX.

45. K. Bow, *History of Corrosion in Power and Communications Cables*, NACE International, Houston, TX, 1993.

46. FM Walker, Marine Technology, pp. 778–782, Dec 1990.

47. J Sedriks, Paper no. 505 in Corrosion 93 Plenary and Keynote Lectures, R Gundry (ed.), NACE International, Houston, TX, 1993.

48. KR Trethewey, *Proceedings of INEC 92*, Institute of Marine Engineers, Plymouth, UK, 1992.

49. *Automotive Corrosion by De-Icing Salts*, R Baboian (ed.), NACE International, Houston, TX, 1981.

50. *Automotive Corrosion and Protection*, R Baboian (ed.), NACE International, Houston, TX, 1992.

51. H McArthur, *Corrosion Prediction and Protection in Motor Vehicles*, Ellis Horwood, Chichester, 1988.

52. EN Soepenberg, Paper 546, Corrosion 93, New Orleans, LA, NACE International, Houston, TX, 1993.

53. H McArthur, *Corrosion Prevention and Control* **28**(3):5–8 (1981).

54. TP Hoar, *Report of the Committee on Corrosion and Protection*, HMSO, London, 1970.

55. *Materials Performance in Waste Incineration Systems*, GY Lai, G Sorrell (eds), NACE International, Houston, TX, 1992.

56. HH Krause, Paper no. 200 in Corrosion 93 Plenary and Keynote Lectures, RD Gundry (ed.), NACE International, Houston, TX, 1993.

57. J Stringer, Paper no. 225 in Corrosion 93 Plenary and Keynote Lectures, RD Gundry (ed.), NACE International, Houston, TX, 1993.

58. CM Anthony, GY Lai, MD Kannair, Paper no. 214 in Corrosion 93, NACE International, Houston, TX, 1993.

59. MJ Szeliga, Stray Current Corrosion: the Past, Present and Future of Rail Transit, *Paper 300*, *Corrosion 94*, Baltimore, MD, NACE International, Houston, TX, 1994.

60. BE Brown, DJ Duquette, Paper no. 486, Corrosion 94, Baltimore, MD, NACE International, Houston, TX, 1994.

INDEX

Challenges in Corrosion: Costs, Causes, Consequences, and Control, First Edition. V. S. Sastri.
© 2015 John Wiley & Sons, Inc. Published 2015 by John Wiley & Sons, Inc.